ELEMENTARY
LINEAR
ALGEBRA

ELEMENTARY
LINEAR
ALGEBRA

A. WAYNE ROBERTS
Macalester College

THE BENJAMIN/CUMMINGS PUBLISHING COMPANY, INC.

Menlo Park, California · Reading, Massachusetts
London · Amsterdam · Don Mills, Ontario · Sydney

To Dolores

Sponsoring Editor: *S. A. Newman*
Production Editor: *Greg Hubit*
Book and Cover Designer: *John Edeen*
Artist: *Ayxa Art*

Copyright © 1982 by The Benjamin/Cummings
Publishing Company, Inc.

Library of Congress Cataloging in Publication Data

Roberts, A. Wayne (Arthur Wayne), 1934–
    Linear algebra.

    Includes index.
    1. Algebra, Linear.   I. Title.
QA184.R6   512′.5   81-18007
ISBN 0-8053-8302-6   AACR2

abcdefghij-HA-898765432

The Benjamin/Cummings Publishing Company, Inc.
2727 Sand Hill Road
Menlo Park, California 94025

CONTENTS

FOREWORD

More times than I care to count, I have taught a one-term course in elementary linear algebra, taken mainly by engineering and science students and based on one of the most popular textbooks. At first, I adhered rigorously to the text and to the official schedule for the course, with the result that the students regarded linear algebra as a sort of ritual symbol pushing, having little relevance to anything else in their experience or to their future careers. That was a shame, for while the intrinsic interest of linear algebra may be a matter of taste, there is no denying that the subject provides some of the most useful tools of applied mathematics and has many important connections with various areas of science and technology. The students, unfortunately, were in no position to appreciate the intrinsic interest of the subject, and the main part of the text provided no indication of applications or of connections with the rest of the world. The official schedule was defensible, however, for the entire term really was needed to cover basic mathematic details, thus precluding a deep discussion of any application.

What could be done to improve the situation? The "vignettes" in *Elementary Linear Algebra* provide an answer. They do not go into much detail and hence do not take much time, but they are cleverly interspersed throughout the main body of the text and thus serve to illustrate a few important connections and perhaps to convince the students of the existence of many more. I have used some of these "vignettes" as supplementary material in my own classes and have found that they greatly increased student interest in the subject.

It would be even better to have more time to cover some applications more deeply. Which areas of application? The natural candidates are linear differential equations, linear programming, and Markov chains.

In a two-term course, they could all be covered. In a one-term course, a choice must be made, depending on the instructor, on the makeup of the class, and on what material is covered in other courses. Wayne Roberts's book contains a chapter on each of these areas of application and thus provides a welcome flexibility in the use of the book.

Victor Klee
Seattle, Washington
November, 1981

PREFACE

A COMMON PROBLEM IN TEACHING LINEAR ALGEBRA

Students in a linear algebra course, perhaps more than in other mathematics courses, want to know how they will be able to use linear algebra after they have completed the course. Their instructors usually have the same nagging feeling, which they have in most courses, namely, that more should be done to relate the topic to its applications.

The solution to this problem seems obvious: Use applications as motivational material throughout the course. Those who have tried know, however, that it's not as easy as it sounds. Students don't get very far into an application before they feel that they're being told more than they really want to know, that they are being led too far afield with something they don't much care about, that this probably won't be on the test anyway, and that they are sorry they asked what the subject was good for. Also, instructors begin to worry about not getting through the standard topics because of the time devoted to applications.

A SOLUTION TO THE PROBLEM

I have tried in this book to respond to the problem in a way that avoids the pitfalls described above.

COVERAGE OF STANDARD TOPICS. The standard topics of a first course in linear algebra are covered in a straightforward manner in Chapters 1 through 6.

VIGNETTES. In every section of the first six chapters, I have included a vignette—a clearly marked short digression that relates the topic of that section to an application or to another area of mathematics. These vignettes are easily distinguishable from the body of the text by the use of a second color and by this symbol ⬭. Their location and distribution throughout the text can easily be seen by flipping through the book and looking for the swatches of second color. The titles and page numbers of these vignettes are completely listed in the table of contents.

Each section (with one clearly indicated exception) is independent of its vignette. This allows instrucfors and students to choose the vignettes that interest them. I hope that the very placement of these vignettes in the shadows of the text will make students want to turn aside and examine them. If they do so, they should find that the vignettes have been written to give them what they want: a lighthearted, sometimes humorous indication of the kinds of things for which linear algebra is useful, without the details that would turn this fun into work. Readers of *Elementary Linear Algebra* may skip them, but I don't think they will.

The vignettes, then, are quite sketchy in their treatment of the topics taken up. They are, however, just one of the responses I have made to the belief that there should be some attention paid to applications.

CHAPTERS 7, 8, AND 9. Each of the last three chapters discusses in some depth a significant application of linear algebra. Although I do not expect that anyone will be able to cover all three chapters, in a semester or two-quarter course there is time to cover at least one of them. Instructors may choose a topic because it is of personal interest, of special interest to the class, or appropriate to a particular curriculum. Taken together, these chapters offer great flexibility to the instructor, and they indicate important areas of application to any student who will peruse them. The page numbers in these chapters are shaded with gray for easy identification.

THE LEVEL
OF THIS BOOK

PROOFS. The applications are in the nature of appetizers, but the text serves up the staples of a first course in linear algebra. Like all writers at this level, I have omitted a good many proofs. Where I have done this, however, I have done it in such a way as to indicate clearly what needs to be proved and why. This effort, carried consistently through the text, can be seen in the treatment of the uniqueness of the row echelon form in Chapter 1. Too often, key ideas like this one are dismissed as obvious, whereas they should be dismissed as important, difficult to prove, but easy to believe after a little experience. To this end I have included numerous exercises contrived to help students discover what they should and should not believe. Problem Set 2-4, together with the discussion in Section 2-5, illustrates both the opportunities and the warnings I have given about discovery through examples. Instructors with a taste for proofs will find, in addition to those usually included at this level, that I have included two that are often omitted: a proof that the eigenvalues of a symmetric matrix are real and a proof of the Cayley-Hamilton Theorem.

THE USE OF CALCULUS. It has become fashionable in the prefaces of linear algebra books to say that the book may be used by students who have not had calculus. That is true of this book as well, in the sense that examples and problems requiring calculus are identified with a **C** in the margin. One of the greatest attributes of linear algebra, however, is that it unifies and illuminates ideas from other areas of mathematics; this fact will be most evident to the student who has had a standard calculus course. I have not made any effort in choosing the examples or writing the vignettes to pretend that this is not so.

FOUR TYPES
OF PROBLEMS

SECTION PROBLEMS. A set of problems appears at the end of each section of the book. I have attempted throughout to cover the same concepts with both even- and odd-numbered problems. I have also gone to great pains to grade these exercises and carefully link them to the text. Answers to odd-numbered problems are at the back of the book.

VIGNETTE PROBLEMS. Most section problem sets include a problem or two marked with a **V** in the margin. Such specially designated problems relate to the vignette in that section.

CALCULUS PROBLEMS. Problems that require the use of calculus are denoted by a **C**.

CHAPTER SELF-TESTS. Each chapter concludes with a self-test. This test is more than a collection of several problems from each of the sections in the chapter. Rather, it is intended to probe the students' understanding of the chapter as a whole.

SUPPLEMENTS

INSTRUCTOR'S GUIDE. The *Instructor's Guide* contains the answers to the even-numbered exercises, together with some suggestions and references that the instructor might find useful.

STUDENT SOLUTIONS MANUAL. The *Student Solutions Manual* contains solutions to selected odd-numbered exercises and to the self-tests.

ACKNOWLEDGMENTS

It was Victor Klee who first suggested to me that I write this book, and I am deeply indebted to him for his continual encouragement, for his ideas on how this course should be taught, and for many specific suggestions for vignette topics. The book would not have been written without his active involvement. I am also grateful to the many people who have reviewed this book in manuscript form, and I wish to mention in particular Donald Albers, Alan Weinstein, Alan Shuchat, and Dale Varberg. A special note of thanks goes to Bruce Edwards, who not only reviewed what I had written, but provided the basis for several vignettes dealing with computer methods in linear algebra.

I cannot ignore the contributions of students in several classes at Macalester College, and among these, Kathy Gretler and Rachael Buhse must be cited in a special way for their efforts in helping me to provide correct answers to problems. Finally, it is a great pleasure to pay tribute to the enthusiasm and determination of Susan Newman, the kind of editor every writer wishes for, but seldom finds.

A. Wayne Roberts

ELEMENTARY
LINEAR
ALGEBRA

SYSTEMS OF LINEAR EQUATIONS 1

INTRODUCTION

Much of what we say in this chapter will not be new to our readers. Some of it may be so old as to have become rusty. It is not new, for example, to visualize the relationship between x_1 and x_2 in the equation

$$2x_1 + x_2 - 5 = 0$$

by plotting on coordinate axes some values of x_1 and x_2 that satisfy this equation (Figure 1-1). Neither is it new to find that the resulting points fall in a straight line and that the equation is consequently called **linear.**

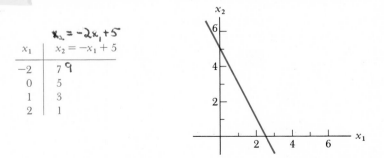

x_1	$x_2 = -x_1 + 5$
-2	7 9
0	5
1	3
2	1

$x_2 = -2x_1 + 5$

Figure 1-1

There is something, however, that will probably be new. It is the insistence on being systematic in our work. Being systematic from the beginning helps to avoid confusion later on when we deal with systems involving more than two variables. Being systematic is the only way to utilize a computer. Being systematic in our approach to specific problems provides a foundation for theoretical concepts to follow.

This desire to emphasize a systematic approach does present a dilemma in writing about a subject. If examples are kept simple, they invite shortcuts around the formality of the systematic methods being developed. But if examples are complicated enough to demand a systematic approach, the principles being illustrated are submerged in a sea of calculations. The way out of this dilemma is to use simple examples preceded by a plea to readers to use these examples as vehicles for learning systematic procedures. This is the plea: *Do learn the systematic methods as they are introduced.*

In this spirit of building on simple examples, let us return to the concept of a linear equation in two variables. If we can find real numbers r_1 and r_2 so that two such equations

(1)

$$a_{11}x_1 + a_{12}x_2 = b_1$$
$$a_{21}x_1 + a_{22}x_2 = b_2$$

are both satisfied when we set $x_1 = r_1$ and $x_2 = r_2$, then the ordered pair (r_1, r_2) is said to be a **solution** to the system. The two straight-line graphs of these equations (Figure 1-2) intersect at the point (r_1, r_2).

Figure 1-2

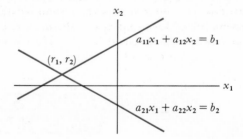

Since it is obvious that two straight lines may be parallel, intersect in a single point, or be coincident, this geometric viewpoint explains why there may be no solutions, exactly one solution, or an infinite number of solutions.

When $x_1 = 0$, $-x_2 + x_3 = 5$
When $x_2 = 0$, $2x_1 + x_3 = 5$
When $x_3 = 0$, $2x_1 - x_2 = 5$

Figure 1-3

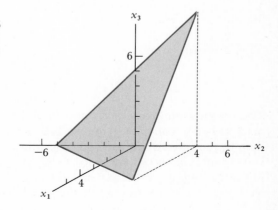

The name *linear* is carried over to first degree equations in three variables such as $2x_1 - x_2 + x_3 = 5$ even though the graph, now requiring three dimensions, is a plane, not a line (Figure 1-3). In general, an equation involving n variables is called **linear** if it can be written in the form

$$a_1x_1 + \cdots + a_nx_n = b$$

where the a_i and b are constants.

In this chapter we consider systems of m linear equations in n variables. Such a system is commonly written

$$
\begin{aligned}
a_{11}x_1 + a_{12}x_2 + \cdots + a_{1n}x_n &= b_1 \\
a_{21}x_1 + a_{22}x_2 + \cdots + a_{2n}x_n &= b_2 \\
\vdots \qquad \vdots \qquad\quad \vdots \qquad \vdots& \\
a_{m1}x_1 + a_{m2}x_2 + \cdots + a_{mn}x_n &= b_m
\end{aligned}
$$

(2)

where a_{ij} is the **coefficient** of x_j in the ith equation.

A **solution** to such a system is an **n-tuple** of numbers $(r_1, r_2, \ldots, r_n)$ such that if we set $x_1 = r_1, x_2 = r_2, \ldots, x_n = r_n$ in (2), all m equations will be satisfied. The set of all solutions to a system is called the **solution set** for that system. We will see that as was the case for $m = n = 2$, the solution set can always be described in one of three ways: empty, a single n-tuple, or an infinite set of n-tuples.

EQUIVALENT SYSTEMS OF EQUATIONS   *1-2*

TWO LINEAR EQUATIONS IN TWO VARIABLES

Much can be learned sometimes by looking carefully at something that has been familiar for a long time. To this end, let us review the solving of two linear equations in two variables as it might be explained in a beginning algebra text:

Find the solution set of

Example A

$$
\begin{aligned}
x_1 + 2x_2 &= 5 \\
2x_1 - 3x_2 &= -4
\end{aligned}
$$

(1)

Our goal is to eliminate a variable from one of the equations. One way to do this is to copy the first equation, then add -2 times this copied equation to the second equation:

$$
\begin{aligned}
x_1 + 2x_2 &= 5 \\
- 7x_2 &= -14
\end{aligned}
$$

(2)

Next, multiply the second equation by $-\frac{1}{7}$ to get

(3)
$$x_1 + 2x_2 = 5$$
$$x_2 = 2$$

Finally, copy the second equation; then add -2 times the second equation to the first; thus obtaining

(4)
$$x_1 \qquad = 1$$
$$x_2 = 2$$

The solution set consists of the single ordered pair $(1, 2)$, a fact you should check by substitution in (1). $\square$

The method used in this example is called **complete elimination.** A second method presents itself in equations (3), where we could take advantage of knowing $x_2 = 2$ by substituting this value back into the first equation. That is, write

(5)
$$x_2 = 2$$
$$x_1 = 5 - 2x_2 = 5 - 4 = 1$$

This second method is called **elimination with back substitution.** Both methods are discussed at greater length in Section 1-3.

Whatever method is used, notice that we have at every step maintained a set of two equations. It seems tedious to copy a set of equations with each step, but this is one of the places where we should discipline ourselves to be systematic even when the simplicity of the problem invites shortcuts.

Now let us comment on the key idea used in Example A which occurs in moving from the system of equations (1)

$$x_1 + 2x_2 = 5$$
$$2x_1 - 3x_2 = -4$$

to the system of equations (2)

$$x_1 + 2x_2 = 5$$
$$- 7x_2 = -14$$

Equations (1) have graphs that are straight lines. To seek the solution to our problem is to seek the point (r_1, r_2) where these lines intersect. Equations (2) which can be written in the form

(6)
$$x_1 + 2x_2 - 5 = 0$$
$$-2(x_1 + 2x_2 - 5) + (2x_1 - 3x_2 + 4) = 0$$

also have graphs that are straight lines. Our method relies on the fact that these two lines also intersect at (r_1, r_2). Why does it work?

Consider any equation of the form

$$k(x_1 + 2x_2 - 5) + (2x_1 - 3x_2 + 4) = 0 \qquad (7)$$

Its graph is a straight line. Now if (1) is satisfied by setting $x_1 = r_1$ and $x_2 = r_2$, then surely (7) will also be satisfied. Thus, for any choice of k, the graph of (7) will pass through (r_1, r_2). In particular, both lines described by (6) will pass through (r_1, r_2).

We say that (7) describes a **family of straight lines.** Some members of the family are drawn in Figure 1-4. Among all possible choices of k in

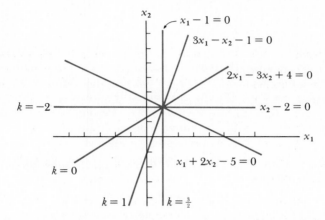

Figure 1-4

(7), the choice $k = -2$ stands out as particularly useful because the corresponding graph is a horizontal line, giving us the x_2 coordinate of the point common to all members of the family. What would be significant about choosing $k = \frac{3}{2}$?

THE ELEMENTARY OPERATIONS

We are about to describe in words the steps used to solve a system of linear equations similar to (1). Our purpose is not to introduce confusion where clarity has prevailed, but to give rules that will guide us when we work with systems of m equations involving n variables. We list three such rules or, as they are commonly called, three elementary operations that may be performed on the system

$$\begin{aligned} a_{11}x_1 + \cdots + a_{1n}x_n &= b_1 \\ \vdots \qquad \qquad \vdots \quad &\ \ \vdots \\ a_{m1}x_1 + \cdots + a_{mn}x_n &= b_m \end{aligned}$$

I. Interchange the positions of two equations in the list.
II. Multiply an equation by a nonzero constant.
III. Replace an equation with the sum of that equation and a multiple of any other equation.

ELEMENTARY OPERATIONS

All of these operations are illustrated in the solution of the following system:

Example B

(8)

$$x_1 + 3x_2 - x_3 = 1$$
$$3x_1 + 4x_2 - 4x_3 = 7$$
$$3x_1 + 6x_2 + 2x_3 = -3$$

After copying the first equation, we use operation III to add -3 times the first equation to both the second and the third equations:

(9)

$$x_1 + 3x_2 - x_3 = 1$$
$$- 5x_2 - x_3 = 4$$
$$- 3x_2 + 5x_3 = -6$$

Note that operation III requires copying the equation that is to be multiplied, then replacing that equation to which the multiplied equation has been added. Having eliminated the variable x_1 from all but one equation, we shift our attention to x_2. Because we prefer working with a coefficient of 1, and because we prefer division by -3 as opposed to division by -5, we use operation II (multiplying the third equation by $-\frac{1}{3}$) and then operation I (interchanging the last two equations) to get

$$x_1 + 3x_2 - x_3 = 1$$
$$x_2 - \tfrac{5}{3}x_3 = 2$$
$$- 5x_2 - x_3 = 4$$

Things are now set up to use operation III twice, first adding -3 times the second equation to the first, then adding 5 times the second equation to the third:

$$x_1 \qquad + 4x_3 = -5$$
$$x_2 - \tfrac{5}{3}x_3 = 2$$
$$- \tfrac{28}{3}x_3 = 14$$

Operation II allows us to multiply the last equation by $-\frac{3}{28}$ to obtain $x_3 = -\frac{3}{2}$. Use of operation III two more times, adding $\frac{5}{3}x_3 = -\frac{5}{2}$ to the second equation and $-4x_3 = 6$ to the first equation, brings us to the solution

$$x_1 \qquad\qquad = 1$$
$$x_2 \qquad = -\tfrac{1}{2}$$
$$x_3 = -\tfrac{3}{2} \quad \square$$

All the elementary operations are reversible. Equations whose positions have been interchanged can be interchanged again. An equation multiplied by the nonzero constant k can then be multiplied by $1/k$. If an equation B has been replaced by an equation C = k(equation A) + (equation B), then equation C can be replaced by equation B = $-k$(equation A) + (equation C). In every case, we can use an elementary operation to return to the system from which we started, a fact which motivates the following important definition.

Two systems of m linear equations in n variables are **equivalent systems** if one can be obtained from the other by a sequence of elementary operations.

All the systems obtained in our solution of Example B are equivalent systems.

It is obvious that if we go from one system of equations to an equivalent system using only elementary operations I and II, then the systems will have the same solution sets; that is, if setting $x_1 = r_1, \ldots, x_n = r_n$ satisfies one system, it will satisfy the other.

Now consider the argument used above to show that any solution of the system (1) will be a solution to (6). This same argument can be used with two equations selected from a system of m equations in n variables. Taken together with the fact that operation III is reversible, this shows that if operation III is used to move from one system to another, the systems will have the same solution set. These observations can be summarized as follows:

Two equivalent systems of linear equations have the same solution set.

Since any sequence of elementary operations can be used to move from one system of equations to an equivalent system, something should be said about our decision in Example B to move from left to right, first eliminating all but one of the x_1 terms, then all but one of the x_2 terms, and so on. The left-to-right procedure is methodical and easy to use in programming a computer. It is not always the most efficient procedure. A person solving Example B by hand calculations would be well advised, upon encountering the system

$$
\begin{aligned}
x_1 + 3x_2 - x_3 &= 1 \\
-5x_2 - x_3 &= 4 \\
-3x_2 + 5x_3 &= -6
\end{aligned}
$$

to avoid fractions by next concentrating on the variable x_3. If the second equation is multiplied by -1 and copied into the third position, that equation can then be used to eliminate the x_3 variable from the other two equations.

$$x_1 + 8x_2 \qquad = -3$$
$$- 28x_2 \qquad = 14$$
$$5x_2 + x_3 = -4$$

(Why would a concern for efficiency cause the same person to ignore the -1 coefficient of x_3 in the system (10)?)

Our goal, of course, is to perform operations that move efficiently toward an equivalent system from which any solutions that exist can be easily obtained. One such process makes use of *pivot operations*.

PIVOTING

Any nonzero coefficient a_{rs} of the system

$$a_{11}x_1 + \cdots + a_{1s}x_s + \cdots + a_{1n}x_n = b_1$$
$$\vdots \qquad \qquad \vdots \qquad \qquad \vdots \qquad \vdots$$
$$a_{r1}x_1 + \cdots + a_{rs}x_s + \cdots + a_{rn}x_n = b_r$$
$$\vdots \qquad \qquad \vdots \qquad \qquad \vdots \qquad \vdots$$
$$a_{m1}x_1 + \cdots + a_{ms}x_s + \cdots + a_{mn}x_n = b_m$$

may be selected as the **pivot element**. To **pivot** on a_{rs}, add multiples of the rth equation to the other equations, choosing the multiples so as to eliminate the x_s term from all the other equations.

Example C

Using the 1 coefficient of x_3 in the second equation as a pivot element, find by pivoting a system equivalent to

$$x_1 + 2x_2 + x_3 + \quad x_4 = \quad 2$$
$$3x_1 + 4x_2 + x_3 + 2x_4 = \quad 5$$
$$2x_1 \qquad + x_3 + 3x_4 = \quad 3$$

$$-2x_1 - 2x_2 \qquad - \quad x_4 = -3 \qquad -1(\text{second eq.}) + (\text{first eq.})$$
$$3x_1 + 4x_2 + x_3 + 2x_4 = \quad 5 \qquad \text{second eq. copied}$$
$$-x_1 - 4x_2 \qquad + \quad x_4 = -2 \qquad -1(\text{second eq.}) + (\text{third eq.}) \quad \square$$

If x_s appears in only the rth equation of a system and has a coefficient of 1 in the rth equation, then x_s is said to be a **basic variable** and the rth equation is called a **basic equation** in that system. The second system of equations in Example C exhibits x_3 as a basic variable; the second equation of that system is a basic equation. A pivot using a_{rs} as the pivot element always leads to a system in which x_s appears as a basic variable.

STEPS IN PIVOTING

1. Select a nonzero pivot element a_{rs}.
2. Multiply the rth equation (if $a_{rs} \neq 1$) by $1/a_{rs}$, thus obtaining what is to be a basic equation in the next step.
3. Add multiples of this basic equation to each of the other equations so as to eliminate all other x_s terms.

A system with m equations always leads to a system with m equations. The *only* exception to this occurs when a pivot operation leads to an equation of the form $0 = 0$. This equation, being redundant, can be dropped from the system. This situation is illustrated in our final example.

Use a sequence of two pivots to find a system displaying two basic variables that is equivalent to

Example D

$$\begin{aligned} x_1 + x_2 - 3x_3 &= -5 \\ -5x_1 - 2x_2 + 3x_3 &= 7 \\ 3x_1 + x_2 - x_3 &= -3 \end{aligned}$$

Suppose we begin by pivoting on the 1 coefficient of x_1 in the first equation.

$$\begin{aligned} x_1 + x_2 - 3x_3 &= -5 & &\text{first eq. copied} \\ 3x_2 - 12x_3 &= -18 & &\text{5(first eq.) + (second eq.)} \\ -2x_2 + 8x_3 &= 12 & &-3\text{(first eq.) + (third eq.)} \end{aligned}$$

(11)

We are now free to pick our next pivot element wherever we please, but it won't please us very much if it is chosen in the first equation. For if we now add multiples of the first equation to the others, we will destroy the zero coefficients of x_1 that were just obtained. Let's choose the -2 coefficient of x_2 in the third equation as our next pivot, making $x_2 - 4x_3 = -6$ our next basic equation.

$$\begin{aligned} x_1 + x_3 &= 1 & &-1\text{(basic eq.) + (first eq.)} \\ 0 &= 0 & &-3\text{(basic eq.) + (second eq.)} \\ x_2 - 4x_3 &= -6 & &\text{basic eq.} \end{aligned}$$

In this system, one equation is redundant and may be dropped. Two variables, x_1 and x_2, appear as basic variables. □

The choice of which two variables would be basic was arbitrary in the previous example. If in (11), for instance, we had chosen the 8 in the third equation as our pivot element, then x_1 and x_3 would have been our final basic variables.

Though we did not set out in Example D to find the solution set to the given system, we have practically done so. From the final system that displays x_1 and x_2 as basic variables, it is clear that

$$\begin{aligned} x_1 &= 1 - x_3 \\ x_2 &= -6 + 4x_3 \end{aligned}$$

We may assign values to x_3 arbitrarily; then x_1 and x_2 are determined.

The solution set consists of all ordered triples of the form

$$(x_1, x_2, x_3) = (1 - x_3, -6 + 4x_3, x_3)$$

Two of the infinite number of solutions are $(1, -6, 0)$ and $(-1, 2, 2)$.

We conclude this section by looking at Example D geometrically. The three given equations have graphs that are planes. It might generally be expected that two of these planes would determine a line, and that this line would then pierce the third plane in a unique point corresponding to the solution of the system. That is not what happens here, however, for this is a special case, first noticed when one of the equations became redundant. In this case, all three planes pass through a common line. Two of the infinite number of points on this line are $(1, -6, 0)$ and $(-1, 2, 2)$.

1. Graph on the same set of axes the two lines corresponding to

$$3x_1 + 4x_2 + 1 = 0 \quad \text{and} \quad x_1 - 2x_2 + 2 = 0$$

   Find the intersection.

2. Graph on the same set of axes the two lines corresponding to

$$2x_1 - 3x_2 + 4 = 0 \quad \text{and} \quad 4x_1 + x_2 + 1 = 0$$

   Find the intersection.

3. Graph, for various values of k, members of the family

$$k(3x_1 + 4x_2 + 1) + (x_1 - 2x_2 + 2) = 0$$

   What value of k gives the horizontal member of this family? The vertical member?

4. Graph, for various values of k, members of the family

$$k(2x_1 - 3x_2 + 4) + (4x_1 + x_2 + 1) = 0$$

   What value of k gives the horizontal member of this family? The vertical member?

5. Find a member of the family described in Problem 3 that passes through $(2, -4)$.

6. Find a member of the family described in Problem 4 that passes through $(3, 4)$.

Problems 7 and 8 refer to the system

$$3x_1 - x_2 + x_3 = 5$$
$$x_1 + 2x_2 - 3x_3 = 4$$
$$4x_1 - 3x_2 - 2x_3 = 5$$

7. Use the indicated coefficient as a pivot to obtain a system equivalent to the given system.
   (a) the -1 coefficient of x_2 (b) the 4 coefficient of x_1

BACK TO BASICS

Since World War II, a technique called **linear programming** has developed that has been used to solve a host of problems ranging from wartime activities (the logistics of getting troops and supplies where they are needed when they are needed) to pastoral pursuits (finding the most economical way to provide proper nutrition for livestock). The method has been particularly useful in management science, and a course explaining the technique is often required in schools of business administration.

The method for solving such problems, called the **simplex method,** is introduced in Chapter 7. There it will be seen that the method prescribes an orderly procedure for going from a system of equations displaying certain basic variables to an equivalent system in which other basic variables are displayed. Thus, in Section 7-4, we use the simplex method to move from the system

$$
\begin{aligned}
-2x_1 + x_2 + x_3 \qquad\qquad &= 2 \\
-3x_1 \qquad + 2x_3 + x_4 \qquad &= 7 \\
5x_1 \qquad - 2x_3 \qquad + x_5 &= 7
\end{aligned}
\qquad
\begin{aligned}
&\text{basic variables} \\
&\quad x_2, x_4, x_5
\end{aligned}
\qquad (i)
$$

to the system

$$
\begin{aligned}
x_2 + \tfrac{1}{5}x_3 \qquad + \tfrac{2}{5}x_5 &= \tfrac{24}{5} \\
\tfrac{4}{5}x_3 + x_4 + \tfrac{3}{5}x_5 &= \tfrac{56}{5} \\
x_1 \qquad - \tfrac{2}{5}x_3 \qquad + \tfrac{1}{5}x_5 &= \tfrac{7}{5}
\end{aligned}
\qquad
\begin{aligned}
&\text{basic variables} \\
&\quad x_1, x_2, x_4
\end{aligned}
\qquad (ii)
$$

Can you see how, beginning with (i), to obtain (ii) by using elementary operations? Find a few solutions to one system by assigning arbitrary values to the nonbasic variables; verify that these solutions also satisfy the other system.

8. Use the indicated coefficient as a pivot to obtain a system equivalent to the given system.
   (a) the 1 coefficient of x_3 (b) the -3 coefficient of x_3

Problems 9–12 refer to the system

$$
\begin{aligned}
2x_1 + 3x_2 - 5x_3 + x_4 &= 8 \\
4x_1 \qquad - 2x_3 \qquad &= 10 \\
3x_1 + 5x_2 - 3x_3 \qquad &= 4 \\
5x_1 - 2x_2 + 7x_3 \qquad &= 5
\end{aligned}
$$

(*Problem set continues on p. 12*)

9. Use the indicated coefficient as a pivot to obtain a system equivalent to the given system.
   (a) the 4 coefficient in row 2, column 1
   (b) the 2 coefficient in row 1, column 1
   (c) the 5 coefficient in row 3, column 2

10. Use the indicated coefficient as a pivot to obtain a system equivalent to the given system.
    (a) the -2 coefficient in row 2, column 3
    (b) the -5 coefficient in row 1, column 3
    (c) the -2 coefficient in row 4, column 2

11. Find a system equivalent to the given system that displays x_3 and x_4 as basic variables.

12. Find a system equivalent to the given system that displays x_2 and x_4 as basic variables.

13. Find all solutions to the system

$$x_1 - 2x_2 + 2x_3 - 4 = 0$$
$$2x_1 + 3x_2 - x_3 + 3 = 0$$

Express your answer three ways, using in turn each of x_1, x_2, and x_3 as the nonbasic variable.

14. Find all solutions to the system

$$2x_1 - 3x_2 + x_3 - 5 = 0$$
$$x_1 + 2x_2 + 2x_3 - 2 = 0$$

Express your answer in three ways.

15. Use back substitution to find the unique solution to the system

$$3x_1 + x_2 - 2x_3 = 3$$
$$5x_2 - 4x_3 = -3$$
$$2x_3 = -1$$

16. Use back substitution to find the unique solution to the system

$$x_1 - 4x_2 - 2x_3 = 6$$
$$-6x_2 + 5x_3 = -2$$
$$x_3 = -1$$

17. The equation $k(x_1 + 2x_2 + x_3) + (3x_1 - x_2 + x_3) = 1$ describes a family of planes. What choice of k gives a member of the family
   (a) perpendicular to the $x_1 x_3$ plane
   (b) passing through $(3, -1, -4)$
   (c) perpendicular to the plane $x_1 + x_2 - x_3 = 5$

18. What condition on a, b, c, and d will guarantee that there will be exactly one solution to the system

$$ax_1 + bx_2 = 1$$
$$cx_1 + dx_2 = 0$$

19. Beginning with the system (i) in the vignette *Back to Basics*, find an equivalent system in which x_2, x_3, and x_5 are basic variables.   **V**

FINDING THE SOLUTION SET   1-3

Two methods for finding the solution set of a system of equations will be presented in this section. The first method, **complete elimination**, is easy to describe and of great theoretical importance. When the first method is understood, the second method, **elimination with back substitution**, is also easy to understand. The second method requires fewer steps than the first, making it especially important to those who program computers to solve large systems.

COMPLETE ELIMINATION

A list of the steps to be taken will be better understood after we have seen an example.

Find the solution set of   *Example A*

$$x_1 + 2x_2 + x_3 + x_4 = 2$$
$$3x_1 + 4x_2 + x_3 + 2x_4 = 5$$
$$2x_1 \qquad + x_3 + 3x_4 = 3$$

It would be natural here to begin by using the 1 coefficient of x_1 in the first equation to eliminate the x_1 terms in the next two equations. This is the way a computer programmed to work from left to right would proceed. In the absence of a compelling reason to act otherwise, it is the way your own hand calculation should proceed, and it is the way we proceed in Example B where this same system is solved again. But to illustrate that there is freedom to begin anywhere, we shall use the fact that beginning with this system in Example C of Section 1-2, a pivot on the 1 coefficient of x_3 in the second equation was shown to lead to the equivalent system

$$-2x_1 - 2x_2 \qquad - x_4 = -3$$
$$3x_1 + 4x_2 + x_3 + 2x_4 = 5$$
$$-x_1 - 4x_2 \qquad + x_4 = -2$$

(*1*)

This system has one basic variable, x_3. We now wish to pivot to another equivalent system in which there are two basic variables, x_3 and one now to be introduced. Note that in selecting our next pivot coefficient, we must avoid the second equation (our basic equation) since its use would destroy the 0 coefficients of x_3 already obtained in the other equations. We will, to continue this illustration, choose the -1 coefficient of x_4 in the first equation as our next pivot.

$$2x_1 + 2x_2 \qquad + x_4 = \quad 3$$
$$-x_1 \qquad + x_3 \qquad = -1$$
$$-3x_1 - 6x_2 \qquad\qquad = -5$$

Our next pivot element must not be selected from the first or second equations (why?), both of which are now basic equations. We can use the third equation to introduce either x_1 or x_2 as the third basic variable. We choose x_1 because we prefer division by -3 to division by -6.

(1)

$$- 2x_2 \qquad + x_4 = -\tfrac{1}{3}$$
$$2x_2 + x_3 \qquad = \quad \tfrac{2}{3}$$
$$x_1 + 2x_2 \qquad\qquad = \quad \tfrac{5}{3}$$

We now have as many basic variables as we have equations, meaning we have carried this as far as possible. Isolating the basic variable in each equation and interchanging the first and third equations in this system, we get

$$x_1 = \quad \tfrac{5}{3} - 2x_2$$
$$x_3 = \quad \tfrac{2}{3} - 2x_2$$
$$x_4 = -\tfrac{1}{3} + 2x_2$$

Values can now be assigned to x_2 arbitrarily, after which the basic variables are determined. Thus, to each value of x_2 there corresponds a unique 4-tuple in the solution set. The correspondence can be emphasized by writing

$$(x_1, x_2, x_3, x_4) = (\tfrac{5}{3} - 2x_2, x_2, \tfrac{2}{3} - 2x_2, -\tfrac{1}{3} + 2x_2)$$

If, for example, we take $x_2 = 0$ and $x_2 = \tfrac{1}{3}$, we find that $(\tfrac{5}{3}, 0, \tfrac{2}{3}, -\tfrac{1}{3})$ and $(1, \tfrac{1}{3}, 0, \tfrac{1}{3})$ are in the infinite set of solutions. □

At each step in Example A, we used a pivot to obtain an equivalent system that had one more basic variable than was present in the previous system. The first step and each succeeding step of the process followed the same pattern.

ELIMINATION

1. Choose a pivot element in a nonbasic equation.
2. Pivot, thus introducing one more basic variable into the system.
3. If there is still a nonbasic equation in the system, return to step 1.

Sometimes the degree of freedom that is allowed in choosing the pivot element can be very useful. It allows us to take advantage of special circumstances as we see them develop in calculations done by hand, and other applications arise in Chapter 7.

There are times, as when one wants to program a computer to solve a system of equations, that we wish to remove all freedom of choice. In such situations, step 1 is modified as follows:

1-A. List all basic equations first in such a way that if the basic variable in equation s is to the right of the basic variable in equation r, then equation s appears below equation r.

1-B. Among the nonbasic equations, identify the leftmost column of coefficients that does not consist entirely of zeros. From this column, choose as the pivot element the uppermost nonzero coefficient.

ELIMINATION WITHOUT FREEDOM

With the pivot element chosen by these more specific rules, continue with steps 2 and 3. We illustrate these modified rules by solving once again the system of Example A.

Find the solution set of

Example B

$$x_1 + 2x_2 + x_3 + x_4 = 2$$
$$3x_1 + 4x_2 + x_3 + 2x_4 = 5$$
$$2x_1 \qquad + x_3 + 3x_4 = 3$$

There being no basic equations yet, we are directed by step 1-B to choose as pivot element the coefficient of x_1 in the first equation (which in an act of foresighted kindness to ourselves happens to be 1).

$$x_1 + 2x_2 + x_3 + x_4 = 2$$
$$- 2x_2 - 2x_3 - x_4 = -1$$
$$- 4x_2 - x_3 + x_4 = -1$$

The next pivot element is the -2 coefficient of x_2 in the second equation.

$$x_1 \qquad - x_3 \qquad = 1$$
$$x_2 + x_3 + \tfrac{1}{2}x_4 = \tfrac{1}{2}$$
$$3x_3 + 3x_4 = 1$$

(2)

The final pivot element is the 3 coefficient of x_3 in the third equation.

$$x_1 \qquad\qquad + x_4 = \tfrac{4}{3}$$
$$x_2 \qquad - \tfrac{1}{2}x_4 = \tfrac{1}{6}$$
$$x_3 + x_4 = \tfrac{1}{3}$$

From these equations we see that

$$x_1 = \tfrac{4}{3} - x_4$$
$$x_2 = \tfrac{1}{6} + \tfrac{1}{2}x_4$$
$$x_3 = \tfrac{1}{3} - x_4$$

so that there is a correspondence between values arbitrarily assigned to x_4 and to 4-tuples in the solution set. This correspondence is emphasized by writing

$$(x_1, x_2, x_3, x_4) = (\tfrac{4}{3} - x_4, \tfrac{1}{6} + \tfrac{1}{2}x_4, \tfrac{1}{3} - x_4, x_4) \quad \square$$

It is important to note that both systems (1) and (2) are equivalent to the same original set of equations, hence to each other. The difference is that the basic variables in (1) and (2) are different; the common property is that both describe the same solution set. Note, for example, that the choices of $x_4 = -\tfrac{1}{3}$ and $x_4 = \tfrac{1}{3}$ in our solution to Example B give the same 4-tuples we found at the conclusion of Example A by setting $x_2 = 0$ and $x_2 = \tfrac{1}{3}$.

These two variations of elimination are often attributed to the mathematicians Karl Friedrich Gauss (1777–1855) and Camille Jordan (1838–1922), but there is no consensus in assigning particular names to particular variations of the methods.

ELIMINATION WITH BACK SUBSTITUTION

The basic step in this method is something less than a full pivot, but we begin our work with steps 1-A and 1-B as if we were going to pivot. The difference is this: after selecting the would-be pivot element a_{rs}, we use it to eliminate only those nonzero coefficients *below* a_{rs}. The contrast will be pointed out as we solve again the system already solved twice in this section.

Example C

Find the solution set of

$$
\begin{aligned}
x_1 + 2x_2 + x_3 + x_4 &= 2 \\
3x_1 + 4x_2 + x_3 + 2x_4 &= 5 \\
2x_1 \phantom{{}+ 4x_2} + x_3 + 3x_4 &= 3
\end{aligned}
$$

The first step is as it was before. We use the 1 coefficient of x_1 in the first equation to eliminate the nonzero coefficients *below* that 1 (which in this case is all of them).

$$
\begin{aligned}
x_1 + 2x_2 + x_3 + x_4 &= 2 \\
-2x_2 - 2x_3 - x_4 &= -1 \\
-4x_2 - x_3 + x_4 &= -1
\end{aligned}
$$

We next select the -2 coefficient of x_2 in the second equation, but this time we have only one coefficient *below* it to be eliminated.

$$
\begin{aligned}
x_1 + 2x_2 + x_3 + \phantom{\tfrac{1}{2}}x_4 &= 2 \\
x_2 + x_3 + \tfrac{1}{2}x_4 &= \tfrac{1}{2} \\
3x_3 + 3x_4 &= 1
\end{aligned}
$$

Selection of the 3 coefficient of x_3 in the third equation leads to nothing more than division by 3 since there are no coefficients below it to eliminate.

$$x_1 + 2x_2 + x_3 + x_4 = 2$$
$$x_2 + x_3 + \tfrac{1}{2}x_4 = \tfrac{1}{2}$$
$$x_3 + x_4 = \tfrac{1}{3}$$

Now comes the **back substitution.** Working from the bottom up, we first write

$$x_3 = \tfrac{1}{3} - x_4$$
$$x_2 = \tfrac{1}{2} - \tfrac{1}{2}x_4 - x_3$$
$$x_1 = 2 - x_4 - x_3 - 2x_2$$

Finally, use the top equation to substitute for x_3 in the second equation, then the top two expressions for x_3 and x_2 in the last equation.

$$x_3 = \tfrac{1}{3} - x_4$$
$$x_2 = \tfrac{1}{2} - \tfrac{1}{2}x_4 - (\tfrac{1}{3} - x_4) = \tfrac{1}{6} + \tfrac{1}{2}x_4$$
$$x_1 = 2 - x_4 - (\tfrac{1}{3} - x_4) - 2(\tfrac{1}{6} + \tfrac{1}{2}x_4) = \tfrac{4}{3} - x_4$$

This leads to the same description of the solution set as was obtained in (2) of Example B.   □

The number of steps saved by using back substitution instead of elimination is not significant for the small problems we are likely to tackle by hand, but the saving becomes important when a computer is programmed for large systems. The name of **Gaussian elimination** is sometimes associated with this procedure.

A system of linear equations is called **underdetermined** if we are free to assign values arbitrarily to one or more of the variables before the values of the others are determined. The system used repeatedly in the examples of this section is an underdetermined system. The last two examples of this section illustrate the back substitution method and emphasize that there are systems of equations that are not underdetermined.

Find the solution set of   *Example D*

$$x_1 - 5x_2 + 4x_3 = -3$$
$$3x_1 + 3x_2 - 2x_3 = 2$$
$$2x_1 - x_2 + x_3 = 1$$

We begin with the lead coefficient of the first equation.

$$x_1 - 5x_2 + 4x_3 = -3$$
$$18x_2 - 14x_3 = 11$$
$$9x_2 - 7x_3 = 7$$

Next, the 18 coefficient of x_2 in the second equation is used to eliminate the 1 coefficient below it.

$$x_1 - 5x_2 + 4x_3 = -3$$
$$x_2 - \tfrac{14}{18}x_3 = \tfrac{11}{18}$$
$$0x_3 = \tfrac{3}{2}$$

It is clear at this point that we can stop, since no choice of x_2 and x_3 can satisfy the third equation. The solution set is empty. □

The system of Example D is said to be **overdetermined** or **inconsistent**. What is left but to be **completely determined?**

Example E | Find all solutions of

$$x_1 + 10x_2 + 4x_3 = 36$$
$$x_1 - 2x_2 \qquad = 0$$
$$-x_1 + 5x_2 + 6x_3 = 24$$

Proceeding in the forward direction leads us in turn to the systems

$$x_1 + 10x_2 + 4x_3 = 36$$
$$-12x_2 - 4x_3 = -36$$
$$15x_2 + 10x_3 = 60$$

$$x_1 + 10x_2 + 4x_3 = 36$$
$$x_2 + \tfrac{1}{3}x_3 = 3$$
$$5x_3 = 15$$

The process of back substitution now gives

$$x_3 = 3$$
$$x_2 = 3 - \tfrac{1}{3}x_3 \qquad = 3 - 1 = 2$$
$$x_1 = 36 - 4x_3 - 10x_2 = 36 - 12 - 20 = 4$$

The unique solution to this system is $(4, 2, 3)$. □

PROBLEM SET 1-3 | *In Problems 1–4, use back substitution to find all solutions to the given systems.*

1. $x_1 + 6x_2 + 4x_3 = -1$
$x_2 + 2x_3 = -\tfrac{10}{3}$
$x_3 = -2$

2. $x_1 + x_2 + x_3 = 2$
$x_2 - 2x_3 = 5$
$x_3 = -2$

3. $x_1 + 2x_2 - x_3 + x_4 = 6$
$x_2 + 3x_3 + 2x_4 = 2$
$x_3 - x_4 = -3$

4. $x_1 - 3x_2 + 2x_3 - x_4 = 0$
$x_2 + 3x_3 + x_4 = -2$
$x_3 + 2x_4 = 1$

(Problem set continues on p. 20)

ROUGH APPROXIMATIONS CAN BECOME LUMPY

Suppose we set out to solve the system

$$0.4000x_1 + 561.6x_2 = 562.0$$
$$73.03x_1 - 43.03x_2 = 30.00$$

with the aid of a calculator that rounds everything off to four significant figures. Multiplication of the first equation by $1/(0.4000)$ proceeds smoothly enough, giving

$$x_1 + 1404x_2 = 1405 \qquad\qquad (i)$$

Elimination of the x_1 in the second equation now calls for multiplying (i) by -73.03 and adding it to the second equation. In doing so, however, remember that our calculator will show the product of $(-73.03)(1404) = -102,534.12$ as $-102,500$, rounded off to four significant figures. Added to -43.03 and rounded off again, we still get $-102,500$. Similar rounding off with $(-73.03)(1405)$ followed by the addition of 30 and still more rounding off gives us the new system

$$x_1 + \qquad 1404x_2 = \qquad 1405$$
$$- 102,500x_2 = -102,600$$

By back substitution, we get

$$x_2 = \qquad 1.001$$

and then, since $(1.001)(1404) = 1405.404$ rounds off to 1405,

$$x_1 = 1405 - 1405 = 0$$

Because of the rounding off in our limited calculator, we would expect some error in our answer, but our answer of $x_1 = 1.001$, $x_2 = 0$ seems to go quite beyond our expectations when compared with the correct answer of $x_1 = x_2 = 1$ (check it).

Admittedly, the calculator we used for this computation is quite limited, but similar examples can be contrived for a calculator or computer capable of handling many more significant digits. You are right, of course, if you point to the word *contrived* and argue against the likelihood of such things happening in a practical problem involving two variables. But we are right in saying that problems for which computers get used involve many more than two variables, often hundreds or even thousands of variables.

Such problems do not have to be contrived in order to have the round-off errors inherent in any computer build up into significant problems.

Difficulties of this sort are treated in numerical analysis courses. You might be interested to know about one elementary technique, called **partial pivoting**, that is often used to contend with the difficulty just described. The idea is to avoid divisions by numbers that have a small absolute value by choosing a pivot of large absolute value. Thus, in the previous example, we begin by interchanging the two equations so as to use 73.03 as a pivot.

$$73.03x_1 - 43.03x_2 = 30.00$$
$$0.4000x_1 + 561.6x_2 = 562.0$$

Division of the first equation by 73.03 gives, using the same practice of rounding 43.03/73.03 to 0.5892 and 30/73.03 to 0.4108,

$$x_1 - 0.5892x_2 = 0.4108 \qquad (ii)$$

Now elimination of the x_1 in the second equation, which calls for multiplying (ii) by -0.4000 and adding it to the second equation, leads after the usual rounding off to

$$561.8x_2 = 561.8$$

Back substitution gives $x_2 = 1.000$ and $x_1 = 1.000$, a much better result.

Find all solutions to the following systems of equations.

5. $\begin{aligned} 2x_1 - x_2 + x_3 &= 1 \\ 3x_1 + 3x_2 - 2x_3 &= 5 \\ x_1 - 5x_2 + 4x_3 &= -3 \end{aligned}$

6. $\begin{aligned} 3x_1 + 2x_2 - x_3 &= 6 \\ 2x_1 - x_2 + x_3 &= 1 \\ -x_1 - 10x_2 + 7x_3 &= -14 \end{aligned}$

7. $\begin{aligned} x_1 - x_2 + 2x_3 &= 1 \\ 3x_1 + 2x_2 - x_3 &= 2 \\ -x_1 - 4x_2 + 5x_3 &= 4 \end{aligned}$

8. $\begin{aligned} 2x_1 + x_2 - 3x_3 &= 2 \\ x_1 - 2x_2 + 4x_3 &= 4 \\ 3x_1 - x_2 + 2x_3 &= 7 \end{aligned}$

9. $\begin{aligned} x_1 - x_2 + 2x_3 &= 6 \\ -2x_1 + 3x_2 - x_3 &= -7 \\ 3x_1 + 2x_2 + 2x_3 &= 5 \end{aligned}$

10. $\begin{aligned} 3x_1 + 2x_2 + x_3 &= 4 \\ 2x_1 + 4x_2 - 2x_3 &= 6 \\ 5x_1 - 2x_2 + 7x_3 &= 0 \end{aligned}$

11. $\begin{aligned} 3x_1 - x_2 + x_3 &= 1 \\ 2x_1 + 3x_2 - 2x_3 &= -2 \\ 4x_1 - 5x_2 + 4x_3 &= 4 \end{aligned}$

12. $\begin{aligned} x_1 + x_2 - x_3 &= 4 \\ 2x_1 - 3x_2 + 2x_3 &= 3 \\ -x_1 + 9x_2 - 7x_3 &= 2 \end{aligned}$

13. The general equation of a circle is $Ax^2 + Ay^2 + Dx + Ey + F = 0$. If a circle passes through $(1, 1)$, $(2, -1)$, and $(2, 3)$, then

$$2A + D + E + F = 0$$
$$5A + 2D - E + F = 0$$
$$13A + 2D + 3E + F = 0$$

Find A, D, E, and F.

14. Follow the procedure outlined in Problem 13 to find the equation of the circle passing through $(2, 3)$, $(-1, 1)$, and $(-2, -1)$.

15. It is desirable in calculus to be able to choose A, B, C, and D so that for all x,

$$\frac{4x^3 + x^2 - 4x + 1}{(x^2 + 1)(x^2 - 1)} = \frac{Ax + B}{x^2 + 1} + \frac{C}{x - 1} + \frac{D}{x + 1}$$

Do so.

16. Choose A, B, C, and D so that for all x,

$$\frac{9x^2 - 16x + 12}{3x^2(x^2 - 5x + 6)} = \frac{A}{x} + \frac{B}{x^2} + \frac{C}{x - 2} + \frac{D}{x - 3}$$

17. Choose A, B, C, and D so that for all x,

$$A \cos 4x + B \sin 2x + C(\sin x + \cos x)^4 + D = 0$$

18. Choose A, B, C, and D so that for all x,

$$A \sin 3x + B \sin 2x + C \sin x + D \cos \frac{3}{2}x \cos x \sin \frac{x}{2} = 0$$

19. Find all solutions to the system

$$\frac{1}{x_1} - \frac{1}{x_2} + \frac{1}{x_3} = 4$$

$$\frac{2}{x_1} - \frac{1}{x_2} - \frac{3}{x_3} = 2$$

$$\frac{1}{x_1} + \frac{2}{x_2} + \frac{2}{x_3} = 2$$

20. Find all solutions to the system

$$\frac{1}{x_1} + \frac{2}{x_2} + \frac{3}{x_3} = 3$$

$$\frac{2}{x_1} + \frac{3}{x_2} + \frac{1}{x_3} = 2$$

$$\frac{3}{x_1} + \frac{1}{x_2} + \frac{2}{x_3} = 7$$

ROW ECHELON FORM 1-4

Good mathematical notation should, among other things, focus our attention on what is important, and it should be easy to read. On these two counts, the subscript-studded notation of Sections 1-2 and 1-3 fails. The coefficients and constants do not stand out with eye-catching bold-ness, and the 1 and 0 coefficients that are so important to computa-tional procedures and goals are often omitted altogether. Subscripts, especially as they get written on a chalkboard, are often difficult to read; and writing the wrong subscripts is an error not confined to teachers.

One visual aid, used without comment in the last two sections, is to write the same variables of a system of equations in a vertical column. If this convention is faithfully observed, it is but a small step to omit the variables altogether. Thus, the system written in Example A of the previous section in the form

(1)

$$
\begin{aligned}
x_1 + 2x_2 + x_3 + \ x_4 &= 2 \\
3x_1 + 4x_2 + x_3 + 2x_4 &= 5 \\
2x_1 \qquad\quad + x_3 + 3x_4 &= 3
\end{aligned}
$$

can be abbreviated to

(2)

$$
\begin{bmatrix}
1 & 2 & 1 & 1 & 2 \\
3 & 4 & 1 & 2 & 5 \\
2 & 0 & 1 & 3 & 3
\end{bmatrix}
$$

MATRICES

The dictionary defines a matrix to be that which encloses or gives form. In mathematics, a **matrix** (plural, **matrices**) is a rectangular array of numbers formed into *rows* and *columns*. The brackets serve to give form. A matrix having m rows and n columns is said to be an $m \times n$ (read "m by n") matrix; matrix (2) is 3×5.

When a matrix is used to represent a system of equations, several conventions are common. The coefficients of 0 and 1, usually not written in elementary algebra (one writes $x_1 - 2x_3$, not $1x_1 + 0x_2 - 2x_3$), are written in the matrix. It is understood that the coefficients of x_j are found in the jth column, and that the constants are placed in the last column.

The matrix (2) is called the **augmented matrix** of the linear system (1). This sets it apart from the matrix

$$
\begin{bmatrix}
1 & 2 & 1 & 1 \\
3 & 4 & 1 & 2 \\
2 & 0 & 1 & 3
\end{bmatrix}
$$

which is called the **coefficient matrix** for the linear system (1). Some writers insert a dotted vertical line in the augmented matrix to separate the coefficients from the column of constants, but we shall not use this convention. (It guards against a confusion that rarely happens.)

Our procedure for solving a system of equations is not altered in any way if we work with the augmented matrices of the equivalent systems. We will find it more natural, however, to describe our actions in terms of operations on rows instead of equations. Compare the following solving of the system (1) with the same work done in Example A of the previous section. This time we begin with the augmented matrix.

$$\begin{bmatrix} 1 & 2 & 1 & 1 & 2 \\ 3 & 4 & 1 & 2 & 5 \\ 2 & 0 & 1 & 3 & 3 \end{bmatrix}$$

$$\begin{bmatrix} -2 & -2 & 0 & -1 & -3 \\ 3 & 4 & 1 & 2 & 5 \\ -1 & -4 & 0 & 1 & -2 \end{bmatrix} \quad \begin{matrix} -1(\text{second row}) + (\text{first row}) \\ \text{second row copied} \\ -1(\text{second row}) + (\text{third row}) \end{matrix}$$

$$\begin{bmatrix} 2 & 2 & 0 & 1 & 3 \\ -1 & 0 & 1 & 0 & -1 \\ -3 & -6 & 0 & 0 & -5 \end{bmatrix} \quad \begin{matrix} -(\text{first row}) \\ 2(\text{first row}) + (\text{second row}) \\ (\text{first row}) + (\text{third row}) \end{matrix}$$

$$\begin{bmatrix} 0 & -2 & 0 & 1 & -\frac{1}{3} \\ 0 & 2 & 1 & 0 & \frac{2}{3} \\ 1 & 2 & 0 & 0 & \frac{5}{3} \end{bmatrix} \quad \begin{matrix} \frac{2}{3}(\text{third row}) + (\text{first row}) \\ -\frac{1}{3}(\text{third row}) + (\text{second row}) \\ -\frac{1}{3}(\text{third row}) \end{matrix}$$

The system of equations corresponding to this matrix has no nonbasic equation, so we have reached the end of the line. You may find it helpful, at least in the beginning, to write out the corresponding system of equations:

$$\begin{aligned} -2x_2 \qquad + x_4 &= -\tfrac{1}{3} \\ 2x_2 + x_3 \qquad &= \tfrac{2}{3} \\ x_1 + 2x_2 \qquad &= \tfrac{5}{3} \end{aligned}$$

As in Example A of Section 1-3, we have a solution which can be summarized by writing

$$(x_1, x_2, x_3, x_4) = (\tfrac{5}{3} - 2x_2, x_2, \tfrac{2}{3} - 2x_2, -\tfrac{1}{3} + 2x_2)$$

ELEMENTARY ROW OPERATIONS

Later in the text, we will find it convenient to work with the columns of a matrix. Since we are now confining our interest to rows, we refer to what we are doing as **row operations.** The elementary row operations on a matrix are precisely those that were identified in Section 1-2 as the elementary operations used in solving systems of equations.

I. Interchange the positions of two rows in the matrix.
II. Multiply a row by a nonzero constant.
III. Replace a row with the sum of that row and a multiple of any other row.

ELEMENTARY ROW OPERATIONS

If we begin with a matrix A and perform a sequence of elementary row operations to obtain B, then it is possible to begin with B and perform a sequence of elementary row operations to obtain A. In such a case, we say that the matrices A and B are **row equivalent;** we write $A \sim B$.

The illustration at the beginning of this section, following the steps of Example A of Section 1-3 with matrix notation, showed that

$$A = \begin{bmatrix} 1 & 2 & 1 & 1 & 2 \\ 3 & 4 & 1 & 2 & 5 \\ 2 & 0 & 1 & 2 & 3 \end{bmatrix} \sim B = \begin{bmatrix} 0 & -2 & 0 & 1 & -\frac{1}{3} \\ 0 & 2 & 1 & 0 & \frac{2}{3} \\ 1 & 2 & 0 & 0 & \frac{5}{3} \end{bmatrix}$$

Since the steps of Example 1-3A were chosen to emphasize that we have freedom in choosing the order in which we proceed, we followed it with Example 1-3B where the same system of equations was solved using the more methodical left-to-right procedure that one would expect to use in a computer. The steps of Example 1-3B applied to matrix A above show that it is equivalent to another matrix C.

$$A = \begin{bmatrix} 1 & 2 & 1 & 1 & 2 \\ 3 & 4 & 1 & 2 & 5 \\ 2 & 0 & 1 & 2 & 3 \end{bmatrix} \sim C = \begin{bmatrix} 1 & 0 & 0 & 1 & \frac{4}{3} \\ 0 & 1 & 0 & -\frac{1}{2} & \frac{1}{6} \\ 0 & 0 & 1 & 1 & \frac{1}{3} \end{bmatrix}$$

The matrices B and C are both row equivalent to the matrix A, but C certainly has a sense of orderliness missing in B. Can we describe this orderliness more specifically?

ROW ECHELON FORM

The dictionary says an **echelon** is a formation of troops, ships, and so on, in which groups are set into parallel lines, each to the right (or each to the left) of the one in front of it, so that the whole presents the appearance of steps. A matrix arranged like the matrix C is said to be an **echelon matrix;** it is characterized as follows:

ROW ECHELON FORM

1. All zero rows (rows consisting entirely of 0 entries) are grouped together at the bottom of the matrix.
2. The first element in any nonzero row is a 1, called a *leading* 1.
3. Each column that contains a leading 1 has all other entries equal to 0.
4. If the leading 1 in row s is in a column to the right of the leading 1 in row r, then row s must appear below row r.

Some examples of matrices in row echelon form:

$$\begin{bmatrix} 1 & 2 & 0 & 0 \\ 0 & 0 & 1 & 0 \\ 0 & 0 & 0 & 1 \end{bmatrix} \quad \begin{bmatrix} 0 & 0 & 1 & 0 \\ 0 & 0 & 0 & 1 \\ 0 & 0 & 0 & 0 \end{bmatrix} \quad \begin{bmatrix} 1 & 0 & 2 & 0 \\ 0 & 1 & -1 & 0 \\ 0 & 0 & 0 & 1 \end{bmatrix}$$

The following matrices are *not* in row echelon form:

$$\begin{bmatrix} 1 & 2 & 0 & 0 \\ 0 & 0 & 0 & 1 \\ 0 & 0 & 1 & 0 \end{bmatrix} \quad \begin{bmatrix} 0 & 0 & 1 & 0 \\ 0 & 0 & 0 & 2 \\ 0 & 0 & -1 & 0 \end{bmatrix} \quad \begin{bmatrix} 1 & 0 & 2 & 0 \\ 0 & 1 & -1 & 0 \\ 0 & 0 & 2 & 1 \end{bmatrix}$$

Given a matrix A, we can always find a row equivalent matrix that is in row echelon form. Simply use the elementary row operations specified by elimination without freedom. We then say that A has been **reduced to row echelon form.**

Reduce to row echelon form the matrix

Example A

$$\begin{bmatrix} \boxed{1} & 2 & 0 & -1 & -1 \\ -1 & -3 & 1 & 2 & 3 \\ 1 & -1 & 3 & 1 & 1 \\ 2 & -3 & 7 & 3 & 4 \end{bmatrix}$$

There is little to comment upon. Every step is specified. In each matrix we have circled the pivot element that is used in moving to the succeeding matrix.

$$\begin{bmatrix} 1 & 2 & 0 & -1 & -1 \\ 0 & \boxed{-1} & 1 & 1 & 2 \\ 0 & -3 & 3 & 2 & 2 \\ 0 & -7 & 7 & 5 & 6 \end{bmatrix}$$

$$\begin{bmatrix} 1 & 0 & 2 & 1 & 3 \\ 0 & 1 & -1 & -1 & -2 \\ 0 & 0 & 0 & \boxed{-1} & -4 \\ 0 & 0 & 0 & -2 & -8 \end{bmatrix}$$

$$\begin{bmatrix} 1 & 0 & 2 & 0 & -1 \\ 0 & 1 & -1 & 0 & 2 \\ 0 & 0 & 0 & 1 & 4 \\ 0 & 0 & 0 & 0 & 0 \end{bmatrix} \quad \square$$

When a given matrix is being reduced to row echelon form using hand calculations, it may still be true that computation can be reduced at certain stages by exercising discretion in the choice of a pivot. That is to say, the process of elimination without freedom is not the only way to reduce a matrix to row echelon form. This possibility of using different methods goes to the heart of a principal property about row echelon form. It is essential to future developments to know that *no matter how we go about reducing a given matrix to row echelon form, we will always wind up with the same matrix.* We rely a great deal on the uniqueness of this matrix, making it the subject therefore of our first theorem concerning matrices.

THEOREM A

Every matrix is row equivalent to a unique row echelon matrix.

It is clear that every matrix is equivalent to a row echelon matrix, and the fact that one matrix cannot be equivalent to two different row echelon matrices is commonly taken for granted in elementary texts. We wish to avoid the tedious digression necessary to prove the uniqueness of the form, but we want to emphasize that it needs to be proved and that the proof is surprisingly involved to write out in detail. (See Thrall, R. M., and Tornheim, L., 1967. *Vector spaces and matrices.* New York: Wiley.)

One way to solve a given system of linear equations is to reduce its augmented matrix to row echelon form. Conversely, reduction of a given matrix to row echelon form is always equivalent to solving a system of linear equations. The work done in Example A is, for example, equivalent to finding solutions to the system

$$\begin{aligned}
x_1 + 2x_2 + \quad\;\; - \;x_4 &= -1 \\
-x_1 - 3x_2 + \;x_3 + 2x_4 &= \quad 3 \\
x_1 - \;x_2 + 3x_3 + \;x_4 &= \quad 1 \\
2x_1 - 3x_2 + 7x_3 + 3x_4 &= \quad 4
\end{aligned}$$

The final matrix of Example A is the augmented matrix for a system that can be written

$$\begin{aligned}
x_1 &= -1 - 2x_3 \\
x_2 &= \quad 2 + \;x_3 \\
x_4 &= \quad 4
\end{aligned}$$

Written as a 4-tuple, we have the solutions

$$(-1 - 2x_3, 2 + x_3, x_3, 4)$$

HOMOGENEOUS SYSTEMS

A **homogeneous system** of equations is a system in which all of the constant terms are 0.

$$\begin{aligned}
a_{11}x_1 + \cdots + a_{1n}x_n &= 0 \\
\vdots \qquad\qquad \vdots \;\; &\;\; \vdots \\
a_{m1}x_1 + \cdots + a_{mn}x_n &= 0
\end{aligned}$$

It is obvious that such a system always has a solution, the so-called **trivial solution,** $x_1 = x_2 = \cdots = x_n = 0$. The interesting question to ask about a homogeneous system, therefore, is this: Does it have any nontrivial solutions?

We will find out that the answer is yes, that there are nontrivial solutions whenever $m < n$, that is, whenever the number of equations is less than the number of variables. Let us consider such an example.

Find the solution set of

Example B

$$2x_1 - 5x_2 - x_3 + 4x_4 = 0$$
$$3x_1 - 5x_2 - 9x_3 + 11x_4 = 0$$
$$3x_1 - 7x_2 - 3x_3 + 7x_4 = 0$$

The augmented matrix for the system

$$\begin{bmatrix} 2 & -5 & -1 & 4 & 0 \\ 3 & -5 & -9 & 11 & 0 \\ 3 & -7 & -3 & 7 & 0 \end{bmatrix}$$

reduces to the row echelon form

$$\begin{bmatrix} 1 & 0 & -8 & 7 & 0 \\ 0 & 1 & -3 & 2 & 0 \\ 0 & 0 & 0 & 0 & 0 \end{bmatrix}$$

The corresponding system of equations, equivalent to the given system is, of course

$$x_1 = 8x_3 - 7x_4$$
$$x_2 = 3x_3 - 2x_4$$

The solution set consists of 4-tuples of the form

$$(8x_3 - 7x_4, 3x_3 - 2x_4, x_3, x_4)$$

Nontrivial solutions are easily obtained by assigning arbitrary nonzero values to the nonbasic variables. ☐

The example illustrates the general situation. If a homogeneous system has m equations, reduction of the corresponding augmented matrix to row echelon form surely results in a matrix with m or fewer nonzero rows. Then the corresponding system has m or fewer basic variables. Since we began with the assumption that the number n of variables exceeded m, the solution set must be expressed in terms of one or more nonbasic variables that can take on arbitrary nonzero values. This proves the following important result:

> *A system of linear homogeneous equations in which the number of variables exceeds the number of equations always has an infinite number of nontrivial solutions.*

THEOREM B

1. Are the following matrices in row echelon form?

(a) $\begin{bmatrix} 1 & 3 & 0 & 0 & 3 \\ 0 & 0 & 1 & 0 & 0 \\ 0 & 0 & 0 & 1 & 1 \\ 0 & 0 & 0 & 0 & 0 \end{bmatrix}$ (b) $\begin{bmatrix} 1 & -2 & 2 & 1 \\ 0 & 1 & 1 & 4 \\ 0 & 0 & 0 & 0 \end{bmatrix}$ (c) $\begin{bmatrix} 1 & 0 & 0 & 2 & 1 \\ 0 & 1 & 0 & 1 & 0 \\ 0 & 0 & 1 & 3 & 1 \end{bmatrix}$

2. Are the following matrices in row echelon form?

(a) $\begin{bmatrix} 1 & 0 & 0 & 4 & 3 \\ 0 & 1 & 0 & 1 & -2 \\ 0 & -1 & 1 & 2 & 0 \end{bmatrix}$ (b) $\begin{bmatrix} 1 & 0 & 5 & 4 \\ 0 & 1 & 2 & -1 \\ 0 & 0 & 0 & 1 \end{bmatrix}$ (c) $\begin{bmatrix} 1 & 0 & 3 \\ 0 & 1 & 4 \\ 0 & 0 & 1 \end{bmatrix}$

In Problems 3–8, reduce the given matrices to an equivalent matrix in row echelon form. Illustrate Theorem A by using several methods on a given matrix to get the unique row echelon form.

3. $\begin{bmatrix} 1 & 0 & 2 & 4 & 0 \\ 1 & 1 & 1 & 5 & 1 \\ 1 & 2 & 0 & 6 & 3 \\ 1 & 1 & 1 & 5 & 0 \end{bmatrix}$

4. $\begin{bmatrix} 1 & 2 & 1 & 3 & 4 \\ -1 & -2 & 0 & -2 & -3 \\ 1 & 2 & -2 & 2 & 5 \\ -2 & -4 & 4 & -2 & -3 \end{bmatrix}$

5. $\begin{bmatrix} 4 & -1 & 2 \\ 1 & 2 & -1 \\ 3 & 0 & 4 \\ -1 & 0 & 2 \end{bmatrix}$

6. $\begin{bmatrix} 2 & 0 & 3 \\ 3 & -1 & 4 \\ 1 & -1 & 2 \\ 0 & 4 & 5 \end{bmatrix}$

7. $\begin{bmatrix} 2 & 5 & 3 & 3 \\ 3 & 2 & 4 & 9 \\ 5 & -3 & -2 & 4 \end{bmatrix}$

8. $\begin{bmatrix} 4 & 3 & 2 & 5 \\ 7 & 2 & 3 & 11 \\ 3 & 2 & 5 & 11 \end{bmatrix}$

9. Let A designate the matrix of Problem 1(a) above. Write an expression giving all solutions to
(a) a homogeneous system of 4 equations in 5 variables having a coefficient matrix that is row equivalent to A.
(b) a nonhomogeneous system of 4 equations in 4 variables having an augmented matrix that is row equivalent to A.

10. Let B designate the matrix of Problem 2(a) above. Write an expression giving all solutions to
(a) a homogeneous system of 3 equations in 5 variables having a coefficient matrix that is row equivalent to B.
(b) a nonhomogeneous system of 3 equations in 4 variables having an augmented matrix that is row equivalent to B.

11. Let C designate the matrix of Problem 1(b) above. Write an expression giving all solutions to
(a) a homogeneous system of 3 equations in 4 variables having a coefficient matrix that is row equivalent to C.

(b) a nonhomogeneous system of 3 equations in 3 variables having an augmented matrix that is row equivalent to C.

12. Let D designate the matrix of Problem 2(b) above. Write an expression giving all solutions to
(a) a homogeneous system of 3 equations in 4 variables having a coefficient matrix that is row equivalent to D.
(b) a nonhomogeneous system of 3 equations in 3 variables having an augmented matrix that is row equivalent to D.

(*Problem set continues on p. 30*)

KIRCHHOFF'S LAWS

Figure 1-A is a schematic diagram of a direct current electrical network. Resistance R in a line is indicated by the symbol $\mathsf{-\!\!\wedge\!\!\wedge\!\!-}$; the power source is indicated by $+\;\dashv\vdash\;-$. If a current of I amperes flows in a positive direction through a resistance of R ohms, the voltage drop E across the resistance is $E = RI$ volts. The current through a power source, which provides a voltage rise, is taken to flow from $-$ to $+$. Current directions between junctions not including a power source are assigned arbitrarily.

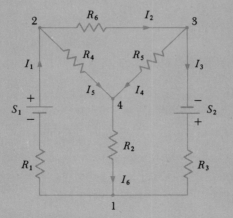

Figure 1-A

Analysis of such an electrical network depends on Kirchhoff's laws and often leads to a rather large system of linear equations. There are two laws. Taking into account the signs of currents and voltages, they are as follows:

(i) The sum of the currents at each junction is 0.
(ii) The sum of the voltage drops around every closed loop is 0.

For the schematic diagram in Figure 1-A, these laws give rise to the following equations:

Law (i)

Junction   *Sum of currents*

$$
\begin{array}{c}
1 \\ 2 \\ 3 \\ 4
\end{array}
\begin{bmatrix}
-1 & 0 & 1 & 0 & 0 & 1 \\
1 & -1 & 0 & 0 & -1 & 0 \\
0 & 1 & -1 & -1 & 0 & 0 \\
0 & 0 & 0 & 1 & 1 & -1
\end{bmatrix}
\begin{bmatrix}
I_1 \\ I_2 \\ I_3 \\ I_4 \\ I_5 \\ I_6
\end{bmatrix}
=
\begin{bmatrix}
0 \\ 0 \\ 0 \\ 0
\end{bmatrix}
$$

Law (ii)

Loop   *Sum of voltage drops*

$$
\begin{array}{c}
124 \\ 134 \\ 123 \\ 234
\end{array}
\begin{bmatrix}
R_1 & 0 & 0 & 0 & R_4 & R_2 \\
0 & 0 & R_3 & -R_5 & 0 & -R_2 \\
R_1 & R_6 & R_3 & 0 & 0 & 0 \\
0 & R_6 & 0 & R_5 & -R_4 & 0
\end{bmatrix}
\begin{bmatrix}
I_1 \\ I_2 \\ I_3 \\ I_4 \\ I_5 \\ I_6
\end{bmatrix}
=
\begin{bmatrix}
S_1 \\ S_2 \\ S_1 + S_2 \\ 0
\end{bmatrix}
$$

Suppose that in this example, $R_1 = R_2 = R_3 = 1$ ohm, $R_4 = R_5 = R_6 = 2$ ohms, $S_1 = 3$ volts, and $S_2 = 6$ volts. Solving the resulting system gives $I_1 = \frac{12}{5}$, $I_2 = \frac{9}{5}$, $I_3 = 3$, $I_4 = -\frac{6}{5}$, $I_5 = \frac{3}{5}$, and $I_6 = -\frac{3}{5}$. Evidently I_4 and I_6 flow in the direction opposite from that which was arbitrarily assigned.

For more about Kirchhoff's laws, see Karni, S. 1966. *Network theory: analysis and synthesis.* Boston: Allyn and Bacon. pp. 67–77.

Find all solutions to the following systems by reducing the associated matrix to row echelon form.

13. $2x_1 - 3x_2 + x_3 - x_4 = 4$
$3x_1 + 2x_2 + x_3 - 3x_4 = 1$
$5x_1 + x_2 - x_3 + x_4 = 3$
$2x_1 - 5x_2 + 4x_3 - 6x_4 = 6$

14. $x_1 + 2x_2 - 3x_3 = 9$
$2x_1 - x_2 + x_3 = 0$
$3x_1 - 2x_2 + 4x_3 = 0$
$4x_1 - x_2 + x_3 = 4$

15. $\begin{aligned} 2x_1 - x_2 + 5x_3 + 2x_4 &= 7 \\ 5x_1 + 2x_2 + 4x_3 + 3x_4 &= 2 \\ -7x_1 - 10x_2 + 8x_3 - x_4 &= 13 \end{aligned}$ 16. $\begin{aligned} x_1 - x_2 + x_3 - 2x_4 &= 3 \\ 3x_1 + x_2 + 4x_3 - x_4 &= 2 \\ 2x_1 + 2x_2 + 3x_3 + x_4 &= -1 \end{aligned}$

17. $\begin{aligned} x_1 - 3x_2 + x_3 &= 3 \\ -x_1 + x_2 + 2x_3 &= 3 \\ 3x_1 - 5x_2 - 3x_3 &= -3 \\ 5x_1 + 2x_2 - x_3 &= 3 \end{aligned}$ 18. $\begin{aligned} 2x_1 - 3x_2 + x_3 + x_4 &= 7 \\ 3x_1 - 2x_2 + 2x_3 - 2x_4 &= 5 \\ 6x_1 + x_2 + 5x_3 - 11x_4 &= 1 \end{aligned}$

19. Describe geometrically the solution set of the homogeneous system

$$\begin{aligned} 3x_1 - x_2 + 2x_3 &= 0 \\ 2x_1 + 2x_2 - x_3 &= 0 \end{aligned}$$

20. What conditions on a, b, c, and d will guarantee that the only solution to the system

$$\begin{aligned} ax_1 + bx_2 &= 0 \\ cx_1 + dx_2 &= 0 \end{aligned}$$

    is the trivial solution?

21. Suppose that the homogeneous system

$$\begin{aligned} a_{11}x_1 + a_{12}x_2 + a_{13}x_3 &= 0 \\ a_{21}x_1 + a_{22}x_2 + a_{23}x_3 &= 0 \\ a_{31}x_1 + a_{32}x_2 + a_{33}x_3 &= 0 \end{aligned}$$

    has only the trivial solution. Then consider the system obtained from the given system by replacing the three zeros on the right with three 1s.
    (a) Must this new system have a solution; that is, can the solution set of the new system be empty?
    (b) Might the new system have more than one solution?

22. Answer the following questions about the system

$$\begin{aligned} x_1 - x_2 + x_3 &= 3 \\ 2x_1 + x_2 - x_3 &= r \\ 4x_1 - x_2 + x_3 &= s \end{aligned}$$

    (a) What conditions on r and s must hold if the system is to have a solution?
    (b) Can r and s be chosen so that the system has exactly one solution?

23. Using 1s and 0s, along with * to designate any number that might not be 0 or 1, indicate the form of all possible row echelon forms of the matrix

$$\begin{bmatrix} a_{11} & a_{12} & a_{13} \\ a_{21} & a_{22} & a_{23} \\ a_{31} & a_{32} & a_{33} \end{bmatrix}$$

The methods of this section can be used to solve systems of equations in which the coefficients and the variables are allowed to have complex values. Solve the following such systems.

24.
$$ix_1 + \quad 2x_2 + (-1 + i)x_3 = -2 + i$$
$$x_1 - (i + 1)x_2 \qquad\qquad = 1 + 2i$$
$$(-2 + i)x_1 \qquad + \qquad ix_3 = -5 + i$$

25.
$$x_1 + \quad ix_2 - \quad (1 - i)x_3 = -2$$
$$ix_1 + \quad x_2 \qquad\qquad = -2$$
$$2ix_2 + \qquad 2x_3 = 2$$

CHAPTER 1
SELF-TEST

1. List the elementary row operations.

2. In solving a homogeneous system of 3 equations in 5 variables, the coefficient matrix has been found to be row equivalent to

$$\begin{bmatrix} 1 & 2 & 0 & 0 & -1 \\ 0 & 0 & 1 & 0 & 1 \\ 0 & 0 & 0 & 1 & 2 \end{bmatrix}$$

Describe the solution set to the system.

3. If the matrix

$$A = \begin{bmatrix} 1 & 0 & 2 \\ 0 & 1 & -1 \end{bmatrix}$$

is the augmented matrix of a nonhomogeneous system, describe the solution set; then describe the solution set if A is the coefficient matrix of a homogeneous system.

4. Reduce to row echelon form the matrix

$$\begin{bmatrix} 2 & -1 & 3 & 1 & 1 \\ -1 & 0 & -2 & 1 & -3 \\ 1 & 2 & -1 & -4 & 3 \\ 3 & 2 & -2 & -3 & -1 \end{bmatrix}$$

5. A system of 4 equations in 4 variables, being solved by back substitution, has been brought to the form

$$2x_1 - x_2 + 4x_3 + 2x_4 = 1$$
$$3x_2 - 8x_3 - 4x_4 = -1$$
$$x_3 + x_4 = 1$$
$$3x_3 + x_4 = -2$$

Finish the solution.

6. By pivoting on the circled 3, find a system equivalent to

$$4x_1 + x_2 - x_3 - 4x_4 = 3$$
$$2x_1 + ③x_2 \qquad - 3x_4 = 2$$
$$5x_1 - 2x_2 + 4x_3 \qquad = 2$$

7. The system below has x_1, x_3, and x_5 as basic variables. Find an equivalent system in which x_2, x_3, and x_5 appear as basic variables.

$$x_1 + x_2 \qquad - 2x_4 \qquad = 1$$
$$x_2 + x_3 - x_4 \qquad = 2$$
$$3x_2 \qquad + x_4 + x_5 = -2$$

8. Describe the solution set for the system

$$x_1 - x_2 + 2x_3 \qquad = -3$$
$$2x_2 + x_3 + 4x_4 = -3$$
$$4x_1 + x_2 + x_3 + 5x_4 = 4$$
$$3x_1 \qquad - 2x_3 + x_4 = 10$$

9. Describe, for various values of k, the solution set to

$$x_1 - x_2 + x_3 = 1$$
$$2x_1 + x_2 - 5x_3 = -1$$
$$7x_1 - x_2 - 7x_3 = k$$

10. The two systems below are equivalent. Determine r, s, and t.

$$-2x_1 + x_2 + rx_3 - 5x_4 = 1 \qquad x_1 \qquad + x_4 = 1$$
$$x_1 + x_2 - x_3 + sx_4 = 4 \qquad x_2 \qquad - 2x_4 = 2$$
$$3x_1 + x_2 + x_3 + 2x_4 = t \qquad x_3 + x_4 = -1$$

THE SOLUTION SET FOR A LINEAR SYSTEM

2

2-1 INTRODUCTION

The linear system

$$
\begin{aligned}
x_1 + 2x_2 + x_3 + x_4 &= 2 \\
3x_1 + 4x_2 + x_3 + 2x_4 &= 5 \\
2x_1 + x_3 + 3x_4 &= 3
\end{aligned}
$$

has the coefficient matrix

$$
A = \begin{bmatrix} 1 & 2 & 1 & 1 \\ 3 & 4 & 1 & 2 \\ 2 & 0 & 1 & 3 \end{bmatrix}
$$

and we saw in Example B of Section 1-3 that its solution set can be written in the form

$$
(x_1, x_2, x_3, x_4) = (\tfrac{4}{3} - x_4, \tfrac{1}{6} + \tfrac{1}{2}x_4, \tfrac{1}{3} - x_4, x_4)
$$

The two goals of this chapter can be stated in reference to this problem.

In the first place, we want to develop notation and language that will enable us to describe the solution set more precisely. The sets of four numbers in the solution set will give way to sets of n numbers when our given system involves n variables, and these ordered sets of numbers will be called **vectors**. Our description of the solution set will be easier when we learn to add these vectors and to multiply them by a real number.

Secondly, in analogy with elementary algebra where $ax = b$ is solved by multiplying both sides by a^{-1} to get $x = a^{-1}b$, we wish to develop an algebra for matrices so that the given system can be solved by multiplying the coefficient matrix with another matrix. This will also contribute to our first goal, enabling us to tell some things about the solution set by studying A.

VECTORS IN R^n | 2-2

In our introduction to Chapter 1, we said that a solution to a system

$$a_{11}x_1 + \cdots + a_{1n}x_n = b_1$$
$$\vdots \qquad \qquad \vdots \quad \vdots$$
$$a_{m1}x_1 + \cdots + a_{mn}x_n = b_m$$

(1)

of m equations in n unknowns is an n-tuple of numbers $(r_1, r_2, \ldots, r_n)$ such that if we set $x_1 = r_1, x_2 = r_2, \ldots, x_n = r_n$ in (1), all m equations will be satisfied. The solution set for (1) is, therefore, a collection of n-tuples. In this and other ways, we shall meet sets in which each member is an n-tuple. Consider two such n-tuples

$$\mathbf{r} = (r_1, r_2, \ldots, r_n) \quad \text{and} \quad \mathbf{s} = (s_1, s_2, \ldots, s_n)$$

It is common to define the sum $\mathbf{r} + \mathbf{s}$ and the product of a real number a with the n-tuple $\mathbf{r}$ as follows:

$$\mathbf{r} + \mathbf{s} = (r_1 + s_1, r_2 + s_2, \ldots, r_n + s_n)$$
$$a\mathbf{r} = (ar_1, ar_2, \ldots, ar_n)$$

For fixed n, R^n designates the set of all n-tuples, and the real number r_i is called the **ith component of the vector r.**

THE MEMBERS OF R^2

The members of R^2 are pairs of numbers (x_1, x_2). It is customary to associate such pairs with a point in the plane, and this is one way to think of the members of R^2. When one begins to talk about the addition of two pairs or the multiplication of a pair by a real number, however, it is more common to picture the members of R^2 as vectors.

Geometrically, a vector is commonly thought of as a directed line segment drawn from an initial point $A(a_1, a_2)$ to a terminal point $B(b_1, b_2)$. It is natural (see Figure 2-1) to write

$$\overrightarrow{AB} = (b_1 - a_1, b_2 - a_2)$$

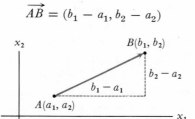

Figure 2-1

but in doing so, we must recognize that distinct directed line segments may correspond to the same number pair. Thus, for both the directed line segment from $A(-1, 2)$ to $B(2, 3)$ and the one from $C(1, -2)$ to $D(4, -1)$, we have

$$\overrightarrow{AB} = (3, 1) \quad \text{and} \quad \overrightarrow{CD} = (3, 1)$$

A little elementary geometry (Figure 2-2) makes it clear that two directed line segments (arrows) drawn in the plane will correspond to the same number pair if and only if the arrows are parallel, have the same length, and have the same sense (point in the same direction).

Figure 2-2

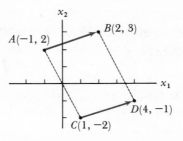

Thus, a vector $\mathbf{x}$ may be thought of analytically as an ordered pair; $\mathbf{x} = (x_1, x_2)$. Or it may be thought of geometrically as an entire class of arrows, any one of which can by parallel displacement be made to coincide with any other. Analytically, $\mathbf{x} = (x_1, x_2)$ and $\mathbf{y} = (y_1, y_2)$ are said to be **equal** if and only if $x_1 = y_1$ and $x_2 = y_2$. Geometrically, they are said to be equal if one can be slid, keeping it parallel to its original position, so as to coincide with the other (Figure 2-2). In particular, note that the vector $\mathbf{x} = (x_1, x_2)$ cannot be assumed to terminate at $P(x_1, x_2)$ unless it is known that $\mathbf{x}$ has its initial point at the origin (Figure 2-3).

Figure 2-3

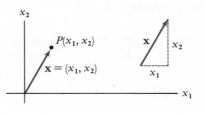

The representation of the same vector by distinct parallel arrows of the same sense and length turns out to be very useful in applications. Consider, for example, the common interpretation of a vector as a force, the direction and magnitude of which are represented by an arrow. If the force is being applied to dislodge a tree stump, it doesn't matter whether we think of pulling or pushing the stump. The arrow may be drawn with either its initial or its terminal point at the stump. Or think of diagrams in books written for the beginning sailor in which the direction and force of the wind may be represented by several parallel vectors in the same picture.

Analytical definitions are given in terms of algebraic operations to be performed on ordered pairs (x_1, x_2). They do not depend on geometric pictures, but the pictures certainly point our thinking in the right direction. Figure 2-3 suggests, for instance, that we define the **magnitude**, also called the **length**, $x = (x_1, x_2)$ to be

$$|x| = \sqrt{x_1^2 + x_2^2}$$

Vector x is called a **unit vector** if $|x| = 1$. Note that if x has magnitude 3, then $y = \frac{1}{3}x$ is a unit vector. More generally, for any vector x having positive magnitude, $y = (1/|x|)x$ is a **unit vector**. The **zero vector** $0 = (0, 0)$ is the only vector with a nonpositive magnitude.

Multiplication of a vector x by a real number a corresponds to lengthening (if $a > 1$), shortening (if $0 < a < 1$), or reversing (if $a < 0$) the sense of x. For a given vector $x = (x_1, x_2)$, $2x = (2x_1, 2x_2)$ and $-\frac{1}{2}x = (-\frac{1}{2}x_1, -\frac{1}{2}x_2)$ are indicated in Figure 2-4.

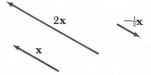

$2x$ $-\frac{1}{2}x$

x

Figure 2-4

The sum of $x = (x_1, x_2)$ and $y = (y_1, y_2)$, $x + y = (x_1 + y_1, x_2 + y_2)$, is obtained geometrically as the diagonal of a parallelogram having x and y as adjacent sides (Figure 2-5a). This process is referred to as *completing the parallelogram*. The same vector $x + y$ can be obtained if the initial point of y is made to coincide with the terminal point of x (Figure 2-5b).

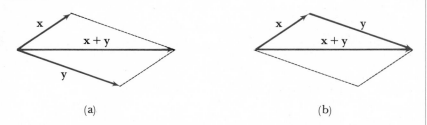

x x y

$x + y$ $x + y$

y

Figure 2-5

(a)   (b)

The vector $x = (3, 2)$ may be represented by the arrow drawn from $A(1, 2)$ to $B(4, 4)$; similarly $y = (2, -5)$ may be represented by an arrow from $B(4, 4)$ to $C(6, -1)$. If this is done, then sure enough

Example A

$x + y = (5, -3)$ is represented by the arrow from $A(1, 2)$ to $C(6, -1)$; see Figure 2-6a.   □

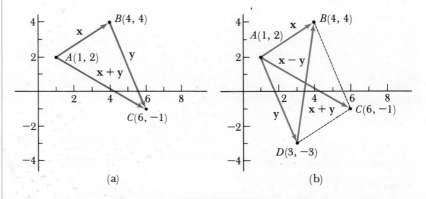

Figure 2-6

(a)   (b)

The difference $x - y = (x_1 - y_1, x_2 - y_2)$ is the "cross diagonal" of the parallelogram having x and y as adjacent sides. The correct direction of the arrow on $x - y$ may be determined by noting that $y + (x - y) = x$.

Refer to Example A again. The vector $y = (2, -5)$ can be represented not only by the arrow from $B(4, 4)$ to $C(6, -1)$, but also by the one from $A(1, 2)$ to $D(3, -3)$. In this representation $x + y$, drawn from A to C, is clearly seen as the diagonal of the parallelogram (Figure 2-6b). And $x - y = (1, 7)$ appears as the cross diagonal from $D(4, -3)$ to $B(4, 4)$. Note in the figure that $y + (x - y) = x.$

Example B

Points $A(-3, 2)$, $B(1, -2)$, and $C(7, 1)$ are given. We wish to find a point D in the first quadrant so that $ABCD$ will form a parallelogram (Figure 2-7).

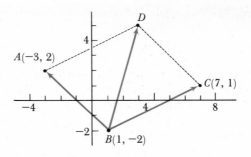

Figure 2-7

It is easy to obtain vectors $\vec{BA} = (-3 - 1, 2 - (-2)) = (-4, 4)$ and $\vec{BC} = (7 - 1, 1 - (-2)) = (6, 3)$, and so it is easy to get

$$\vec{BD} = \vec{BA} + \vec{BC} = (-4, 4) + (6, 3) = (2, 7)$$

We wish to emphasize, however, that the coordinates of D are not $(2, 7)$. The reason is that $\overrightarrow{BD}$ does not have its initial point at the origin. We need the vector $\overrightarrow{OD}$. One way to get $\overrightarrow{OD}$, since we went to the work of finding $\overrightarrow{BD}$, is to note that

$$\overrightarrow{OD} = \overrightarrow{OB} + \overrightarrow{BD} = (1, -2) + (2, 7) = (3, 5)$$

Another way to get it is to write

$$\overrightarrow{OD} = \overrightarrow{OA} + \overrightarrow{AD}$$

We don't know $\overrightarrow{AD}$, of course, but since the line segment from A to D is parallel to and has the same length and direction as the segment from B to C, then $\overrightarrow{AD} = \overrightarrow{BC}$. Hence,

$$\overrightarrow{OD} = \overrightarrow{OA} + \overrightarrow{BC} = (-3, 2) + (6, 3) = (3, 5)$$

You might try getting the same result, beginning with the vector equation $\overrightarrow{OD} = \overrightarrow{OC} + \overrightarrow{CD}$. In any of these ways, we get the point $D(3, 5)$. $\square$

It is frequently of interest, given a vector **a**, to know whether it can be expressed in terms of two other given vectors, say **n** and **t**. This comes up in physics, for example, if a moving body (think of a car) is accelerating while on a curved path (road). Part of the acceleration **a** is directed along a unit normal **n** to the curve, causing the change in direction. The rest is directed along a unit tangent **t** to the curve, changing the speed (see Figure 2-8). With **a, n,** and **t** all known, the goal is to find real numbers a_1 and a_2 such that

$$\mathbf{a} = a_1\mathbf{n} + a_2\mathbf{t} \tag{2}$$

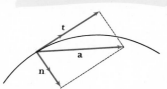

Figure 2-8

When this can be done, we say that **a** is a **linear combination** of **n** and **t**. We also say that the vectors **a, n,** and **t** are a **linearly dependent** set. In order to get away from one vector appearing distinctive as does **a** in (2), it is more common to say that the vectors **a, n,** and **t** are **linearly dependent** if there exist real numbers b_1, b_2, and b_3, not all zero, such that

$$b_1\mathbf{a} + b_2\mathbf{n} + b_3\mathbf{t} = 0$$

Otherwise the vectors are said to be **linearly independent.**

Since linear independence turns out to be a critically important concept in linear algebra, it is worth your time to reflect a moment on its geometric interpretation in R^2. Think about it until you see that two vectors are linearly independent if and only if they are not parallel.

THE MEMBERS OF R^3

Members of R^3 are ordered triples (x_1, x_2, x_3). The connection with the usual analytic geometry of 3-dimensional space is secured by identifying the directed arrow from $A(a_1, a_2, a_3)$ to $B(b_1, b_2, b_3)$ with the triple

$$\overrightarrow{AB} = (b_1 - a_1, b_2 - a_2, b_3 - a_3)$$

Those who have encountered the three mutually perpendicular unit vectors $\mathbf{i}$, $\mathbf{j}$, and $\mathbf{k}$ used in calculus courses will see (Figure 2-9) that they correspond to the triples $(1, 0, 0)$, $(0, 1, 0)$, and $(0, 0, 1)$, respectively. As in R^2, so in R^3, vector $\mathbf{x} = (x_1, x_2, x_3)$ is represented by an entire class of directed line segments, any two of which can be made to coincide under parallel translation.

Figure 2-9

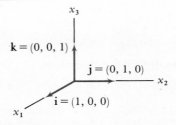

Two vectors $\mathbf{r}$ and $\mathbf{s}$, drawn with the same initial point, lie in a plane, so their sum can still be thought of as the completion of a parallelogram (Figure 2-10a), and the sum of three vectors $\mathbf{r}$, $\mathbf{s}$, and $\mathbf{t}$ can be found by

Figure 2-10

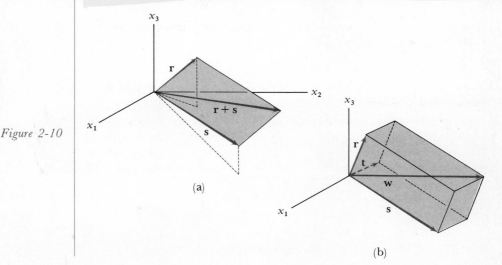

(a)

(b)

completing a parallelepiped (Figure 2-10b). If $\mathbf{w} = \mathbf{r} + \mathbf{s} + \mathbf{t}$ or, more generally, if

$$\mathbf{w} = a_1\mathbf{r} + a_2\mathbf{s} + a_3\mathbf{t}$$

we say that $\mathbf{w}$ is a linear combination of $\mathbf{r}$, $\mathbf{s}$, and $\mathbf{t}$.

VECTORS IN R^n

A vector in R^n is an n-tuple $\mathbf{r} = (r_1, \ldots, r_n)$ in which the real number r_i is called the ith component, and the **zero vector** is the vector having all 0 components. For two vectors $\mathbf{r} = (r_1, \ldots, r_n)$ and $\mathbf{s} = (s_1, \ldots, s_n)$, and for any real number a, we have already defined

$$\mathbf{r} + \mathbf{s} = (r_1 + s_1, \ldots, r_n + s_n)$$
$$a\mathbf{r} = (ar_1, \ldots, ar_n)$$

Although geometric intuition deserts us for $n > 3$, it seems perfectly natural to retain the analytic definitions and even the terminology that is by now familiar. The magnitude of $\mathbf{r}$ is defined, therefore, by

$$|\mathbf{r}| = \sqrt{r_1^2 + \cdots + r_n^2}$$

And, as before, $\mathbf{r}$ is a unit vector if its magnitude is 1.

If $\mathbf{r}_1, \mathbf{r}_2, \ldots, \mathbf{r}_k$ are vectors in R^n and $a_1, a_2, \ldots, a_k$ are real numbers, then

$$\mathbf{w} = a_1\mathbf{r}_1 + a_2\mathbf{r}_2 + \cdots + a_k\mathbf{r}_k$$

is called a **linear combination.** If there exist real numbers $a_1, a_2, \ldots, a_k$, not all zero, such that

$$a_1\mathbf{r}_1 + a_2\mathbf{r}_2 + \cdots + a_k\mathbf{r}_k = 0$$

then the set $\mathbf{r}_1, \mathbf{r}_2, \ldots, \mathbf{r}_k$ is said to be **linearly dependent.** If, on the other hand, we can only have

$$a_1\mathbf{r}_1 + a_2\mathbf{r}_2 + \cdots + a_k\mathbf{r}_k = 0$$

when $a_1 = a_2 = \cdots = a_k = 0$, then the set $\mathbf{r}_1, \mathbf{r}_2, \ldots, \mathbf{r}_k$ is said to be **linearly independent.**

Let $S = \{\mathbf{r}_1, \mathbf{r}_2, \mathbf{r}_3, \mathbf{r}_4\}$ be the set of four vectors in R^5 defined below. Decide whether S is linearly independent, and whether $\mathbf{W} = (5, 7, 10, -1, 1)$ can be written as a linear combination of vectors in S. *Example C*

$$\mathbf{r}_1 = (1, 1, 0, -1, 3)$$
$$\mathbf{r}_2 = (-1, 1, 1, 0, -1)$$
$$\mathbf{r}_3 = (0, -2, 1, 1, 0)$$
$$\mathbf{r}_4 = (-1, 1, 3, 0, 1)$$

To decide whether or not the set is linearly independent, we must decide whether it is possible to choose real numbers a_1, a_2, a_3, and a_4, not all zero, such that

(3)
$$a_1\mathbf{r}_1 + a_2\mathbf{r}_2 + a_3\mathbf{r}_3 + a_4\mathbf{r}_4 = \mathbf{0}$$

or

$$a_1(1, 1, 0, -1, 3) + a_2(-1, 1, 1, 0, -1) +$$
$$a_3(0, -2, 1, 1, 0) + a_4(-1, 1, 3, 0, 1) = (0, 0, 0, 0, 0)$$

This is equivalent to asking whether there is a nontrivial solution to the system

(4)
$$
\begin{aligned}
a_1 - a_2 \quad\quad\ - a_4 &= 0 \\
a_1 + a_2 - 2a_3 + a_4 &= 0 \\
a_2 + a_3 + 3a_4 &= 0 \\
-a_1 \quad\quad + a_3 \quad\quad &= 0 \\
3a_1 - a_2 \quad\quad + a_4 &= 0
\end{aligned}
$$

The methods of Chapter 1 lead us to conclude that there are indeed many nontrivial solutions; for example, $(a_1, a_2, a_3, a_4) = (1, 2, 1, -1)$ is a solution. You should verify that

$$\mathbf{r}_1 + 2\mathbf{r}_2 + \mathbf{r}_3 - \mathbf{r}_4 = \mathbf{0}$$

We conclude that S is a linearly dependent set.

To determine whether or not $\mathbf{w}$ can be written as a linear combination of vectors in S, we ask whether or not we can find a_1, a_2, a_3, and a_4 to satisfy (3) if the $\mathbf{0}$ vector on the right-hand side is replaced by $\mathbf{w}$. That leads to a system similar to (4) except that the constants on the right-hand side will be the components of $\mathbf{w}$. Attempts to solve this system lead to the conclusion that $\mathbf{w} = 2\mathbf{r}_2 - \mathbf{r}_3 + 3\mathbf{r}_4$; $\mathbf{w}$ can indeed be written as a linear combination of vectors in S. □

*PROBLEM
SET 2-2*

In Problems 1–4, draw vectors $\mathbf{x}$, $\mathbf{y}$, $\mathbf{x} + \mathbf{y}$, $\mathbf{x} - \mathbf{y}$, and the linear combination of $\mathbf{w} = a\mathbf{x} + b\mathbf{y}$.

1. $\mathbf{x} = (4, 3)$, $\mathbf{y} = (-1, 2)$; $a = \frac{1}{3}$, $b = -\frac{2}{3}$
2. $\mathbf{x} = (-1, -3)$, $\mathbf{y} = (-2, 5)$; $a = -2$, $b = \frac{1}{2}$
3. $\mathbf{x} = (-3, 2)$, $\mathbf{y} = (-2, -4)$; $a = 2$, $b = -\frac{3}{2}$
4. $\mathbf{x} = (4, -6)$, $\mathbf{y} = (-2, 4)$; $a = \frac{1}{4}$, $b = -2$
5. Verify for each of the given vectors $\mathbf{x}$ that $\mathbf{y} = (1/|\mathbf{x}|)\mathbf{x}$ is a unit vector; then find a vector in the direction of $\mathbf{x}$ having a magnitude of 3.
   (a) $\mathbf{x} = (3, 4)$ (c) $\mathbf{x} = (1, -1, 1)$
   (b) $\mathbf{x} = (\frac{1}{2}, -\frac{1}{2})$ (d) $\mathbf{x} = (1, -2, 2)$

A GEOMETRIC PROOF

Geometric representation of vectors can be used to prove theorems from plane geometry, as we now illustrate.

The medians of a triangle are concurrent at a point located on each median $\frac{2}{3}$ of the way from the vertex to the opposite midpoint.

Proof: Given an arbitrary triangle, we select two medians and impose a coordinate system and coordinates for the vertices as shown in Figure 2-A. The midpoints of OA

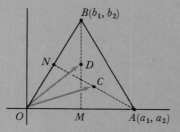

Figure 2-A

and OB are $M(\frac{1}{2}a_1, O)$ and $N(\frac{1}{2}b_1, \frac{1}{2}b_2)$. Let C and D be the points on the two medians AN and BM which are $\frac{2}{3}$ of the way from the vertices to the opposite midpoints.

Expressed as vector equations,

$$\overrightarrow{OC} = \overrightarrow{OA} + \tfrac{2}{3}\overrightarrow{AN}, \qquad \overrightarrow{OD} = \overrightarrow{OB} + \tfrac{2}{3}\overrightarrow{BM}$$

Now shifting to the analytic representation,

$$\overrightarrow{OC} = (a_1, a_2) + \tfrac{2}{3}(\tfrac{1}{2}b_1 - a_1, \tfrac{1}{2}b_2 - a_2)$$
$$= (\tfrac{1}{3}(a_1 + b_1), \tfrac{1}{3}(a_2 + b_2))$$
$$\overrightarrow{OD} = (b_1, b_2) + \tfrac{2}{3}(\tfrac{1}{2}a_1 - b_1, \tfrac{1}{2}a_2 - b_2)$$
$$= (\tfrac{1}{3}(b_1 + a_1), \tfrac{1}{3}(b_2 + a_2))$$

Since $\overrightarrow{OC}$ and $\overrightarrow{OD}$ both have initial points at the origin, both terminate at $\frac{1}{3}(a_1 + b_1), \frac{1}{3}(a_2 + b_2)$, so C and D are in fact the same point. □

6. Verify for each of the given vectors **x** that $\mathbf{y} = (1/|\mathbf{x}|)\mathbf{x}$ is a unit vector; then find a vector in the direction of **x** having a magnitude of 3.
   (a) $\mathbf{x} = (5, -12)$
   (b) $\mathbf{x} = (-\tfrac{1}{3}, \tfrac{1}{4})$
   (c) $\mathbf{x} = (3, 0, -3)$
   (d) $\mathbf{x} = (2, 1, -2)$

(Problem set continues on p. 44)

A regular hexagon ABCDEF is centered at 0 (Figure 2-11). Let **r** *and* **s** *designate the vectors* $\overrightarrow{OD}$ *and* $\overrightarrow{OE}$.

Figure 2-11

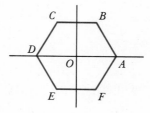

7. **Express the following vectors in terms of r and s.**

(a) $\overrightarrow{ED}$

(b) $\overrightarrow{DE}$

(c) $\overrightarrow{OA}$

(d) $\overrightarrow{AB}$

(e) $\overrightarrow{BE}$

(f) $\overrightarrow{FD}$

8. **Express the following vectors in terms of r and s.**

(a) $\overrightarrow{OB}$

(b) $\overrightarrow{OA}$

(c) $\overrightarrow{AB}$

(d) $\overrightarrow{BA}$

(e) $\overrightarrow{DA}$

(f) $\overrightarrow{DB}$

In Problems 9–12, express vector **w** *as a linear combination of the given vectors* **u** *and* **v;** *that is, find a and b so that* **w** = a**u** + b**v.** *Draw a picture showing vectors* r**u,** s**v,** *and* **w.**

9. $\mathbf{u} = (-2, -3)$, $\mathbf{v} = (-3, 1)$, $\mathbf{w} = (5, 2)$

10. $\mathbf{u} = (2, -3)$, $\mathbf{v} = (-1, -5)$, $\mathbf{w} = (2, 4)$

11. $\mathbf{u} = (-2, -6)$, $\mathbf{v} = (5, 3)$, $\mathbf{w} = (1, 3)$

12. $\mathbf{u} = (2, 5)$, $\mathbf{v} = (-2, 4)$, $\mathbf{w} = (-1, 2)$

In Problems 13–18, decide whether the given sets of vectors in R^4 *are linearly independent or linearly dependent.*

13. $\mathbf{x}_1 = (1, -1, 2, 0)$
$\mathbf{x}_2 = (2, 0, -3, 1)$
$\mathbf{x}_3 = (2, -2, 0, 4)$
$\mathbf{x}_4 = (2, -1, 0, 1)$

14. $\mathbf{x}_1 = (1, -1, 0, 1)$
$\mathbf{x}_2 = (-1, 0, 2, 0)$
$\mathbf{x}_3 = (0, 3, 1, -1)$
$\mathbf{x}_4 = (2, 1, -1, 3)$

15. $\mathbf{x}_1 = (2, 0, 1, 1)$
$\mathbf{x}_2 = (0, 1, 0, -1)$
$\mathbf{x}_3 = (3, 1, -1, 2)$
$\mathbf{x}_4 = (2, -2, 3, -1)$

16. $\mathbf{x}_1 = (-1, 0, 1, 0)$
$\mathbf{x}_2 = (3, -6, 1, 0)$
$\mathbf{x}_3 = (1, -4, -2, 1)$
$\mathbf{x}_4 = (2, -8, 7, -1)$

17. $\mathbf{x}_1 = (3, 5, -2, -4)$
$\mathbf{x}_2 = (-2, -3, 1, 5)$
$\mathbf{x}_3 = (0, 1, -1, 7)$

18. $\mathbf{x}_1 = (1, 0, -1, 3)$
$\mathbf{x}_2 = (-1, 1, 2, -1)$
$\mathbf{x}_3 = (-4, 11, 15, 10)$

19. Show that $\mathbf{w} = (1, 2, 5, 2)$ can be written as a linear combination of

$\mathbf{x}_1 = (1, 0, -1, 2)$, $\mathbf{x}_2 = (-1, 2, 3, -1)$, and $\mathbf{x}_3 = (0, -1, 1, -1)$, but that $\mathbf{v} = (1, 1, 8, 4)$ cannot.

20. Show that $\mathbf{w} = (-3, 5, -4, 9)$ can be written as a linear combination of $\mathbf{x}_1 = (-1, 1, 0, 1)$, $\mathbf{x}_2 = (-3, 2, 2, 0)$, and $\mathbf{x}_3 = (2, -1, -2, 1)$, but that $\mathbf{v} = (1, -1, -1, 3)$ cannot.

21. Show that both $\mathbf{v}$ and $\mathbf{w}$ of Problem 19 can be written as linear combinations of $\mathbf{x}_1 = (2, 1, -1, 2)$, $\mathbf{x}_2 = (-1, -1, 2, 0)$, and $\mathbf{x}_3 = (0, 3, 1, -2)$.

22. Show that both $\mathbf{v}$ and $\mathbf{w}$ of Problem 20 can be written as linear combinations of $\mathbf{x}_1 = (-2, 4, -1, 5)$, $\mathbf{x}_2 = (-2, 2, -2, 1)$, and $\mathbf{x}_3 = (7, -11, 3, -8)$.

23. Given points $A(1, 0, -1)$, $B(2, -1, -4)$, and $C(4, 3, 1)$, find the coordinates of a point D so that $ABCD$ is a parallelogram with AB opposite CD. Then find the coordinates of the point R where the diagonals intersect.

24. Follow the directions for Problem 23, using the points $A(3, 1, -1)$, $B(-1, 4, 2)$, and $C(1, 3, 3)$.

25. If points A, B, and C of Problem 23 are vertices of a triangle, find the coordinates of the point where the medians intersect.

26. If points A, B, and C of Problem 24 are vertices of a triangle, find the coordinates of the point where the medians intersect.

DOT PRODUCTS AND PROJECTIONS | *2-3*

We shall continue to use the geometric pictures available in R^2 to motivate definitions we shall find important in any of the vector spaces R^n. Consider, therefore, the problem of finding the area of a triangle for which you have been given the coordinates of the three vertices $A(a_1, a_2)$, $B(b_1, b_2)$, and $C(c_1, c_2)$. If we choose the segment from A to B as the base and drop a perpendicular from C to the point D on AB (Figure 2-12), then

$$\text{area} = \tfrac{1}{2}bh = \tfrac{1}{2}|\overrightarrow{AB}||\overrightarrow{DC}|$$

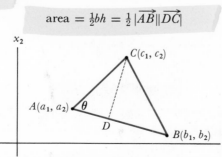

Figure 2-12

Since the coordinates of A and B are known, it is easy to find $\overrightarrow{AB}$. Our problem is to find $\overrightarrow{DC}$.

Two methods suggest themselves, and both are worth investigation. The first method builds on the observation that if we knew the angle θ at vertex A, then we could find the length of $\overrightarrow{DC}$ from the fact that $|\overrightarrow{DC}| = |\overrightarrow{AC}|\sin\theta$. The second method makes use of the vector equation $\overrightarrow{OD} = \overrightarrow{OA} + \overrightarrow{AD}$ which would give us the coordinates of D. This directs our attention to finding vector $\overrightarrow{AD}$ which, for obvious reasons, is called the **projection** of $\overrightarrow{AC}$ onto $\overrightarrow{AB}$, written $\text{proj}_{\overrightarrow{AB}} \overrightarrow{AC}$.

THE ANGLE BETWEEN TWO VECTORS IN R²

It is easy enough, given a protractor, to find the angle θ between **r** and **s** from their geometric representation (Figure 2-13a). To find the angle

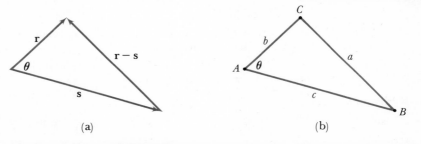

Figure 2-13

(a)　　　　　　　　　　　　(b)

from the analytic representation of $\mathbf{r} = (r_1, r_2)$ and $\mathbf{s} = (s_1, s_2)$ requires appeal to a fact from trigonometry. Recall that in a triangle with side lengths a, b, and c opposite from angles A, B, and C (Figure 2-13b) the *law of cosines* says that

$$\cos A = \frac{b^2 + c^2 - a^2}{2bc}$$

With $A = \theta$, $b = |\mathbf{r}|$, $c = |\mathbf{s}|$, and $a = |\mathbf{r} - \mathbf{s}|$, we see that the numerator is

$$(r_1^2 + r_2^2) + (s_1^2 + s_2^2) - [(r_1 - s_1)^2 + (r_2 - s_2)^2]$$

which simplifies to $2(r_1 s_1 + r_2 s_2)$. Thus,

$$\cos A = \frac{r_1 s_1 + r_2 s_2}{|\mathbf{r}|\|\mathbf{s}|}$$

The resulting formula is simplified if we define the **dot product** of $\mathbf{r} = (r_1, r_2)$ and $s = (s_1, s_2)$ to be

$$\mathbf{r} \bullet \mathbf{s} = r_1 s_1 + r_2 s_2$$

We then have the important formula for the angle θ between two non-zero vectors **r** and **s**,

$$\cos \theta = \frac{\mathbf{r} \cdot \mathbf{s}}{|\mathbf{r}||\mathbf{s}|}$$

Note in particular that two nonzero vectors will be perpendicular, also called **orthogonal,** if and only if $\mathbf{r} \cdot \mathbf{s} = 0$.

The perpendicular lines (Figure 2-14)

Example A

$$L_1: x_2 - 2 = -\tfrac{1}{2}(x_1 - 3), \qquad L_2: x_2 - 2 = 2(x_1 - 3)$$

intersect at $R(3, 2)$. Point $P_1(1, 3)$ lies on L_1, and point $P_2(5, 6)$ lies on L_2.

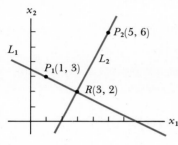

Figure 2-14

Verify that vectors $\overrightarrow{RP_1}$ and $\overrightarrow{RP_2}$ are orthogonal.

We simply write $\overrightarrow{RP_1} = (-2, 1)$ and $\overrightarrow{RP_2} = (2, 4)$, and then note that

$$\overrightarrow{RP_1} \cdot \overrightarrow{RP_2} = -2(2) + 1(4) = 0 \quad \square$$

Given a triangle with vertices $A(2, 4)$, $B(8, 7)$, and $C(1, 11)$, as shown in Figure 2-15, find $\angle BAC$ and the area of the triangle.

Example B

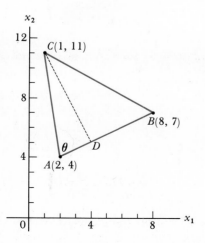

Figure 2-15

We begin by writing $\overrightarrow{AB} = (6, 3)$ and $\overrightarrow{AC} = (-1, 7)$. Then with

$\theta = \angle BAC,$

$$\cos\theta = \frac{\overrightarrow{AB} \cdot \overrightarrow{AC}}{|\overrightarrow{AB}||\overrightarrow{AC}|} = \frac{-6 + 21}{\sqrt{45}\,\sqrt{50}} = \frac{1}{\sqrt{10}}$$

Since $\sin\theta = \sqrt{1 - \cos^2\theta} = 3/\sqrt{10}$,

$$|\overrightarrow{DC}| = |\overrightarrow{AC}|\sin\theta = \sqrt{50}\,\frac{3}{\sqrt{10}} = 3\sqrt{5}$$

and we have

$$\text{area} = \tfrac{1}{2}|\overrightarrow{AB}||\overrightarrow{DC}| = \tfrac{1}{2}\sqrt{45}\,3\sqrt{5} = \tfrac{45}{2} \quad \square$$

PROJECTIONS IN R²

In the introduction to this section, referring to a picture similar to Figure 2-16, we called $\mathbf{p} = \overrightarrow{AD}$ the projection of $\mathbf{r} = \overrightarrow{AC}$ onto $\mathbf{s} = \overrightarrow{AB}$.

Figure 2-16

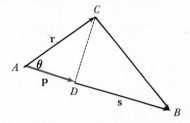

Since $(1/|\mathbf{s}|)\mathbf{s}$ is a unit vector in the direction of $\mathbf{s}$, the desired projection can be written in the form

(1)
$$\mathbf{p} = a\frac{\mathbf{s}}{|\mathbf{s}|}$$

where the real number a is the length of $\mathbf{p}$. Clearly, however, the length of $\mathbf{p}$ is given by $|\mathbf{p}| = |\mathbf{r}|\cos\theta$, which together with the formula already obtained for $\cos\theta$ gives

$$a = |\mathbf{p}| = |\mathbf{r}|\,\frac{\mathbf{r}\cdot\mathbf{s}}{|\mathbf{r}||\mathbf{s}|} = \frac{\mathbf{r}\cdot\mathbf{s}}{|\mathbf{s}|}$$

Substituting this expression for a and $\text{proj}_\mathbf{s}\mathbf{r}$ for $\mathbf{p}$ in (1) gives us our desired formula

$$\text{proj}_\mathbf{s}\mathbf{r} = \frac{\mathbf{r}\cdot\mathbf{s}}{|\mathbf{s}|^2}\mathbf{s}$$

Example C

We again find the area of the triangle with vertices $A(2, 4)$, $B(8, 7)$, and $C(1, 11)$, as shown in Figure 2-17, this time making use of the vector equation

$$\overrightarrow{OD} = \overrightarrow{OA} + \overrightarrow{AD} = \overrightarrow{OA} + \text{proj}_{\overrightarrow{AB}}\overrightarrow{AC}$$

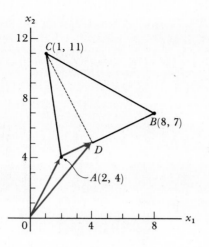

Figure 2-17

Since $\overrightarrow{AB} = (6, 3)$ and $\overrightarrow{AC} = (-1, 7)$,

$$\text{proj}_{\overrightarrow{AB}}\overrightarrow{AC} = \frac{\overrightarrow{AC} \cdot \overrightarrow{AB}}{|\overrightarrow{AB}|^2} \overrightarrow{AB} = \frac{15}{45}(6, 3) = (2, 1)$$

Then $\overrightarrow{OD} = (2, 4) + (2, 1) = (4, 5)$. This locates $D(4, 5)$, enabling us to find $\overrightarrow{DC} = (-3, 6)$. Hence, $|\overrightarrow{DC}| = \sqrt{45}$, and then

$$\text{area} = \tfrac{1}{2}|\overrightarrow{AB}||\overrightarrow{DC}| = \tfrac{1}{2}\sqrt{45}\,\sqrt{45} = \tfrac{45}{2} \quad \square$$

The method just used does show how to find the coordinates of D, but this was an unnecessary step to solve the problem. We could have used the vector equation $\overrightarrow{DC} = \overrightarrow{AC} - \overrightarrow{AD}$, found $\overrightarrow{AD} = \text{proj}_{\overrightarrow{AB}}\overrightarrow{AC}$ as we did in the example, and then found

$$\overrightarrow{DC} = (-1, 7) - (2, 1) = (-3, 6)$$

without ever knowing the coordinates of D.

ANGLES AND PROJECTIONS IN R³

If the dot product of $\mathbf{r} = (r_1, r_2, r_3)$ and $\mathbf{s} = (s_1, s_2, s_3)$ is defined to be

$$\mathbf{r} \bullet \mathbf{s} = r_1 s_1 + r_2 s_2 + r_3 s_3$$

the same argument made for two dimensions will show that for the angle θ between $\mathbf{r}$ and $\mathbf{s}$,

$$\cos \theta = \frac{\mathbf{r} \bullet \mathbf{s}}{|\mathbf{r}||\mathbf{s}|}$$

Orthogonal vectors, that is nonzero vectors having a dot product of 0, do indeed correspond to our geometric notion of being perpendicular.

The dot product test can be used to establish that if $\mathbf{s} = (s_1, s_2, s_3)$ and $\mathbf{t} = (t_1, t_2, t_3)$, then a vector $\mathbf{u}$ orthogonal to both $\mathbf{s}$ and $\mathbf{t}$ is given by

(2) $$\mathbf{u} = (s_2 t_3 - s_3 t_2, \ s_3 t_1 - s_1 t_3, \ s_1 t_2 - s_2 t_1)$$

It will be useful later in the text to know that when $\mathbf{u}$ is defined by (2), then

(3) $$|\mathbf{u}|^2 = |\mathbf{s}|^2 |\mathbf{t}|^2 - (\mathbf{s} \cdot \mathbf{t})^2$$

Verification of (3) is straightforward and is left as an exercise (Problem 19).

Finally, since we have the same formula in R^3 as we had in R^2 for the cosine of the angle formed by two vectors $\mathbf{r}$ and $\mathbf{s}$, the projection of $\mathbf{r}$ onto $\mathbf{s}$ is

$$\text{proj}_{\mathbf{s}} \mathbf{r} = \frac{\mathbf{r} \cdot \mathbf{s}}{|\mathbf{s}|^2} \mathbf{s}$$

Example D

Given the points $P(-1, 3, 2)$, $Q(5, 5, -2)$, and $R(3, -1, 0)$, find a fourth point S so that the points $PQRS$ form a parallelogram, and then find its area.

Since all four points must lie in a common plane, we may visualize the situation as indicated in Figure 2-18. The vector equation

$$\overrightarrow{OS} = \overrightarrow{OR} + \overrightarrow{RS} = \overrightarrow{OR} + \overrightarrow{QP}$$

Figure 2-18

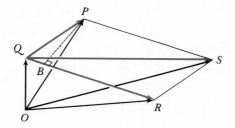

gives $\overrightarrow{OS} = (3, -1, 0) + (-6, -2, 4) = (-3, -3, 4)$. This locates $S(-3, -3, 4)$.

The desired area is the product of the lengths of the base $|\overrightarrow{QR}|$ and the height $|\overrightarrow{BP}|$. We already know $\overrightarrow{QR}$, and

$$\overrightarrow{BP} = \overrightarrow{QP} - \overrightarrow{QB} = \overrightarrow{QP} - \text{proj}_{\overrightarrow{QR}} \overrightarrow{QP}$$

$$\text{proj}_{\overrightarrow{QR}} \overrightarrow{QP} = \frac{\overrightarrow{QP} \cdot \overrightarrow{QR}}{|\overrightarrow{QR}|^2} \overrightarrow{QR} = \frac{12 + 12 + 8}{44}(-2, -6, 2)$$

$$\overrightarrow{BP} = (-6, -2, 4) - \tfrac{8}{11}(-2, -6, 2) = \tfrac{1}{11}(-50, 26, 28)$$

Therefore,

$$\text{area} = |\overrightarrow{QR}||\overrightarrow{BP}| = \sqrt{44}\,\tfrac{1}{11}\sqrt{50^2 + 26^2 + 28^2}$$
$$= \tfrac{1}{11}\sqrt{44}\,\sqrt{9 \cdot 44 \cdot 10} = 12\sqrt{10} \quad \square$$

DOT PRODUCTS AND PROJECTIONS IN R^n

Once again we extend to R^n in an obvious way the analytic definitions found to be useful in R^2 and R^3. For two vectors

$$\mathbf{r} = (r_1, \ldots, r_n), \qquad \mathbf{s} = (s_1, \ldots, s_n)$$

we define the **dot product** by

$$\mathbf{r} \cdot \mathbf{s} = r_1 s_1 + \cdots + r_n s_n$$

Though we have not yet had need of it, it is easy to verify (and is left for you to do in Problem 17) that $\mathbf{r} \cdot \mathbf{s} = \mathbf{s} \cdot \mathbf{r}$ and that for any three vectors $\mathbf{r}$, $\mathbf{s}$, and $\mathbf{t}$, $\mathbf{r} \cdot (\mathbf{s} + \mathbf{t}) = \mathbf{r} \cdot \mathbf{s} + \mathbf{r} \cdot \mathbf{t}$. Watch out though; $\mathbf{r} \cdot (\mathbf{s} \cdot \mathbf{t})$ doesn't even make sense, so neither will an attempt to prove any "obvious" statements about it.

We also define, for any two vectors $\mathbf{r}$ and $\mathbf{s}$ in R^n, the *projection* of $\mathbf{r}$ onto $\mathbf{s}$ by

$$\text{proj}_{\mathbf{s}} \mathbf{r} = \frac{\mathbf{r} \cdot \mathbf{s}}{|\mathbf{s}|^2} \mathbf{s}$$

We shall continue, even for $n > 3$, to call $\mathbf{r}$ and $\mathbf{s}$ *orthogonal* if and only if $\mathbf{r} \cdot \mathbf{s} = 0$.

Although geometric intuition deserts us in R^n for $n > 3$, a virtue of our terminology is that it continues to suggest things that are as true in any space R^n as they were in R^2 and R^3.

Let $\mathbf{r} = (3, -1, 2, 4)$ and $\mathbf{s} = (-1, 2, 0, -1)$. Show that, as in R^2, $\mathbf{r} - \text{proj}_{\mathbf{s}} \mathbf{r}$ is orthogonal to $\mathbf{s}$ (Figure 2-19).

Example E

Figure 2-19

We begin by finding

$$\text{proj}_{\mathbf{s}} \mathbf{r} = \frac{\mathbf{r} \cdot \mathbf{s}}{|\mathbf{s}|^2} \mathbf{s} = -\tfrac{9}{6}(-1, 2, 0, -1)$$

and then

$$\mathbf{r} - \text{proj}_{\mathbf{s}} \mathbf{r} = (3, -1, 2, 4) - (\tfrac{3}{2}, -3, 0, \tfrac{3}{2})$$
$$= (\tfrac{3}{2}, 2, 2, \tfrac{5}{2})$$

You should verify that the dot product of this vector with $\mathbf{s}$ is 0, hence that the vectors are orthogonal. □

In Problems 1–4 find the angle between **r** *and* **s**. *Draw diagrams.*

1. $\mathbf{r} = (2, -1)$ and $\mathbf{s} = (3, 1)$

2. $\mathbf{r} = (1, 3)$ and $\mathbf{s} = (-1, 2)$

3. $\mathbf{r} = (4, 2)$ and $\mathbf{s} = (-1, 3)$

4. $\mathbf{r} = (-2, 4)$ and $\mathbf{s} = (-1, -3)$

5. Using vectors **r** and **s** from Problem 1, find $\text{proj}_{\mathbf{r}}\mathbf{s}$ and $\text{proj}_{\mathbf{s}}\mathbf{r}$.

6. Using vectors **r** and **s** from Problem 2, find $\text{proj}_{\mathbf{r}}\mathbf{s}$ and $\text{proj}_{\mathbf{s}}\mathbf{r}$.

7. Find the area of the triangle having vertices $A(1, 2)$, $B(7, -2)$, and $C(7, \frac{20}{3})$.

8. Find the area of the triangle having vertices $A(-3, 1)$, $B(3, 4)$, and $C(-4, 8)$.

9. For the points $A(3, 1, 1)$, $B(4, 2, 0)$, and $C(0, 4, 3)$, answer the following where O designates the origin:

   (a) What is vector $\overrightarrow{AB}$? $\overrightarrow{BC}$? $\overrightarrow{OA}$?
   (b) What is the distance from A to B?
   (c) What angle is formed by $\overrightarrow{OB}$ and $\overrightarrow{OC}$?
   (d) Suppose line segments OA, OB, and OC are three adjacent edges of a parallelepiped. What is the length of the longest diagonal of the parallelepiped?

10. Answer the questions asked in Problem 9, using the points $A(2, 4, 3)$, $B(2, 0, 4)$, and $C(3, 4, 0)$.

11. The dimensions of a rectangular box are 3, 4, and 5. Find the angle that the longest diagonal of the box makes with the shortest side.

12. Solve Problem 11 for a box with dimensions of 2, 3, and 6.

13. The vertices of a triangle in space are $A(1, -1, 2)$, $B(3, 1, 6)$, and $C(-\frac{5}{2}, \frac{11}{2}, 2)$. A perpendicular is dropped from C to a point D on the base AB. Find the coordinates of D.

14. Solve Problem 13 using the points $A(-1, -3, -2)$, $B(5, 3, 10)$, and $C(7, -5, 7)$.

15. Given $\mathbf{s} = (3, 0, 4)$ and $\mathbf{t} = (2, 3, 6)$, use (2) of this section to form a vector **u** orthogonal to both **s** and **t**. Verify that $\mathbf{s} \cdot \mathbf{u} = \mathbf{t} \cdot \mathbf{u} = 0$, and that relation (3) holds.

16. Follow the instructions of Problem 15, using the vectors $\mathbf{s} = (1, 1, 2)$ and $\mathbf{t} = (3, 1, -5)$.

17. Using $\mathbf{r} = (r_1, r_2, r_3)$, $\mathbf{s} = (s_1, s_2, s_3)$, and $\mathbf{t} = (t_1, t_2, t_3)$, write out expressions for $\mathbf{r} \cdot \mathbf{s}$, $\mathbf{s} \cdot \mathbf{r}$, $\mathbf{r} \cdot (\mathbf{s} + \mathbf{t})$, and $\mathbf{r} \cdot \mathbf{t}$. Verify that $\mathbf{r} \cdot \mathbf{s} = \mathbf{s} \cdot \mathbf{r}$ and that $\mathbf{r} \cdot (\mathbf{s} + \mathbf{t}) = \mathbf{r} \cdot \mathbf{s} + \mathbf{r} \cdot \mathbf{t}$.

18. Given $\mathbf{s} = (s_1, s_2, s_3)$ and $\mathbf{t} = (t_1, t_2, t_3)$, define as in (2) of this section the vector

$$\mathbf{u} = (s_2 t_3 - s_3 t_2, \, s_3 t_1 - s_1 t_3, \, s_1 t_2 - s_2 t_1)$$

Show that $\mathbf{s} \cdot \mathbf{u} = \mathbf{t} \cdot \mathbf{u} = 0$.

WHO NEEDS n DIMENSIONS?

To describe the motion of an airplane in flight, the coordinates of the center of mass at time t don't tell the whole story. One needs to know how the airplane is moving about its axes: the roll L, the pitch M, and the yaw N (Figure 2-B). That's six coordinates to get our study off the ground. It is not uncommon in books on aeronautical engineering to find 12 or 15 variables being used to analyze the motion of a plane in flight.

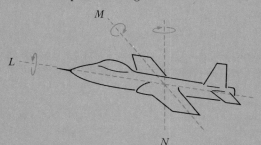

Figure 2-B

In games of chance where n outcomes are possible in a given turn, in studies involving n hereditary characteristics, in analyzing the performance of a time-sharing computer system with n remote terminals, and in many other applications where n outcomes are possible, one is often interested in a probability p_i of each of the possible outcomes. The probabilities are entered into a vector called, not surprisingly, a *probability vector*. Such a vector $(p_1, \ldots, p_n)$ is characterized by the conditions that $p_i \geq 0$ for all i, and that $p_1 + \cdots + p_n = 1$. This turns out to be an important concept in Chapter 8.

19. For the vectors **s**, **t**, and **u** of Problem 18, verify (3) by showing that

$$|\mathbf{u}|^2 = |\mathbf{s}|^2 |\mathbf{t}|^2 - (\mathbf{s} \cdot \mathbf{t})^2$$

20. Starting with $\mathbf{r} = (2, -1, 3, 1, -2)$ and $\mathbf{s} = (4, 2, 5, -1, 1)$ in R^5, proceed as in Example E to verify that $\mathbf{r} - \text{proj}_{\mathbf{s}}\mathbf{r}$ is orthogonal to **s**.

21. Using the definition of $\text{proj}_{\mathbf{s}}\mathbf{r}$, prove that in any space R^n, $\mathbf{r} - \text{proj}_{\mathbf{s}}\mathbf{r}$ will be orthogonal to **s**.

22. Note that for $A(1, -1, 0, 2)$, $B(2, 3, 1, 4)$, and $C(5, 0, 1, 2)$, the vectors $\overrightarrow{BA}$ and $\overrightarrow{BC}$ have the same length. By analogy with R^2, BA and BC might be thought of as adjacent sides of a rhombus. Prove that, sure enough, the diagonals of the rhombus are orthogonal.

23. Given three points A, B, and C in R^n with the property that $|\overrightarrow{BA}| = |\overrightarrow{BC}|$, find the point D such that $ABCD$ might reasonably be called a rhombus. Then prove that $\overrightarrow{AC}$ and $\overrightarrow{BD}$ are orthogonal.

2-4   SUBSPACES OF R^n

Suppose that a teacher gives to a beginning class in linear algebra the assignment of solving the homogeneous system

$$
\begin{aligned}
2x_1 + x_2 - x_3 + 2x_4 - x_5 &= 0 \\
3x_1 - x_2 - 2x_3 \qquad\quad + 3x_5 &= 0 \\
-2x_1 \qquad + x_3 - x_4 \qquad\quad &= 0 \\
5x_1 - x_2 - 4x_3 \qquad\quad + 7x_5 &= 0
\end{aligned}
$$

Students who elect to use x_2, x_3, and x_4 as their first three basic variables find that the given system is equivalent to the system

$$
\begin{aligned}
-x_1 + x_2 \qquad\qquad\quad + x_5 &= 0 \\
-x_1 \qquad + x_3 \qquad\quad - 2x_5 &= 0 \\
x_1 \qquad\qquad + x_4 - 2x_5 &= 0
\end{aligned}
$$

These students will describe the solution set by writing

(1) $(x_1, x_2, x_3, x_4, x_5) = (x_1, x_1 - x_5, x_1 + 2x_5, -x_1 + 2x_5, x_5)$

Other students may decide, however, to use (for instance) x_1, x_3, and x_5 as basic variables, leading them to the equivalent system

$$
\begin{aligned}
x_1 - 2x_2 \qquad\quad - x_4 \qquad\quad &= 0 \\
-4x_2 + x_3 - 3x_4 \qquad\quad &= 0 \\
-x_2 \qquad\quad - x_4 + x_5 &= 0
\end{aligned}
$$

The corresponding description of the solution set is

(2) $(x_1, x_2, x_3, x_4, x_5) = (2x_2 + x_4, x_2, 4x_2 + 3x_4, x_4, x_2 + x_4)$

Consider now the teacher's problem. Since the form of the answer is not unique, how can the papers be corrected with a minimum of work (a goal that teachers share with other sensible folk)?

It will pay us to restate the question. The dependence of solutions (1) upon arbitrary choices of x_1 and x_5 can be emphasized if we rewrite (1) as the sum of two vectors, one involving only x_1, the other involving only x_5.

(3) $(x_1, x_2, x_3, x_4, x_5) = (x_1, x_1, x_1, -x_1, 0) + (0, -x_5, 2x_5, 2x_5, x_5)$
$$= x_1(1, 1, 1, -1, 0) + x_5(0, -1, 2, 2, 1)$$

In the same way, the dependence of the solutions (2) on x_2 and x_4 can be emphasized by writing

$$(x_1, x_2, x_3, x_4, x_5) = (2x_2 + x_4, x_2, 4x_2 + 3x_4, x_4, x_2 + x_4)$$
$$= x_2(2, 1, 4, 0, 1) + x_4(1, 0, 3, 1, 1)$$

The question of equivalence of the sets described in (3) and (4) comes down to this: If arbitrary choices of x_1 and x_5 in (3) produce the vector **r,** can we find choices of x_2 and x_4 in (4) that will also produce **r**? And conversely, can every vector expressible in (4) be obtained from (3)? Some new terminology and methods will help us in addressing this problem.

THE SPAN OF A SET

Beginning with a set of k vectors

$$S = \{\mathbf{r}_1, \mathbf{r}_2, \ldots, \mathbf{r}_k\}$$

consider the set of all vectors

$$\mathbf{r} = x_1\mathbf{r}_1 + x_2\mathbf{r}_2 + \cdots + x_k\mathbf{r}_k$$

that can be formed as linear combinations of vectors in S. These vectors constitute what is called the **span** of the set S.

We specifically mean to allow $x_1 = x_2 \cdots = x_k = 0$, meaning that the span of any set always includes the zero vector.

Two different sets may have the same span. Consider, for instance, the sets

$$S_1 = \{\mathbf{e}_1 = (1, 0), \mathbf{e}_2 = (0, 1)\}, \qquad S_2 = \{\mathbf{r}_1 = (1, 1), \mathbf{r}_2 = (-1, 1)\}$$

From the observation that

$$\mathbf{e}_1 = \tfrac{1}{2}(\mathbf{r}_1 - \mathbf{r}_2) \quad \text{and} \quad \mathbf{e}_2 = \tfrac{1}{2}(\mathbf{r}_1 + \mathbf{r}_2)$$

it is clear that any vector **t** in the span of S_1 can be written as

$$\mathbf{t} = a_1\mathbf{e}_1 + a_2\mathbf{e}_2 = \frac{a_1}{2}(\mathbf{r}_1 - \mathbf{r}_2) + \frac{a_2}{2}(\mathbf{r}_1 + \mathbf{r}_2)$$

$$= \frac{a_1 + a_2}{2}\mathbf{r}_1 + \frac{-a_1 + a_2}{2}\mathbf{r}_2$$

the latter sum showing that **t** is in the span of S_2. Conversely, if **u** is in the span of S_2,

$$\mathbf{u} = b_1\mathbf{r}_1 + b_2\mathbf{r}_2 = b_1(\mathbf{e}_1 + \mathbf{e}_2) + b_2(-\mathbf{e}_1 + \mathbf{e}_2)$$
$$= (b_1 - b_2)\mathbf{e}_1 + (b_1 + b_2)\mathbf{e}_2$$

showing that **u** is in the span of S_1. See Figure 2-20.

It may even be that two sets with different numbers of vectors will have the same span. Consider in R^3 the sets

$$S_1 = \{\mathbf{r}_1 = (1, 0, 1), \mathbf{r}_2 = (0, 1, -1)\}$$
$$S_2 = \{\mathbf{t}_1 = (1, 1, 0), \mathbf{t}_2 = (1, -1, 2), \mathbf{t}_3 = (2, 3, -1)\}$$

Figure 2-20

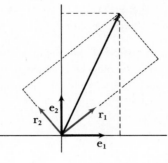

From the observations that

$$t_1 = r_1 + r_2, \qquad t_2 = r_1 - r_2, \qquad t_3 = 2r_1 + 3r_2$$

and

$$r_1 = \tfrac{1}{2}(t_1 + t_2), \qquad r_2 = \tfrac{1}{2}(t_1 - t_2)$$

it is easily seen that any vector that can be expressed as a linear combination of vectors in S_1 can be expressed as a linear combination of vectors in S_2, and vice versa. The span of S_1 is identical to the span of S_2.

Besides the obvious fact that S_2 contains one more vector than does S_1, there is another distinction that is important to note. You can discover with just a little calculation that $5t_1 - t_2 - 2t_3 = 0$, meaning that S_2 is a linearly dependent set. Even less calculation is needed to see that S_1 is linearly independent, since the only way that $a_1(1, 0, 1) + a_2(0, 1, -1)$ can equal $(0, 0, 0)$ is for $a_1 = a_2 = 0$.

In general, if S_1 is a linearly independent set of k vectors, and if S_2 is a set of more than k vectors in the span of S_1, then S_2 is linearly dependent. We can minimize notational problems and still see why this is so if we consider a set of three vectors, say t_1, t_2, and t_3 in the span of a set $S = \{r_1, r_2\}$ of two vectors. Since the three vectors are in the span of S, each can be expressed as a linear combination of the vectors in S.

$$t_1 = a_{11}r_1 + a_{21}r_2$$
$$t_2 = a_{12}r_1 + a_{22}r_2$$
$$t_3 = a_{13}r_1 + a_{23}r_2$$

We now wish to show that three numbers x_1, x_2, and x_3, not all zero, can be chosen so that

$$x_1 t_1 + x_2 t_2 + x_3 t_3 = 0$$

Substituting for the t vectors, this gives

$$x_1(a_{11}r_1 + a_{21}r_2) + x_2(a_{12}r_1 + a_{22}r_2) + x_3(a_{13}r_1 + a_{23}r_2) = 0$$

Collecting coefficients of r_1 and r_2 gives

$$(a_{11}x_1 + a_{12}x_2 + a_{13}x_3)r_1 + (a_{21}x_1 + a_{22}x_2 + a_{23}x_3)r_2 = 0$$

Since $\mathbf{r}_1$ and $\mathbf{r}_2$ are linearly independent, this can only be accomplished if we can find a nontrivial solution to

$$a_{11}x_1 + a_{12}x_2 + a_{13}x_3 = 0$$
$$a_{21}x_1 + a_{22}x_2 + a_{23}x_3 = 0$$

But the existence of a nontrivial solution to this homogeneous system of 2 equations in 3 variables is exactly what is guaranteed by Theorem B of Section 1-4. The same reasoning enables us to prove the following important fact:

> *If the vectors of the set $S = \{\mathbf{r}_1, \ldots, \mathbf{r}_k\}$ are linearly independent, then any set of m vectors in the span of S must be linearly dependent if $m > k$.*

THEOREM A

SUBSPACES

Two properties of the span of a given set $S = \{\mathbf{r}_1, \ldots, \mathbf{r}_k\}$ are worth emphasizing. Note that if $\mathbf{x}$ and $\mathbf{y}$ are both in the span of S, that is, if

$$\mathbf{x} = x_1\mathbf{r}_1 + \cdots + x_k\mathbf{r}_k$$
$$\mathbf{y} = y_1\mathbf{r}_1 + \cdots + y_k\mathbf{r}_k$$

then

$$\mathbf{x} + \mathbf{y} = (x_1 + y_1)\mathbf{r}_1 + \cdots + (x_k + y_k)\mathbf{r}_k$$

and

$$a\mathbf{x} = ax_1\mathbf{r}_1 + \cdots + ax_k\mathbf{r}_k$$

are also in the span of S. These are the two properties that define the important concept of a subspace. A **subspace of** R^n is a collection $\mathcal{S}$ of vectors in R^n having the properties

1. If $\mathbf{x}$ and $\mathbf{y}$ are in $\mathcal{S}$, then so is $\mathbf{x} + \mathbf{y}$.
2. If $\mathbf{x}$ is in $\mathcal{S}$ and a is any real number, then $a\mathbf{x}$ is also in $\mathcal{S}$.

For any n, R^n always contains two obvious subspaces. The $\mathbf{0}$ element taken by itself consists of what we shall call the **zero subspace,** and R^n is a subspace of R^n. This latter observation is an important one, meaning that anything we prove about a subspace of R^n will be true of R^n itself. Any subspace other than $\{\mathbf{0}\}$ or R^n is called a **proper subspace.**

If a vector $\mathbf{x}$ is in a subspace $\mathcal{S}$ of R^n then of course every real multiple $a\mathbf{x}$ is also in $\mathcal{S}$. In particular, $0\mathbf{x} = \mathbf{0}$ must be in $\mathcal{S}$, so any subspace of R^n shares a common zero vector with R^n. Some subspaces and non-subspaces of R^2 are indicated in Figure 2-21. Proper subspaces of R^3 may be visualized as either lines or planes passing through the origin.

Two examples of subspaces of R^n will be of particular interest to us.

The span of a finite set $S = \{\mathbf{r}_1, \ldots, \mathbf{r}_k\}$ forms a subspace $\mathcal{S}$ of R^n. This example is hardly a surprise. Our definition of a subspace was in

Example A

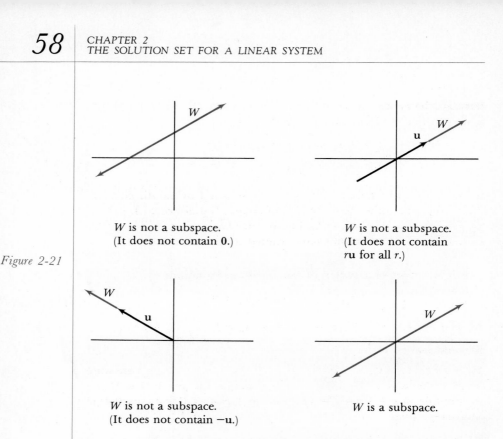

Figure 2-21

W is not a subspace.
(It does not contain **0**.)

W is not a subspace.
(It does not contain
r**u** for all r.)

W is not a subspace.
(It does not contain −**u**.)

W is a subspace.

fact motivated by observing that if **x** and **y** are in the span of a set S, then so are **x** + **y** and a**x** for any real number a.) □

We again have special interest in the case where the vectors of S are linearly independent. In this case we say that they form a **basis** for the subspace spanned by S. Since different sets can span the same subspace, the basis is not unique. This much is true, however: Every basis of a subspace has the same number of vectors. For suppose that a subspace S has the two bases

$$S_1 = \{\mathbf{r}_1, \mathbf{r}_2, \ldots, \mathbf{r}_k\}$$
$$S_2 = \{\mathbf{s}_1, \mathbf{s}_2, \ldots, \mathbf{s}_m\}$$

By definition of a basis, the vectors of S_1 must be linearly independent. Thus, if $m > k$, Theorem A would imply that the set S_2 was linearly dependent, contradicting the fact that S_2 is also a basis. This means that $m \leq k$. But the same argument applied to S_2 would show that $k \leq m$. We have proved another important fact:

THEOREM B

Every basis of a subspace S contains the same number of vectors.)

The number of vectors in a basis of a subspace S is said to be the **dimension** of the space. A commonly used basis of R^n is the so-called

standard basis of

$$e_1 = (1, 0, 0, \ldots, 0)$$
$$e_2 = (0, 1, 0, \ldots, 0)$$
$$\vdots$$
$$e_n = (0, 0, 0, \ldots, 1)$$

in which the only nonzero component of e_i is a 1 in the ith position. There are many other bases of R^n, of course, but Theorem B tells us that every basis, like the standard basis, must have n vectors. Thus, R^n is n-dimensional. We wouldn't have wanted it any other way.

The solution set to a system of homogeneous equations   *Example B*

$$a_{11}x_1 + \cdots + a_{1n}x_n = 0$$
$$\vdots$$
$$a_{m1}x_1 + \cdots + a_{mn}x_n = 0$$

is a subspace of R^n.

We need to show that if $\mathbf{r} = (r_1, \ldots, r_n)$ and $\mathbf{s} = (s_1, \ldots, s_n)$ are solutions to the system, then so are $\mathbf{r} + \mathbf{s}$ and $a\mathbf{r}$ for any real a. Direct substitution of $\mathbf{r} + \mathbf{s} = (r_1 + s_1, \ldots, r_n + s_n)$ into the first equation of the system gives

$$a_{11}(r_1 + s_1) + \cdots + a_{1n}(r_n + s_n)$$
$$= [a_{11}r_1 + \cdots + a_{1n}r_n] + [a_{11}s_1 + \cdots + a_{1n}s_n]$$

The first bracketed term is zero because $\mathbf{r}$ is a solution to the given system, and the second bracketed term is zero because $\mathbf{s}$ is a solution. The same reasoning applied to each of the m equations shows that $\mathbf{r} + \mathbf{s}$ does indeed satisfy the given system. It is similarly easy to show that $a\mathbf{r}$ is a solution. □

Because the solution set to a homogeneous system forms a subspace, it is often called the **solution space** for the system.

We are finally ready to ask the principal question of this section in the context where we want to address it. We now know that the set of solutions to the homogeneous system

$$2x_1 + x_2 - x_3 + 2x_4 - x_5 = 0$$
$$3x_1 - x_2 - 2x_3 + 3x_5 = 0$$
$$-2x_1 + x_3 - x_4 = 0$$
$$5x_1 - x_2 - 4x_3 + 7x_5 = 0$$

is a subspace of R^5. One method of solving the system led to a description of the solution space as the span of the set

$$S_1 = \{\mathbf{r}_1 = (1, 1, 1, -1, 0), \mathbf{r}_2 = (0, -1, 2, 2, 1)\}$$

A second method led to describing the solution space as the span of

$$S_2 = \{s_1 = (2, 1, 4, 0, 1), \; s_2 = (1, 0, 3, 1, 1)\}$$

A general method for verifying that these really do describe the same subspace proceeds as follows:

Given a set S of vectors

$$\mathbf{r}_1 = (r_{11}, r_{12}, \ldots, r_{1n})$$
$$\vdots \qquad \vdots \quad \vdots \qquad \vdots$$
$$\mathbf{r}_k = (r_{k1}, r_{k2}, \ldots, r_{kn})$$

form the $k \times n$ matrix

$$M = \begin{bmatrix} r_{11} & r_{12} & \cdots & r_{1n} \\ \vdots & \vdots & & \vdots \\ r_{k1} & r_{k2} & \cdots & r_{kn} \end{bmatrix}$$

using the vectors of S as the rows of M. It is natural to refer to $\mathbf{r}_1, \ldots, \mathbf{r}_k$ as **row vectors** of M, and to the subspace that they span as the **row space** of M.

Recall now that matrix M is said to be row equivalent to a matrix N, written $M \sim N$, if we can obtain

$$N = \begin{bmatrix} s_{11} & s_{12} & \cdots & s_{1n} \\ \vdots & \vdots & & \vdots \\ s_{k1} & s_{k2} & \cdots & s_{kn} \end{bmatrix}$$

by performing a sequence of elementary row operations on M. Thus, if $M \sim N$, then any row of N must be obtainable by some combination of the following operations:

    I. Interchange rows of M.
    II. Multiply rows of M by a nonzero constant.
    III. Add to some row of M a multiple of any other row.

Given that the rows of M are the vectors $\mathbf{r}_1, \ldots, \mathbf{r}_k$, it follows that any row vector of N must be of the form

$$a_1\mathbf{r}_1 + \cdots + a_k\mathbf{r}_k$$

That is, any row vector of N is in the row space of M. Moreover, since row equivalence goes two ways, it also follows that any row vector of M is in the row space of N. This critical observation is worth stating formally.

THEOREM C

> If the $k \times n$ matrices M and N are row equivalent, their row vectors span the same subspace of R^n.

The uniqueness of the row echelon form of a matrix now provides a practical method to determine whether or not two different sets S_1 and

S_2 span the same subspace of R^n. We use the vectors of S_1 as row vectors of one matrix M and the vectors of S_2 as row vectors of another matrix N. If necessary we can add rows of zeros to one matrix so that they both have k rows. We then reduce M and N to their unique echelon form. Their row spaces are then seen to be equivalent if and only if they have the same row echelon form.

The teacher checking solutions to the homogeneous system studied in this section needs to know whether the same subspace of R^5 is spanned by the sets

Example C

$$S_1 = \{\mathbf{r}_1 = (1, 1, 1, -1, 0),\ \mathbf{r}_2 = (0, -1, 2, 2, 1)\}$$
$$S_2 = \{\mathbf{s}_1 = (2, 1, 4, 0, 1),\ \mathbf{s}_2 = (1, 0, 3, 1, 1)\}$$

We reduce to row echelon form the matrices

$$M = \begin{bmatrix} 1 & 1 & 1 & -1 & 0 \\ 0 & -1 & 2 & 2 & 1 \end{bmatrix}$$

$$\sim \begin{bmatrix} 1 & 0 & 3 & 1 & 1 \\ 0 & 1 & -2 & -2 & -1 \end{bmatrix} \quad \begin{array}{l} \text{(first row) + (second row)} \\ -1 \text{ (second row)} \end{array}$$

$$N = \begin{bmatrix} 2 & 1 & 4 & 0 & 1 \\ 1 & 0 & 3 & 1 & 1 \end{bmatrix}$$

$$\sim \begin{bmatrix} 1 & 0 & 3 & 1 & 1 \\ 0 & 1 & -2 & -2 & -1 \end{bmatrix} \quad \begin{array}{l} \text{second row copied} \\ -2 \text{ (second row) + (first row)} \end{array}$$

Since these matrices are row equivalent, the question is resolved by Theorem C. $\square$

Determine whether or not the same subspace of R^4 is spanned by the sets

$$S_1 = \{\mathbf{r}_1 = (1, -2, 4, -7),\ \mathbf{r}_2 = (3, -2, 8, -13),$$
$$\mathbf{r}_3 = (2, 5, -1, 4)\}$$
$$S_2 = \{\mathbf{s}_1 = (3, -2, 8, -13),\ \mathbf{s}_2 = (7, 6, 8, -9)\}$$

The first matrix to be reduced to row echelon form is

$$M = \begin{bmatrix} 1 & -2 & 4 & -7 \\ 3 & -2 & 8 & -13 \\ 2 & 5 & -1 & 4 \end{bmatrix} \sim \begin{bmatrix} 1 & -2 & 4 & -7 \\ 0 & 4 & -4 & 8 \\ 0 & 9 & -9 & 18 \end{bmatrix}$$

$$\sim \begin{bmatrix} 1 & 0 & 2 & -3 \\ 0 & 1 & -1 & 2 \\ 0 & 0 & 0 & 0 \end{bmatrix}$$

The second matrix to be reduced is formed, to keep the sizes compatible, by using a row of zeros.

$$N = \begin{bmatrix} 3 & -2 & 8 & -13 \\ 7 & 6 & 8 & -9 \\ 0 & 0 & 0 & 0 \end{bmatrix} \sim \begin{bmatrix} 1 & -\frac{2}{3} & \frac{8}{3} & -\frac{13}{3} \\ 0 & \frac{32}{3} & -\frac{32}{3} & \frac{64}{3} \\ 0 & 0 & 0 & 0 \end{bmatrix}$$

$$\sim \begin{bmatrix} 1 & 0 & 2 & -3 \\ 0 & 1 & -1 & 2 \\ 0 & 0 & 0 & 0 \end{bmatrix} \quad \square$$

Finally, note that since the nonzero row vectors of a row echelon matrix are always linearly independent, their number gives the dimension of the space spanned by the original set. In Example D, sets S_1 and S_2 both span a 2-dimensional subspace.

PROBLEM
SET 2-4

In Problems 1–6, decide whether the two sets $\mathbf{r}_1$, $\mathbf{r}_2$, $\cdots$ and $\mathbf{s}_1$, $\mathbf{s}_2$, $\cdots$ span the same subspace.

1. $\mathbf{r}_1 = (1, -1, 2, 0)$ $\mathbf{s}_1 = (1, 2, 1, 2)$
   $\mathbf{r}_2 = (0, 3, -1, 2)$ $\mathbf{s}_2 = (2, -2, 4, 0)$
   $\mathbf{r}_3 = (2, -5, 5, -2)$

2. $\mathbf{r}_1 = (-1, 1, 0, 2)$ $\mathbf{s}_1 = (0, 3, 0, 3)$
   $\mathbf{r}_2 = (1, 2, 0, 1)$ $\mathbf{s}_2 = (0, 0, 1, 0)$
   $\mathbf{r}_3 = (1, 0, 0, -1)$ $\mathbf{s}_3 = (2, 2, 0, 0)$

3. $\mathbf{r}_1 = (-1, 2, 0, 2, 1)$ $\mathbf{s}_1 = (-1, 3, -1, 5, 1)$
   $\mathbf{r}_2 = (0, 1, -1, 3, 0)$ $\mathbf{s}_2 = (-2, 6, -1, -1, 2)$

4. $\mathbf{r}_1 = (1, -1, 3)$ $\mathbf{s}_1 = (0, 1, 1)$
   $\mathbf{r}_2 = (0, -2, 1)$ $\mathbf{s}_2 = (1, 0, 1)$
   $\mathbf{r}_3 = (1, 0, -1)$ $\mathbf{s}_3 = (1, 1, 0)$

5. $\mathbf{r}_1 = (-1, 1, 0, 2, 0)$ $\mathbf{s}_1 = (-2, 3, 0, 7, -1)$
   $\mathbf{r}_2 = (0, 1, 0, 3, -1)$ $\mathbf{s}_2 = (4, -2, 1, -2, 2)$
   $\mathbf{r}_3 = (4, -1, 1, 1, 1)$ $\mathbf{s}_3 = (1, 0, 1, 0, 1)$
   $\mathbf{r}_4 = (5, -3, 1, -4, 2)$

6. $\mathbf{r}_1 = (1, 1, 0, 0, -1)$ $\mathbf{s}_1 = (1, -1, -1, 0, 0)$
   $\mathbf{r}_2 = (-1, 1, 1, 0, 0)$ $\mathbf{s}_2 = (0, 1, -1, -1, 0)$
   $\mathbf{r}_3 = (0, -1, 1, 1, 0)$ $\mathbf{s}_3 = (0, 0, 1, -1, -1)$

Given four vectors $\mathbf{r}_1$, $\mathbf{r}_2$, $\mathbf{r}_3$, $\mathbf{r}_4$ in R^3, it should always be possible, according to Theorem A, to find real numbers x_1, x_2, x_3, x_4, not all zero, so that $x_1\mathbf{r}_1 + x_2\mathbf{r}_2 + x_3\mathbf{r}_3 + x_4\mathbf{r}_4 = \mathbf{0}$. Do so in Problems 7–10.

7. $\mathbf{r}_1 = (4, 1, 1)$ 8. $\mathbf{r}_1 = (3, 0, 5)$
   $\mathbf{r}_2 = (-1, 1, 3)$ $\mathbf{r}_2 = (7, -3, 10)$
   $\mathbf{r}_3 = (3, 2, -2)$ $\mathbf{r}_3 = (-4, 1, 3)$
   $\mathbf{r}_4 = (-2, -3, 2)$ $\mathbf{r}_4 = (1, -2, -4)$

9. $\mathbf{r}_1 = (4, 3, -2)$

    $\mathbf{r}_2 = (-1, 1, 3)$

    $\mathbf{r}_3 = (-3, -2, 9)$

    $\mathbf{r}_4 = (-2, -4, 1)$

10. $\mathbf{r}_1 = (3, -2, 5)$

    $\mathbf{r}_2 = (7, -6, 3)$

    $\mathbf{r}_3 = (-6, 5, -7)$

    $\mathbf{r}_4 = (-5, 5, -8)$

In this section we repeatedly returned to vectors of the forms

$$x_1(1, 1, 1, -1, 0) + x_5(0, -1, 2, 2, 1)$$
$$x_2(2, 1, 4, 0, 1) + x_4(1, 0, 3, 1, 1)$$

and claimed that vectors expressible in one form can be expressed in the other. Problems 11 and 12 ask you to show how to do this.

11. Give formulas telling, for given values of x_1 and x_5, how to choose x_2 and x_4.

12. Give formulas telling, for given values of x_2 and x_4, how to choose x_1 and x_5.

A BALANCING ACT

A 1-dimensional solution space to a homogeneous system of equations does not give us specific values for the variables involved, but merely indicates the proportions that must exist between them. Sometimes the nature of a problem is such that this is exactly the kind of answer to be expected. Consider, for example, the study of chemical reactions in which we wish to ascertain the relative amounts of the compounds that enter into a reaction. This is called **balancing** the chemical equation. Thus, it may be known that

    Cl_2 and KOH react to form KCl, $KClO_3$, and H_2O

Balancing the equation requires that we find coefficients x_i so that in the equation

$$x_1Cl_2 + x_2KOH = x_3KCl + x_4KClO_3 + x_5H_2O$$

we have the same number of atoms of each element on each side of the equation. Thus, we require for the elements present

Chlorine	(Cl)	$2x_1 = x_3 + x_4$
Potassium	(K)	$x_2 = x_3 + x_4$
Oxygen	(O)	$x_2 = 3x_4 + x_5$
Hydrogen	(H)	$x_2 = 2x_5$

This system of 4 equations in 5 variables can be solved by the methods described in this section. The solution space is spanned by $(3, 6, 5, 1, 3)$, a vector that gives the relative amounts of each compound entering into the reaction.

13. Given

$$S = \{\mathbf{r}_1 = (1, -1, 0, 2), \mathbf{r}_2 = (0, 2, -1, 3), \mathbf{r}_3 = (2, -4, 1, 1)\}$$

note that the vector $\mathbf{r} = (3, -5, 1, 3)$ can be expressed in two ways: $\mathbf{r} = \mathbf{r}_1 + \mathbf{r}_3$ and $\mathbf{r} = 3\mathbf{r}_1 - \mathbf{r}_2$. Show that this could not have happened if S had been linearly independent.

14. Suppose that $S = \{\mathbf{r}_1, \cdots, \mathbf{r}_k\}$ is a linearly dependent set. Prove that there is at least one vector $\mathbf{r}$ in the span of S that can be represented in the following two distinct ways:

$$\mathbf{r} = a_1\mathbf{r}_1 + \cdots + a_k\mathbf{r}_k \quad \text{and} \quad \mathbf{r} = b_1\mathbf{r}_1 + \cdots + b_k\mathbf{r}_k$$

15. Let $S_1 = \{\mathbf{r}_1, \mathbf{r}_2\}$ and $S_2 = \{\mathbf{s}_1, \mathbf{s}_2, \mathbf{s}_3\}$ be sets in R^5. Suppose the span of S_1 and the span of S_2 have only the vector $\mathbf{0}$ in common. Must $\{\mathbf{r}_1, \mathbf{r}_2, \mathbf{s}_1, \mathbf{s}_2, \mathbf{s}_3\}$ span R^5?

16. Refer to the sets S_1 and S_2 in Problem 15. Suppose that $\{\mathbf{r}_1, \mathbf{r}_2, \mathbf{s}_1, \mathbf{s}_2, \mathbf{s}_3\}$ spans R^5. Can there be a nonzero vector common to the span of S_1 and the span of S_2?

17. Suppose the set $\{\mathbf{r}_1, \mathbf{r}_2, \mathbf{r}_3, \mathbf{r}_4\}$ spans R^4. Must the set be linearly independent?

18. Suppose that all the vectors of the set $\{\mathbf{r}_1, \mathbf{r}_2, \mathbf{r}_3, \mathbf{r}_4\}$ are in R^4, and that they are linearly independent. Must they form a basis for R^4?

19. Prove that the row vectors of a row echelon matrix are linearly independent.

20. Suppose the vectors of the set $S = \{\mathbf{r}_1, \cdots, \mathbf{r}_k\}$ are mutually orthogonal; that is, $\mathbf{r}_i \cdot \mathbf{r}_j = 0$ when $i \neq j$. Must S be a linearly independent set?

V   21. Find the relative amounts of each compound needed to balance the chemical equation

$$x_1\text{Pb}(\text{N}_3)_2 + x_2\text{Cr}(\text{MnO}_4)_2 = x_3\text{Cr}_2\text{O}_3 + x_4\text{MnO}_2 + x_5\text{Pb}_3\text{O}_4 + x_6\text{NO}$$

2-5   MULTIPLICATION OF MATRICES

The matrix

$$A = \begin{bmatrix} a_{11} & a_{12} & \cdots & a_{1n} \\ a_{21} & a_{22} & \cdots & a_{2n} \\ \vdots & \vdots & & \vdots \\ a_{m1} & a_{m2} & \cdots & a_{mn} \end{bmatrix}$$

having m rows and n columns is said to be an $m \times n$ matrix. We have already referred to the m rows of A as row vectors; they are vectors in R^n. Similarly, the n columns of A may be thought of as vectors in R^m. They are the **column vectors** of A.

If $m = 1$ so that A consists of just one row vector, A is called a **row matrix**. And if $n = 1$, then A is called a **column matrix**.

A SPECIAL MATRIX PRODUCT

With the announced goal of writing the linear system

$$
\begin{aligned}
a_{11}x_1 + \cdots + a_{1n}x_n &= b_1 \\
a_{21}x_1 + \cdots + a_{2n}x_n &= b_2 \\
&\ \vdots \\
a_{m1}x_1 + \cdots + a_{mn}x_n &= b_m
\end{aligned}
$$

(1)

in the form $A\mathbf{x} = \mathbf{b}$, it is natural to define the product of an $m \times n$ matrix A with an $n \times 1$ column matrix $\mathbf{x}$ by

$$
A\mathbf{x} = \begin{bmatrix} a_{11} & a_{12} & \cdots & a_{1n} \\ \vdots & \vdots & & \vdots \\ a_{m1} & a_{m2} & \cdots & a_{mn} \end{bmatrix} \begin{bmatrix} x_1 \\ x_2 \\ \vdots \\ x_n \end{bmatrix} = \begin{bmatrix} a_{11}x_1 + a_{12}x_2 + \cdots + a_{1n}x_n \\ \vdots & \vdots & \vdots \\ a_{m1}x_1 + a_{m2}x_2 + \cdots + a_{mn}x_n \end{bmatrix}
$$

Note that the product on the right is an $m \times 1$ matrix, another column matrix. If we now agree that two matrices will be called **equal** if and only if all their corresponding entries are equal, the system (1) can indeed be written in the form $A\mathbf{x} = \mathbf{b}$, where $\mathbf{b}$ represents the obvious $m \times 1$ column matrix of constants.

Before going on, be certain that you understand how to obtain the two products illustrated below.

$$
\begin{bmatrix} -1 & 2 & 1 \\ 1 & -1 & 0 \\ 0 & 1 & 3 \\ 2 & -1 & -1 \end{bmatrix} \begin{bmatrix} 1 \\ -1 \\ -2 \end{bmatrix} = \begin{bmatrix} -5 \\ 2 \\ -7 \\ 5 \end{bmatrix}, \quad \begin{bmatrix} -1 & 2 & 1 \\ 1 & -1 & 0 \\ 0 & 1 & 3 \\ 2 & -1 & -1 \end{bmatrix} \begin{bmatrix} 2 \\ -3 \\ 1 \end{bmatrix} = \begin{bmatrix} -7 \\ 5 \\ 0 \\ 6 \end{bmatrix} \qquad \square
$$

Example A

A PRODUCT OF TWO MATRICES

Suppose that the ith row vector of the $m \times k$ matrix A is designated by $\mathbf{A}_{i\cdot}$, and that the jth column vector of the $k \times n$ matrix B is designated by $\mathbf{B}_{\cdot j}$.

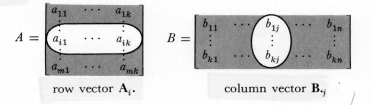

$$
A = \begin{bmatrix} a_{11} & \cdots & a_{1k} \\ & \vdots & \\ a_{i1} & \cdots & a_{ik} \\ & \vdots & \\ a_{m1} & \cdots & a_{mk} \end{bmatrix} \qquad B = \begin{bmatrix} b_{11} & \cdots & b_{1j} & \cdots & b_{1n} \\ \vdots & & \vdots & & \vdots \\ b_{k1} & \cdots & b_{kj} & \cdots & b_{kn} \end{bmatrix}
$$

row vector $\mathbf{A}_{i\cdot}$.   column vector $\mathbf{B}_{\cdot j}$

Then the products

$$AB._1, AB._2, \cdots, AB._n$$

are all defined, and they point us to the correct way to multiply the $m \times k$ matrix A and the $k \times n$ matrix B having $\mathbf{B}._1, \mathbf{B}._2, \cdots, \mathbf{B}._n$ as column vectors. Again, simply multiply A on the right by $\mathbf{B}._1$, then $\mathbf{B}._2$, and so on, setting the resulting column vectors into a matrix in the same sequence.

$$AB = A[\mathbf{B}._1 \quad \mathbf{B}._2 \quad \cdots \quad \mathbf{B}._n] = [A\mathbf{B}._1 \quad A\mathbf{B}._2 \quad \cdots \quad A\mathbf{B}._n]$$

It is important, if the product AB is to be formed, that the number of columns in A and the number of rows in B be equal. Note carefully the dimensions of the product (2).

(2)

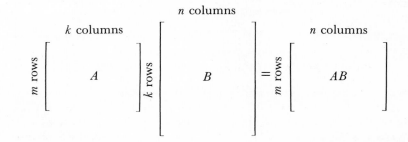

Example B

Find the product of

$$A = \begin{bmatrix} 2 & 0 & -1 & 1 \\ 0 & 3 & 1 & -2 \\ 1 & 0 & 2 & 3 \end{bmatrix}, \quad B = \begin{bmatrix} 1 & -3 \\ 2 & 1 \\ 0 & -2 \\ -1 & 2 \end{bmatrix}$$

The product AB will be 3×2. We begin by finding the first column of AB.

$$\begin{bmatrix} 2 & 0 & -1 & 1 \\ 0 & 3 & 1 & -2 \\ 1 & 0 & 2 & 3 \end{bmatrix} \begin{bmatrix} 1 & -3 \\ 2 & 1 \\ 0 & -2 \\ -1 & 2 \end{bmatrix} = \begin{bmatrix} 1 \\ 8 \\ -2 \end{bmatrix}$$

With that completed, we move on to the second column.

$$\begin{bmatrix} 2 & 0 & -1 & 1 \\ 0 & 3 & 1 & -2 \\ 1 & 0 & 2 & 3 \end{bmatrix} \begin{bmatrix} 1 & -3 \\ 2 & 1 \\ 0 & -2 \\ -1 & 2 \end{bmatrix} = \begin{bmatrix} 1 & -2 \\ 8 & -3 \\ -2 & -1 \end{bmatrix} \quad \square$$

A formal definition of matrix multiplication is most easily stated in terms of the dot product of vectors. Let $A_i.$ be the ith row vector of an $m \times k$ matrix A, and let $B_{.j}$ be the jth column vector of a $k \times n$ matrix B. Then the product AB is an $m \times n$ matrix C having as its ith row, jth column the entry

$$c_{ij} = A_i. \cdot B_{.j}$$

For instance, in Example B, the entry in the first row, second column of the product is

$$A_1. \cdot B_{.2} = [2 \quad 0 \quad -1 \quad 1] \begin{bmatrix} -3 \\ 1 \\ -2 \\ 2 \end{bmatrix} = -2$$

NOTATION

We have written $A_1.$ as a row matrix and $B_{.2}$ as a column matrix to emphasize that the dot product of these two vectors may also be thought of as the product of a 1×4 matrix and a 4×1 matrix (in that order). The distinction between a vector $(2, 0, -1, 1)$ and a row matrix $[2 \quad 0 \quad -1 \quad 1]$ is artificial, and whether we write the components in a row or a column is a matter of convenience.

When we wish to think of a vector $\mathbf{x}$ as a matrix, the context usually makes clear whether we are thinking of a row matrix or a column matrix. When there is doubt, we specifically say that $\mathbf{x}$ is a row or a column matrix. In cases where it is convenient to represent $\mathbf{x}$ both as a row matrix and a column matrix in the same problem, we refer to the row matrix $\mathbf{x}$ and its **transpose** $\mathbf{x}^t$.

$$\mathbf{x} = [x_1 \quad x_2 \quad \cdots \quad x_n], \qquad \mathbf{x}^t = \begin{bmatrix} x_1 \\ x_2 \\ \vdots \\ x_n \end{bmatrix}$$

Note that for the row vector $\mathbf{x}$, $\mathbf{x} \cdot \mathbf{x}^t = |\mathbf{x}|^2$.

The transpose of a row matrix is a special case of a more general concept. The **transpose** of an $m \times n$ matrix A is the $n \times m$ matrix A^t obtained by writing all the row vectors of A as the corresponding column vectors of A^t.)

If $\qquad A = \begin{bmatrix} 2 & 1 & -3 \\ 4 & -1 & 5 \end{bmatrix}, \qquad A^t = \begin{bmatrix} 2 & 4 \\ 1 & -1 \\ -3 & 5 \end{bmatrix}$

If $\qquad A = \begin{bmatrix} 4 & 1 & 3 \\ 1 & 0 & 2 \\ 2 & -1 & 5 \end{bmatrix}, \qquad A^t = \begin{bmatrix} 4 & 1 & 2 \\ 1 & 0 & -1 \\ 3 & 2 & 5 \end{bmatrix}$

(Obviously, $(A^t)^t = A$, and for the row vector $\mathbf{x}$ in particular, $(\mathbf{x}^t)^t = \mathbf{x}$. If A is square, A^t is the reflection of A in its main diagonal.)

Products of transposes behave—or you may prefer to say that they misbehave—in an unexpected way. You can anticipate the general rule by observing that if

$$A = \begin{bmatrix} 1 & -2 \\ 3 & -1 \end{bmatrix} \quad \text{and} \quad B = \begin{bmatrix} 0 & 2 \\ 1 & -2 \end{bmatrix}$$

then

$$AB = \begin{bmatrix} 1 & -2 \\ 3 & -1 \end{bmatrix}\begin{bmatrix} 0 & 2 \\ 1 & -2 \end{bmatrix} = \begin{bmatrix} -2 & 6 \\ -1 & 8 \end{bmatrix}$$

$$A^t B^t = \begin{bmatrix} 1 & 3 \\ -2 & -1 \end{bmatrix}\begin{bmatrix} 0 & 1 \\ 2 & -2 \end{bmatrix} = \begin{bmatrix} 6 & -5 \\ -2 & 0 \end{bmatrix}$$

$$B^t A^t = \begin{bmatrix} 0 & 1 \\ 2 & -2 \end{bmatrix}\begin{bmatrix} 1 & 3 \\ -2 & -1 \end{bmatrix} = \begin{bmatrix} -2 & -1 \\ 6 & 8 \end{bmatrix}$$

Which product gives $(AB)^t$?

THEOREM A

(*If matrices A and B can be multiplied to form AB, then $(AB)^t = B^t A^t$.*

Proof: Let $C = AB$. Then the element in row i, column j of C^t is the element in row j, column i of C. That is,

$$\left(\begin{array}{c} \text{entry in row } i, \text{ column } j \\ \text{of } C^t \end{array}\right) = \left(\begin{array}{c} \text{entry in row } j, \text{ column } i \\ \text{of } C \end{array}\right)$$

$$= [\text{row } j \text{ of } A] \cdot [\text{column } i \text{ of } B]$$
$$= [\text{column } i \text{ of } B] \cdot [\text{row } j \text{ of } A]$$
$$= [\text{row } i \text{ of } B^t] \cdot [\text{column } j \text{ of } A^t]$$
$$= \left(\begin{array}{c} \text{entry in row } i, \text{ column } j \\ \text{of } B^t A^t \end{array}\right)$$

We conclude therefore that $C^t = B^t A^t$.) □

Consider the product of the 3×4 column matrix A and the 4×1 *Example C*
column matrix $\mathbf{b}$ where

$$A\mathbf{b} = \begin{bmatrix} 1 & -1 & 0 & 2 \\ 0 & 3 & 1 & -1 \\ 2 & 1 & -1 & 4 \end{bmatrix} \begin{bmatrix} 1 \\ 1 \\ 2 \\ -1 \end{bmatrix} = \begin{bmatrix} -2 \\ 6 \\ -3 \end{bmatrix} = \mathbf{c}$$

Then

$$\mathbf{b}^t A^t = \begin{bmatrix} 1 & 1 & 2 & -1 \end{bmatrix} \begin{bmatrix} 1 & 0 & 2 \\ -1 & 3 & 1 \\ 0 & 1 & -1 \\ 2 & -1 & 4 \end{bmatrix} = \begin{bmatrix} -2 & 6 & -3 \end{bmatrix} = \mathbf{c}^t \quad \square$$

Other properties of matrix multiplication will be developed in the next
section. Many of them can be anticipated by careful observation as you
work the exercises in Problem Set 2-5.

PETRIFIED MATRICES

Before archaeological finds in a region can be placed on a time-
scale, or even when they cannot be so placed, it is often helpful to
order them according to which came earlier in time, a practice
called sequence-dating or **seriation.** Flinders Petrie proposed in
1899 that matrices could be used as a tool in such an effort.

Let the rows $i = 1, \cdots, g$ correspond to g graves opened in a certain region. Let the columns $j = 1, \cdots, c$ represent c characteristics (varieties of pottery, methods of interment, and so on) observed in the graves. If characteristic j is observed in grave i, place
a 1 in position ij; otherwise, place a 0. The resulting matrix is
called an **incidence matrix.** For example, A and B are incidence
matrices:

$$A = \begin{bmatrix} 0 & 1 & 0 & 0 & 1 & 1 & 1 \\ 1 & 1 & 0 & 1 & 0 & 1 & 0 \\ 0 & 0 & 0 & 0 & 1 & 1 & 0 \\ 1 & 0 & 0 & 1 & 0 & 0 & 0 \\ 0 & 1 & 0 & 1 & 1 & 1 & 1 \end{bmatrix} \qquad B = \begin{bmatrix} 1 & 0 & 0 & 1 & 0 & 0 & 0 \\ 1 & 1 & 0 & 1 & 0 & 1 & 0 \\ 0 & 1 & 0 & 1 & 1 & 1 & 1 \\ 0 & 1 & 0 & 0 & 1 & 1 & 1 \\ 0 & 0 & 0 & 0 & 1 & 1 & 0 \end{bmatrix}$$

Notice that in matrix B, the 1s in each column are bunched. Such
a matrix is called a **Petrie matrix.** The idea is that if an incidence

matrix can be brought into this form, then graves with the same characteristics are bunched and it might be that they have been listed in something approximating correct temporal order. Given an incidence matrix A, the question is this: Is there a permutation of the rows of A that will give a Petrie matrix? More succinctly, can A be petrified? Matrix B above is a petrified form of matrix A. (We say a petrified form, since a petrified form is not generally unique.)

Form the square matrix $C = A^t A$; the dimension of C is the number of characteristics, and the entry c_{ij} tells us the number of graves that exhibit both characteristics i and j. Form $G = AA^t$; the dimension of this square matrix is the number of graves, and g_{ij} tells us how many characteristics are held in common by graves i and j. There is a good deal more buried in these matrices. The interested reader should consult Kendall, D. G. 1969. Some problems and methods in statistical archaeology. *World Archaeology* 1:68–76.

It is a delicate transition from graves, but much of what we know today about Petrie matrices was motivated by the study of the genetics of microorganisms. (See Fulkerson, D. R., and Gross, O. A. 1965. Incidence matrices and interval graphs. *Pacific J. Math.* 15:835–55.)

PROBLEM SET 2-5

Problems 1–18 all refer to the 2×2 matrices

$$A = \begin{bmatrix} 9 & 13 \\ 2 & 3 \end{bmatrix} \qquad B = \begin{bmatrix} 7 & 5 \\ -3 & -2 \end{bmatrix} \qquad C = \begin{bmatrix} 5 & -8 \\ 2 & -3 \end{bmatrix} \qquad D = \begin{bmatrix} 5 & 3 \\ 3 & 2 \end{bmatrix}$$

1. Compare AB and BA.

2. Compare CD and DC.

3. Compare AC and CA.

4. Compare BD and DB.

5. Compare $(AB)^t$, $A^t B^t$, and $B^t A^t$.

6. Compare $(CD)^t$, $C^t D^t$, and $D^t C^t$.

7. Compare $(BA)^t$, $A^t B^t$, and $B^t A^t$.

8. Compare $(DC)^t$, $C^t D^t$, and $D^t C^t$.

9. Find all of the following products:
   (a) $E = AB$ (b) $F = BC$ (c) EC (d) AF

10. Find all of the following products:
    (a) $G = BC$ (b) $H = CD$ (c) GD (d) BH

11. Find all of the following products:
    (a) $R = AC$ (b) $S = CD$ (c) $T = RD$ (d) $U = AS$

12. Find all of the following products:
    (a) $K = BD$ (b) $L = DA$ (c) $M = KA$ (d) $N = BL$

13. For $I = \begin{bmatrix} 1 & 0 \\ 0 & 1 \end{bmatrix}$, find AI and IA.

14. For $I = \begin{bmatrix} 1 & 0 \\ 0 & 1 \end{bmatrix}$, find BI and IB.

15. Find LA where $L = \begin{bmatrix} 3 & -13 \\ -2 & 9 \end{bmatrix}$

16. Find NB where $N = \begin{bmatrix} -2 & -5 \\ 3 & 7 \end{bmatrix}$

17. Find PC where $P = \begin{bmatrix} -3 & 8 \\ -2 & 5 \end{bmatrix}$

18. Find QD were $Q = \begin{bmatrix} 2 & -3 \\ -3 & 5 \end{bmatrix}$

19. Find $M = AB$, $N = BC$, MC, and AN if

$$A = \begin{bmatrix} 3 & -1 \\ 0 & 2 \\ 1 & -1 \end{bmatrix} \qquad B = \begin{bmatrix} 2 & -1 & 0 & 1 \\ 3 & 0 & 1 & 2 \end{bmatrix} \qquad C = \begin{bmatrix} 1 & -1 & 1 \\ 2 & 0 & -1 \\ 0 & 2 & 1 \\ -1 & 0 & 2 \end{bmatrix}$$

20. Find $M = AB$, $N = BC$, MC, and AN if

$$A = \begin{bmatrix} -1 & 1 & 4 & 1 \\ 3 & 0 & 1 & 2 \end{bmatrix} \qquad B = \begin{bmatrix} -1 & 1 & 0 \\ 2 & 0 & 1 \\ -1 & 1 & 0 \\ 0 & 3 & -1 \end{bmatrix} \qquad C = \begin{bmatrix} 3 & -1 \\ 1 & 0 \\ -1 & 1 \end{bmatrix}$$

21. Using the matrices of Problem 19, find B^tA^t in two ways.

22. Using the matrices of Problem 20, find B^tA^t in two ways.

On the basis of your experience in Problems 1–22, answer the following questions about the multiplication of matrices.

23. Do you think that the multiplication of two matrices R and S is generally commutative? That is, will it generally be true that $RS = SR$?

24. Do you think that the multiplication of three matrices R, S, and T is generally associative? That is, will it generally be true that $(RS)T = R(ST)$?

25. What special properties does I possess?

26. Given a 2×2 matrix J, can you generally find another matrix W so that $JW = I$?

27. Given two 2×2 matrices J and W such that $JW = I$, is there anything noteworthy about WJ?

V 28. Six children from the same day camp are taken one day to a nearby fair. Each child carries an identical leather pouch that had been made as a craft project and, when they return from the fair, each is asked to tell about the contents of their pouch. Items mentioned by various children were:

B—Bike trail map distributed at the 4H booth.
C—Coin stamped with the child's name and address.
F—Fortune from a fortune cookie.
J—Jackknife sold at a stand.
K—Key ring given away by an exhibitor.
M—Small bag of marbles sold at a stand.
P—Prize at the ring toss game.
R—Raincheck from the grandstand show.
S—Stick from a pronto pup.
T—Unused token from one of the carnival rides.

The contents of individual pouches are as indicated:

Child	Contents of Pouch
1	CFKPT
2	BFKR
3	CFJKP
4	BJKMT
5	CMPT
6	BKMRS

Form an incidence matrix A with rows corresponding to children, columns corresponding to items. Can this matrix be petrified? Verify the properties of $C = A^t A$ and $G = A A^t$ as described in the vignette *Petrified Matrices*. If told that the children traveled in pairs, which ones would you guess traveled together?

*29. In the matrix product (2) pictured in this section, suppose that numbers m', k', n' are chosen to satisfy $0 < m' < m$, $0 < k' < k$, $0 < n' < n$, respectively. Now suppose that matrices A and B are **partitioned** as indicated,

$$m'\left\{\begin{bmatrix} \overbrace{A_1}^{k'} & A_2 \\ \hline A_3 & A_4 \end{bmatrix}\right. \quad k'\left\{\begin{bmatrix} \overbrace{B_1}^{n'} & B_2 \\ \hline B_3 & B_4 \end{bmatrix}\right. = m'\left\{\begin{bmatrix} \overbrace{C_1}^{n'} & C_2 \\ \hline C_3 & C_4 \end{bmatrix}\right.$$

each consisting of four submatrices. Try some examples to convince yourself that the product matrix $C = AB$ can then be represented as a partitioned matrix in which $C_1 = A_1B_1 + A_2B_3$, $C_2 = A_1B_2 + A_2B_4$, $C_3 = A_3B_1 + A_4B_3$, $C_4 = A_3B_2 + A_4B_4$. Can you prove that the result always holds?

THE ALGEBRA OF MATRICES   2-6

Two matrices of the same dimension

$$A = \begin{bmatrix} a_{11} & \cdots & a_{1n} \\ \vdots & & \vdots \\ a_{m1} & \cdots & a_{mn} \end{bmatrix} \qquad B = \begin{bmatrix} b_{11} & \cdots & b_{1n} \\ \vdots & & \vdots \\ b_{m1} & \cdots & b_{mn} \end{bmatrix}$$

are said to be equal if all corresponding entries are equal; that is, if $a_{ij} = b_{ij}$ for all $i = 1, \cdots, m, j = 1, \cdots, n$. With equality thus defined, there are three operations involved in the algebra of matrices. Addition is defined between matrices having the same dimensions, and is accomplished in the straightforward way of adding corresponding terms. Multiplication of a matrix A by a real number a (called **scalar multiplication** for reasons to be discussed in Section 4-2) is accomplished by multiplying each term of A by a. And multiplication of two matrices was defined in the last section.

addition
$$A + B = \begin{bmatrix} a_{11} + b_{11} & \cdots & a_{1n} + b_{1n} \\ \vdots & & \vdots \\ a_{m1} + b_{m1} & \cdots & a_{mn} + b_{mn} \end{bmatrix}$$

scalar multiplication
$$rA = \begin{bmatrix} ra_{11} & \cdots & ra_{1n} \\ \vdots & & \vdots \\ ra_{m1} & \cdots & ra_{mn} \end{bmatrix}$$

multiplication of matrices
$$AB = C \quad \text{where} \quad c_{ij} = a_{i1}b_{1j} + \cdots + a_{in}b_{nj}$$

Our goal in this section is to point out which of the properties familiar from elementary algebra can be carried over to the algebra of matrices.

The roles of examples and counterexamples are sharply distinguished in mathematics. As a ray of sun breaking through leaden skies, an example may cause a certain property to dawn upon us. Examples only suggest, however, and even many examples will not drive away the clouds of doubt that must surround the property until it is proved. A counterexample to a property, on the other hand, strikes out like lightning and thunders, "No!"

This is the way we must regard the problems at the end of Section 2-5. They were meant to be, in their own way, suggestive. Most of them suggest things that are true about matrix algebra. A few, however, issue clear messages. The clearest is that in matrix algebra, the familiar property $AB = BA$ is *false*—we have seen counterexamples. Thus,

PROPERTY 1

Matrix multiplication is not commutative.

On a more positive note, Problems 9 through 12 in Problem Set 2-5 suggest that matrix multiplication is associative. A general proof requires that beginning with the matrices

$$A = \begin{bmatrix} a_{11} & \cdots & a_{1r} \\ \vdots & & \vdots \\ a_{m1} & \cdots & a_{mr} \end{bmatrix} \quad B = \begin{bmatrix} b_{11} & \cdots & b_{1s} \\ \vdots & & \vdots \\ b_{r1} & \cdots & b_{rs} \end{bmatrix} \quad C = \begin{bmatrix} c_{11} & \cdots & c_{1n} \\ \vdots & & \vdots \\ c_{s1} & \cdots & c_{sn} \end{bmatrix}$$

we show that the entry in the ith row, jth column of $(AB)C$ is the same as the similarly placed element in $A(BC)$. This is not hard, but it requires either more room or more notation than we wish to use. Thus, we shall state without proof

PROPERTY 2

Matrix multiplication is associative.

Another fact about matrix multiplication that is easier to believe than to establish by writing out all the details is that the distributive laws hold. Note that because of the noncommutativity of multiplication, there are two distributive laws to write down.

$$A(B + C) = AB + AC \quad \text{and} \quad (B + C)A = BA + CA$$

By way of example, if

$$A = \begin{bmatrix} 1 & -1 \\ 0 & 2 \end{bmatrix} \quad B = \begin{bmatrix} -2 & 0 \\ 1 & -1 \end{bmatrix} \quad C = \begin{bmatrix} 0 & -1 \\ 2 & 1 \end{bmatrix}$$

then

$$A(B + C) = \begin{bmatrix} 1 & -1 \\ 0 & 2 \end{bmatrix} \begin{bmatrix} -2 & -1 \\ 3 & 0 \end{bmatrix} = \begin{bmatrix} -5 & -1 \\ 6 & 0 \end{bmatrix}$$

and

$$AB + AC = \begin{bmatrix} -3 & 1 \\ 2 & -2 \end{bmatrix} + \begin{bmatrix} -2 & -2 \\ 4 & 2 \end{bmatrix} = \begin{bmatrix} -5 & -1 \\ 6 & 0 \end{bmatrix}$$

In Problems 3–6 at the end of this section you will find further examples intended to throw light on the truth.

PROPERTY 3

Matrix multiplication is distributive over addition.

The **main diagonal** of an $n \times n$ matrix is the diagonal connecting the upper left and lower right corners. The matrix having 1's on the main diagonal as its only nonzero entries is called the **multiplicative identity**

matrix, usually designated by I.

$$I = \begin{bmatrix} 1 & & \\ & 1 & 0 \\ & & \ddots & \\ & 0 & & 1 \end{bmatrix}$$

It is distinguished by the fact that if A is any other $n \times n$ matrix, then

INFORMATION PLEASE

In 1878, the telephone directory in Washington, D.C., consisted of a single page. In 1971, the last year in which a complete list of numbers in that metropolitan area went to every subscriber, the directory was published in four volumes weighing sixteen pounds. Since 1971, each subscriber has been provided with a directory covering only his or her area of the city.

When the decision is made to provide different directories in different parts of a city, the goal is to provide at each phone the numbers most likely to be called from that phone. To aid in deciding how to partition a city, a count is made of the number of calls made between subareas (called units and defined in ways convenient to the telephone company, such as an area having a certain exchange or serviced by a local office). Depending on the size of the city, the number of units may range between $n = 30$ and $n = 120$. It is natural to record this information in an $n \times n$ matrix.

$$\begin{array}{c} \\ 1 \\ 2 \\ \vdots \\ n \end{array} \begin{array}{c} 1, 2, \ldots, n \\ \begin{bmatrix} r_{ij} \\ \text{(number of calls} \\ \text{from unit } i \text{ to} \\ \text{unit } j) \end{bmatrix} \end{array}$$

An analysis of this matrix provides the principal guidance as to which units should be grouped in the same book. Bell Telephone Laboratories developed a computer program to aid in this analysis. On the basis of the first test of this program made in 1977 in Knoxville, Tennessee, projected savings to the Bell System nationwide were conservatively placed at $5.4 million annually. (See Chen and McCallum. 1977. The application of management science to the design of telephone directories. *Interfaces* 8:58–69.)

$AI = IA = A$. The matrix I plays in matrix multiplication the role of 1 in ordinary multiplication.

The multiplicative property of the zero matrix $\mathcal{O}$ should also be noted. If A is any $n \times n$ matrix and $\mathcal{O}$ is the $n \times n$ matrix having all entries equal to 0, then $A\mathcal{O} = \mathcal{O}A = \mathcal{O}$.

PROPERTY 4

The matrices I and $\mathcal{O}$ behave in matrix multiplication as do the numbers 1 and 0 in ordinary multiplication.

PRODUCTS INVOLVING ROW OR COLUMN MATRICES

The case in which an $m \times n$ matrix A is multiplied on the right by an $n \times 1$ column matrix **x** can be written in an interesting way. We use our definitions for the addition and scalar multiplication of matrices.

$$
\begin{bmatrix} a_{11} & a_{12} & \cdots & a_{1n} \\ a_{21} & a_{22} & \cdots & a_{2n} \\ \vdots & \vdots & & \vdots \\ a_{m1} & a_{m2} & \cdots & a_{mn} \end{bmatrix}
\begin{bmatrix} x_1 \\ x_2 \\ \vdots \\ x_n \end{bmatrix}
$$

$$
= \begin{bmatrix} a_{11}x_1 + a_{12}x_2 + \cdots + a_{1n}x_n \\ a_{21}x_1 + a_{22}x_2 + \cdots + a_{2n}x_n \\ \vdots \\ a_{m1}x_1 + a_{m2}x_2 + \cdots + a_{mn}x_n \end{bmatrix}
$$

$$
= x_1 \begin{bmatrix} a_{11} \\ a_{21} \\ \vdots \\ a_{m1} \end{bmatrix} + x_2 \begin{bmatrix} a_{12} \\ a_{22} \\ \vdots \\ a_{m2} \end{bmatrix} + \cdots + x_n \begin{bmatrix} a_{1n} \\ a_{2n} \\ \vdots \\ a_{mn} \end{bmatrix}
$$

$$
= x_1 \mathbf{A}_{\cdot 1} + x_2 \mathbf{A}_{\cdot 2} + \cdots + x_n \mathbf{A}_{\cdot n}
$$

PROPERTY 5

*If A is an $m \times n$ matrix and **x** is an $n \times 1$ column matrix, then $A\mathbf{x}$ is a linear combination of the column vectors of A.*

A similar result is obtained if the $m \times n$ matrix A is multiplied on the left by a $1 \times m$ row matrix **y**.

$$
\begin{bmatrix} y_1 & y_2 & \cdots & y_m \end{bmatrix}
\begin{bmatrix} a_{11} & a_{12} & \cdots & a_{1n} \\ a_{21} & a_{22} & \cdots & a_{2n} \\ \vdots & & & \vdots \\ a_{m1} & a_{m2} & \cdots & a_{mn} \end{bmatrix}
$$

$$= [a_{11}y_1 + a_{21}y_2 + \cdots + a_{m1}y_m, \cdots, a_{1n}y_1 + a_{2n}y_2 + \cdots + a_{mn}y_m]$$
$$= [a_{11}, a_{12}, \cdots, a_{1n}]y_1 + \cdots + [a_{m1}, a_{m2}, \cdots, a_{mn}]y_m$$
$$= y_1 \mathbf{A}_1. + \cdots + y_m \mathbf{A}_m.$$

If A is an $m \times n$ matrix and $\mathbf{y}$ is a $1 \times m$ row matrix, then $\mathbf{y}A$ is a linear combination of the row vectors of A.

PROPERTY 6

According to Property 5,

Example A

$$\begin{bmatrix} 4 & -2 \\ 2 & 5 \\ -1 & 3 \end{bmatrix} \begin{bmatrix} -2 \\ 3 \end{bmatrix} = -2 \begin{bmatrix} 4 \\ 2 \\ -1 \end{bmatrix} + 3 \begin{bmatrix} -2 \\ 5 \\ 3 \end{bmatrix} = \begin{bmatrix} -8-6 \\ -4+15 \\ 2+9 \end{bmatrix} = \begin{bmatrix} -14 \\ 11 \\ 11 \end{bmatrix}$$

This is easily verified by direct multiplication. Similarly, one can verify the following illustration of Property 6.

$$[2 \quad -1 \quad 3] \begin{bmatrix} 4 & -2 \\ 2 & 5 \\ -1 & 3 \end{bmatrix}$$

$$= 2[4 \quad -2] - [2 \quad 5] + 3[-1 \quad 3]$$
$$= [3 \quad 0] \quad \square$$

ELEMENTARY MATRICES

In our discussion of how to find solutions of systems of equations (Section 1-4), we listed three elementary row operations that might be used on the augmented matrix of the system. When an $n \times n$ matrix is obtained from the identity matrix I by just one elementary operation, that matrix is called an **elementary matrix.** Thus, an elementary matrix is obtained from I in one of three ways.

1. The ith and jth rows are interchanged.

$$\begin{bmatrix} 1 & 0 & 0 & & \cdots & & 0 \\ 0 & 1 & 0 & & \cdots & & 0 \\ 0 & 0 & 0 & \cdots & 1 & \cdots & 0 \\ \vdots & \vdots & \vdots & & \vdots & & \vdots \\ 0 & 0 & 1 & \cdots & 0 & \cdots & 0 \\ \vdots & \vdots & \vdots & & \vdots & & \vdots \\ 0 & 0 & 0 & \cdots & 0 & \cdots & 1 \end{bmatrix}$$

2. The ith row is multiplied by $r \neq 0$.

$$\begin{bmatrix} 1 & 0 & 0 & & \cdots & & 0 \\ 0 & 1 & 0 & & \cdots & & 0 \\ \vdots & \vdots & \vdots & & \ddots & & \vdots \\ 0 & 0 & 0 & \cdots & r & \cdots & 0 \\ & & & & & \ddots & \\ 0 & 0 & 0 & & \cdots & & 1 \end{bmatrix}$$

3. We add a constant multiple of row i to row j.

$$\begin{bmatrix} 1 & & 0 & & \cdots & & 0 \\ 0 & & 1 & & \cdots & & 0 \\ 0 & & \cdots & 1 & & \cdots & 0 \\ 0 & \cdots & r & \cdots & 1 & \cdots & 0 \\ 0 & & & \cdots & & & 1 \end{bmatrix}$$

Examples of each kind of elementary matrix for the case $n = 3$ follow:

1. Rows 2 and 3 are interchanged.

$$E_1 = \begin{bmatrix} 1 & 0 & 0 \\ 0 & 0 & 1 \\ 0 & 1 & 0 \end{bmatrix}$$

2. Row 3 is multiplied by r.

$$E_2 = \begin{bmatrix} 1 & 0 & 0 \\ 0 & 1 & 0 \\ 0 & 0 & r \end{bmatrix}$$

3. To row 1 we add r times row 3.

$$E_3 = \begin{bmatrix} 1 & 0 & r \\ 0 & 1 & 0 \\ 0 & 0 & 1 \end{bmatrix}$$

Our interest in elementary matrices stems from the effect of multiplying an arbitrary matrix on the left by an elementary matrix. We illustrate the general principles that apply to $n \times n$ matrices by continuing to examine the case $n = 3$.

$$E_1 A = \begin{bmatrix} 1 & 0 & 0 \\ 0 & 0 & 1 \\ 0 & 1 & 0 \end{bmatrix} \begin{bmatrix} a_{11} & a_{12} & a_{13} \\ a_{21} & a_{22} & a_{23} \\ a_{31} & a_{32} & a_{33} \end{bmatrix} = \begin{bmatrix} a_{11} & a_{12} & a_{13} \\ a_{31} & a_{32} & a_{33} \\ a_{21} & a_{22} & a_{23} \end{bmatrix}$$

$$E_2 A = \begin{bmatrix} 1 & 0 & 0 \\ 0 & 1 & 0 \\ 0 & 0 & r \end{bmatrix} \begin{bmatrix} a_{11} & a_{12} & a_{13} \\ a_{21} & a_{22} & a_{23} \\ a_{31} & a_{32} & a_{33} \end{bmatrix} = \begin{bmatrix} a_{11} & a_{12} & a_{13} \\ a_{21} & a_{22} & a_{23} \\ ra_{31} & ra_{32} & ra_{33} \end{bmatrix}$$

$$E_3 A = \begin{bmatrix} 1 & 0 & r \\ 0 & 1 & 0 \\ 0 & 0 & 1 \end{bmatrix} \begin{bmatrix} a_{11} & a_{12} & a_{13} \\ a_{21} & a_{22} & a_{23} \\ a_{31} & a_{32} & a_{33} \end{bmatrix}$$

$$= \begin{bmatrix} a_{11} + ra_{31} & a_{12} + ra_{32} & a_{13} + ra_{33} \\ a_{21} & a_{22} & a_{23} \\ a_{31} & a_{32} & a_{33} \end{bmatrix}$$

Multiplication of an arbitrary matrix A on the left by an elementary matrix E changes the rows of A in the same way that the rows of I are changed to obtain E.

PROPERTY 7

This secures the connection between elementary operations on a system of equations and elementary matrices. If a system of equations is written in the form $A\mathbf{x} = \mathbf{b}$, then any of the three elementary row operations may be accomplished by multiplication by an appropriate elementary matrix.

One might naturally wonder about the effect of multiplying on the right by elementary matrices. The easiest way to find out is to try it. We use the same E_1, E_2, and E_3 as we used above.

$$A E_1 = \begin{bmatrix} a_{11} & a_{12} & a_{13} \\ a_{21} & a_{22} & a_{23} \\ a_{31} & a_{32} & a_{33} \end{bmatrix} \begin{bmatrix} 1 & 0 & 0 \\ 0 & 0 & 1 \\ 0 & 1 & 0 \end{bmatrix} = \begin{bmatrix} a_{11} & a_{13} & a_{12} \\ a_{21} & a_{23} & a_{22} \\ a_{31} & a_{33} & a_{32} \end{bmatrix}$$

$$A E_2 = \begin{bmatrix} a_{11} & a_{12} & a_{13} \\ a_{21} & a_{22} & a_{23} \\ a_{31} & a_{32} & a_{33} \end{bmatrix} \begin{bmatrix} 1 & 0 & 0 \\ 0 & 1 & 0 \\ 0 & 0 & r \end{bmatrix} = \begin{bmatrix} a_{11} & a_{12} & ra_{13} \\ a_{21} & a_{22} & ra_{23} \\ a_{31} & a_{32} & ra_{33} \end{bmatrix}$$

$$A E_3 = \begin{bmatrix} a_{11} & a_{12} & a_{13} \\ a_{21} & a_{22} & a_{23} \\ a_{31} & a_{32} & a_{33} \end{bmatrix} \begin{bmatrix} 1 & 0 & r \\ 0 & 1 & 0 \\ 0 & 0 & 1 \end{bmatrix} = \begin{bmatrix} a_{11} & a_{12} & ra_{11} + a_{13} \\ a_{21} & a_{22} & ra_{21} + a_{23} \\ a_{31} & a_{32} & ra_{31} + a_{33} \end{bmatrix}$$

This time it is the columns that are affected. How are they affected? In AE_1, the second and third columns have been interchanged. Now note something. Though we actually obtained E_1 by interchanging the second and third rows of I, who is to say that E_1 was not obtained by interchanging the second and third columns of I? Similarly, E_2 might have been obtained by multiplying the third column of I by r, and E_3 can be thought of as having been obtained from I by adding to the third column r times the first column.

PROPERTY 8

Multiplication of an arbitrary matrix A on the right by an elementary matrix E changes the columns of A in the same way that the columns of I are changed to obtain E.

PROBLEM
SET 2-6

Associativity of matrix multiplication does not depend on the matrices being square. In Problems 1 and 2, show that $(AB)C = A(BC)$.

1. $A = \begin{bmatrix} 1 & -1 & 0 \\ 2 & 1 & 3 \end{bmatrix}$, $\quad B = \begin{bmatrix} 1 & 2 & 3 \\ -1 & 1 & 0 \\ 2 & 0 & -1 \end{bmatrix}$, $\quad C = \begin{bmatrix} 5 & 3 \\ 1 & 1 \\ -2 & 2 \end{bmatrix}$

2. $A = \begin{bmatrix} -1 & 2 \\ 0 & 4 \\ 3 & 1 \end{bmatrix}$, $\quad B = \begin{bmatrix} 1 & 0 & 3 & 1 \\ -1 & 2 & 1 & -2 \end{bmatrix}$, $\quad C = \begin{bmatrix} -1 & 0 \\ 0 & 1 \\ 2 & -1 \\ -3 & 2 \end{bmatrix}$

The distributive law also holds for nonsquare matrices. In Problems 3 and 4, verify that $A(B + C) = AB + AC$.

3. $A = \begin{bmatrix} 2 & -1 & 3 \\ -1 & 2 & 4 \end{bmatrix}$, $\quad B = \begin{bmatrix} 1 & -2 & 0 \\ 0 & 1 & 3 \\ 1 & 0 & -2 \end{bmatrix}$, $\quad C = \begin{bmatrix} -2 & 1 & 1 \\ -1 & 1 & -1 \\ 0 & 1 & 1 \end{bmatrix}$

4. $A = \begin{bmatrix} -1 & 2 \\ 0 & -2 \\ 2 & 3 \end{bmatrix}$, $\quad B = \begin{bmatrix} 2 & 1 & 2 & 1 \\ 3 & 4 & -2 & -3 \end{bmatrix}$, $\quad C = \begin{bmatrix} 1 & 1 & -2 & 1 \\ -1 & -2 & 1 & 2 \end{bmatrix}$

In Problems 5 and 6, verify that $(B + C)A = BA + CA$.

5. $A = \begin{bmatrix} 1 & -1 & 0 \\ -2 & 2 & 3 \end{bmatrix}$, $\quad B = \begin{bmatrix} -1 & 0 \\ 1 & 3 \\ 0 & -1 \\ 2 & 1 \end{bmatrix}$, $\quad C = \begin{bmatrix} 3 & 1 \\ -2 & -4 \\ 1 & 1 \\ 0 & -2 \end{bmatrix}$

6. $A = \begin{bmatrix} 1 \\ 3 \\ 5 \end{bmatrix}$, $\quad B = \begin{bmatrix} 1 & -1 & 2 \\ 4 & -1 & -1 \end{bmatrix}$, $\quad \begin{bmatrix} C = & 2 & 0 & -1 \\ & 1 & 3 & -2 \end{bmatrix}$

7. Add $3\begin{bmatrix} 1 \\ 2 \\ -1 \end{bmatrix} + 4\begin{bmatrix} 2 \\ -1 \\ 3 \end{bmatrix} - 2\begin{bmatrix} 1 \\ -2 \\ 1 \end{bmatrix}$. Then multiply $\begin{bmatrix} 1 & 2 & 1 \\ 2 & -1 & -2 \\ -1 & 3 & 1 \end{bmatrix}\begin{bmatrix} 3 \\ 4 \\ -2 \end{bmatrix}$.

NOT THE ONLY WAY TO FLY

Suppose a small airline flies to five cities, providing one-way service or two-way service as indicated by the directed paths (Figure 2-C). We may describe this same picture in a way that a computer

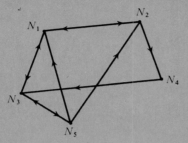

Figure 2-C

can read. Form a 5×5 matrix A in which we enter a 1 in position a_{ij} if there is direct service *from* city N_i to city N_j; otherwise, enter a 0.

$$A = \begin{bmatrix} 0 & 1 & 1 & 0 & 0 \\ 1 & 0 & 0 & 1 & 0 \\ 1 & 0 & 0 & 0 & 1 \\ 0 & 0 & 1 & 0 & 0 \\ 1 & 1 & 1 & 0 & 0 \end{bmatrix}$$

The powers of A now have this interpretation. The entry $a_{ij}^{(k)}$ of the kth power of A tells us how many k-stop trips can be set up from N_i to N_j.

$$A^2 = \begin{bmatrix} 2 & 0 & 0 & 1 & 1 \\ 0 & 1 & 2 & 0 & 0 \\ 1 & 2 & 2 & 0 & 0 \\ 1 & 0 & 0 & 0 & 1 \\ 2 & 1 & 1 & 1 & 1 \end{bmatrix}, \qquad A^3 = \begin{bmatrix} 1 & 3 & 4 & 0 & 0 \\ 3 & 0 & 0 & 1 & 2 \\ 4 & 1 & 1 & 2 & 2 \\ 1 & 2 & 2 & 0 & 0 \\ 3 & 3 & 4 & 1 & 1 \end{bmatrix}$$

From A^3, for instance, we see that we can arrange a 3-stop flight from N_2 to N_4 in one way (2–1–2–4 is the way, found by inspection) and from N_5 to N_1 in three ways, one of which is 5–3–5–1. (Can you find two others?) No 3-stop flight goes from N_4 to N_5, but that trip can (according to A^2) be arranged in one way as a 2-stop flight.

Problems of this sort are a part of that field of mathematics called **graph theory** (a new use of the word graph). The "cities" in our picture are called **nodes,** the paths are **edges,** and the matrix A is an **adjacency matrix.** Applications of graph theory occur in electrical circuit theory, the analysis of a communications network, traffic management, job scheduling, many parts of computer science, and other places.

8. Add $2\begin{bmatrix} -1 \\ 2 \\ 3 \end{bmatrix} - 3\begin{bmatrix} 2 \\ 1 \\ 0 \end{bmatrix} + 4\begin{bmatrix} 2 \\ 1 \\ -1 \end{bmatrix}$. Then multiply $\begin{bmatrix} -1 & 2 & 2 \\ 2 & 1 & 1 \\ 3 & 0 & -1 \end{bmatrix}\begin{bmatrix} 2 \\ -3 \\ 4 \end{bmatrix}$.

9. Add $2\begin{bmatrix} 1 & -1 & 0 \end{bmatrix} + 3\begin{bmatrix} 2 & 4 & 1 \end{bmatrix} - 5\begin{bmatrix} 1 & 0 & -1 \end{bmatrix}$.

   Then multiply $\begin{bmatrix} 2 & 3 & -5 \end{bmatrix}\begin{bmatrix} 1 & -1 & 0 \\ 2 & 4 & 1 \\ 1 & 0 & -1 \end{bmatrix}$.

10. Add $3\begin{bmatrix} 1 & 0 & -2 \end{bmatrix} + 2\begin{bmatrix} -2 & 1 & 1 \end{bmatrix} - 2\begin{bmatrix} 1 & 2 & 0 \end{bmatrix}$.

    Then multiply $\begin{bmatrix} 3 & 2 & -2 \end{bmatrix}\begin{bmatrix} 1 & 0 & -2 \\ -2 & 1 & 1 \\ 1 & 2 & 0 \end{bmatrix}$.

11. With as little calculation as possible, find the entry in the third row, second column of the product

$$\begin{bmatrix} 3 & -1 & 2 & 3 \\ 0 & 2 & 1 & -2 \\ 1 & 0 & -1 & 2 \\ 5 & 4 & -1 & 6 \end{bmatrix}\begin{bmatrix} 1 & -3 \\ -2 & 0 \\ 1 & 1 \\ 3 & 2 \end{bmatrix}\begin{bmatrix} 1 & -1 & 5 & -1 & 4 \\ 6 & 2 & -3 & 1 & 8 \end{bmatrix}$$

12. With as little calculation as possible, find the entry in the second row, fourth column of the product in Problem 11.

In Problems 13–20, you are to determine the elementary matrix E, given the matrix

$$A = \begin{bmatrix} 4 & 3 & -2 & 4 \\ 5 & 1 & 2 & -1 \\ 6 & 0 & 1 & 3 \end{bmatrix}$$

13. $EA = \begin{bmatrix} 6 & 0 & 1 & 3 \\ 5 & 1 & 2 & -1 \\ 4 & 3 & -2 & 4 \end{bmatrix}$ 14. $EA = \begin{bmatrix} -6 & 1 & -6 & 6 \\ 5 & 1 & 2 & -1 \\ 6 & 0 & 1 & 3 \end{bmatrix}$

15. $EA = \begin{bmatrix} 4 & 3 & -2 & 4 \\ -7 & 1 & 0 & -7 \\ 6 & 0 & 1 & 3 \end{bmatrix}$ 16. $EA = \begin{bmatrix} 5 & 1 & 2 & -1 \\ 4 & 3 & -2 & 4 \\ 6 & 0 & 1 & 3 \end{bmatrix}$

17. $AE = \begin{bmatrix} 4 & 3 & 4 & -2 \\ 5 & 1 & -1 & 2 \\ 6 & 0 & 3 & 1 \end{bmatrix}$ 18. $AE = \begin{bmatrix} -2 & 3 & -2 & 4 \\ 3 & 1 & 2 & -1 \\ 6 & 0 & 1 & 3 \end{bmatrix}$

19. $AE = \begin{bmatrix} 4 & 3 & -2 & 2 \\ 5 & 1 & 2 & -\frac{7}{2} \\ 6 & 0 & 1 & 0 \end{bmatrix}$ 20. $AE = \begin{bmatrix} 4 & -6 & -2 & 4 \\ 5 & -2 & 2 & -1 \\ 6 & 0 & 1 & 3 \end{bmatrix}$

21. Find examples that illustrate the following peculiar properties of matrix multiplication:
    (a) It may be that $AB = \mathcal{O}$ even though $A \neq \mathcal{O}$ and $B \neq \mathcal{O}$.
    (b) It may be that $A^2 = \mathcal{O}$ even though $A \neq \mathcal{O}$.
    (c) It is possible for $A^3 = \mathcal{O}$ while $A \neq \mathcal{O}$ and $A^2 \neq \mathcal{O}$.
    (d) For A different from both $\mathcal{O}$ and I, it is still possible that $A^2 = A$.
    (e) For A different from both I and $-I$, it is still possible that $A^2 = I$.

22. For $n \times n$ matrices A and B, does $A^2 - B^2 = (A - B)(A + B)$?

23. For $n \times n$ matrices A and B, does $(A + B)^2 = A^2 + 2AB + B^2$?

24. Let $\mathbf{x}$ be a 3×1 column matrix, $\mathbf{y}$ be a 4×1 column matrix, and A a 3×4 matrix. Will it always be true that $\mathbf{x}^t A \mathbf{y} = \mathbf{y}^t A^t \mathbf{x}$?

25. **A diagonal matrix** is a matrix having its only nonzero entries on the main diagonal. A 3×3 diagonal matrix takes the form

$$D = \begin{bmatrix} a & 0 & 0 \\ 0 & b & 0 \\ 0 & 0 & c \end{bmatrix}$$

   (a) Describe the products DM and MD if M is an arbitrary 3×3 matrix.
   (b) Describe the product of two diagonal matrices.

THE INVERSE OF A MATRIX   2-7

In elementary algebra, that number b for which $ab = ba = 1$ is called the **multiplicative inverse** (or, more carelessly, the **inverse**) of a. It is designated by either $1/a$ or a^{-1}. The same terminology (and the same carelessness) is employed in matrix algebra. That matrix B, if there is one, for which $AB = BA = I$, is called the **inverse** of A.

Those intentionally suggestive exercises in Problem Set 2-5 included the multiplication of

$$LA = \begin{bmatrix} 3 & -13 \\ -2 & 9 \end{bmatrix} \begin{bmatrix} 9 & 13 \\ 2 & 3 \end{bmatrix} = \begin{bmatrix} 1 & 0 \\ 0 & 1 \end{bmatrix} = I$$

The matrix L is, in the terminology just introduced, the inverse of A.

It may seem to you, since matrix multiplication is not commutative, that the matrix L should be called a *left* inverse, allowing the possibility that some other matrix R may serve as a *right* inverse. That can't happen, however, because the associative law tells us that $(LA)R = L(AR)$. Thus, if $LA = I$ and $AR = I$, we have $IR = LI$ or $R = L$. In the example above,

$$AL = \begin{bmatrix} 9 & 13 \\ 2 & 3 \end{bmatrix} \begin{bmatrix} 3 & -13 \\ -2 & 9 \end{bmatrix} = \begin{bmatrix} 1 & 0 \\ 0 & 1 \end{bmatrix} = I$$

We shall see in this section that if an $n \times n$ matrix A has an inverse that works on one side, it will always work on the other. Some things turn out more simply than we have any right to expect. If A has such an inverse, it is always designated by A^{-1}.

INVERSES OF ELEMENTARY MATRICES

We noted in the very first section of this book that the effect of any elementary operation could be reversed by an elementary operation of the same type. In the last section we saw that an elementary operation on the rows of a matrix can be effected by multiplying on the left by an elementary matrix E. It follows that for every elementary matrix E, there is another elementary matrix E^{-1} of the same type that will reverse the effect of E. This principle may be observed in the following products:

$$E_1^{-1}E_1 = \begin{bmatrix} 1 & 0 & 0 \\ 0 & 0 & 1 \\ 0 & 1 & 0 \end{bmatrix} \begin{bmatrix} 1 & 0 & 0 \\ 0 & 0 & 1 \\ 0 & 1 & 0 \end{bmatrix} = I$$

$$E_2^{-1}E_2 = \begin{bmatrix} 1 & 0 & 0 \\ 0 & 1 & 0 \\ 0 & 0 & 1/r \end{bmatrix} \begin{bmatrix} 1 & 0 & 0 \\ 0 & 1 & 0 \\ 0 & 0 & r \end{bmatrix} = I$$

$$E_3^{-1}E_3 = \begin{bmatrix} 1 & 0 & -r \\ 0 & 1 & 0 \\ 0 & 0 & 1 \end{bmatrix} \begin{bmatrix} 1 & 0 & r \\ 0 & 1 & 0 \\ 0 & 0 & 1 \end{bmatrix} = I$$

Thus, given an elementary matrix E, we may always act on the rows of E with another elementary matrix E^{-1} so as to restore order; $E^{-1}E = I$. Now recall that an elementary matrix can be thought of as having resulted from an elementary operation performed on the rows of I, or from the same operation having been performed on the columns of I. Thus, the same restoration that works on the rows of E will work on the columns of E, leading to the observation that $EE^{-1} = I$. Evidently, then, the inverse of an elementary matrix E is two-sided; $E^{-1}E = EE^{-1} = I$. You should verify for the examples above that $E_1E_1^{-1} = E_2E_2^{-1} = E_3E_3^{-1} = I$.

EQUIVALENT MATRICES

Two linear systems of equations were said to be *equivalent* in Section 1-2 if one of the systems was obtainable from the other by a sequence of elementary row operations. This terminology was carried over to the matrices introduced in Section 1-4 to keep track of calculations; matrices A and B were said to be *row equivalent* if one was obtainable from the other by a sequence of elementary row operations. This can now be rephrased to say that A *is row equivalent to* B *if there exists a sequence of elementary matrices* $E_1, \ldots, E_k$ *such that* $B = E_k \ldots E_2 E_1 A$.

Since every matrix A is row equivalent to a unique matrix R in row echelon form, it follows from the previous paragraph that, for every matrix A, we can find elementary matrices such that

$$E_k \ldots E_2 E_1 A = R$$

is in row echelon form. Let us restate this observation more formally.

> *If A is an arbitrary $m \times n$ matrix, then there exists a product of $m \times m$ elementary matrices $P = E_k \ldots E_2 E_1$ such that PA is in row echelon form.*

THEOREM A

There is a practical way to determine the matrix P that will be of future use to us. Observe that since $IA = A$, and since multiplication is associative, (1) may be written in the form

$$(E_k \ldots E_2 E_1 I)A = R$$

The matrix P can therefore be determined if we perform on I the same sequence of elementary row operations used to reduce A to row echelon form. The most common way to do this in practice is to write I to the right of A, and then reduce this matrix to row echelon form.

Find a matrix P such that PA will be in row echelon form, where

Example A

$$A = \begin{bmatrix} 2 & -1 & -3 & 1 \\ 1 & 0 & -2 & 1 \\ -3 & 1 & 1 & 2 \end{bmatrix}$$

We begin by copying the 3×3 matrix I to the right of A.

$$\begin{bmatrix} 2 & -1 & -3 & \vdots & 1 & 1 & 0 & 0 \\ 1 & 0 & -2 & \vdots & 1 & 0 & 1 & 0 \\ -3 & 1 & 1 & \vdots & 2 & 0 & 0 & 1 \end{bmatrix}$$

$$\begin{bmatrix} 1 & 0 & -2 & \vdots & 1 & 0 & 1 & 0 \\ 0 & -1 & 1 & \vdots & -1 & 1 & -2 & 0 \\ 0 & 1 & -5 & \vdots & 5 & 0 & 3 & 1 \end{bmatrix} \quad \begin{array}{l} \text{second row copied} \\ -2(\text{second row}) + (\text{first row}) \\ 3(\text{second row}) + (\text{third row}) \end{array}$$

$$\begin{bmatrix} 1 & 0 & -2 & \vline & 1 & 0 & 1 & 0 \\ 0 & 1 & -1 & \vline & 1 & -1 & 2 & 0 \\ 0 & 0 & -4 & \vline & 4 & 1 & 1 & 1 \end{bmatrix}$$

first row copied
-1(second row)
(second row) + (third row)

$$\begin{bmatrix} 1 & 0 & 0 & \vline & -1 & -\frac{2}{4} & \frac{2}{4} & -\frac{2}{4} \\ 0 & 1 & 0 & \vline & 0 & -\frac{5}{4} & \frac{7}{4} & -\frac{1}{4} \\ 0 & 0 & 1 & \vline & -1 & -\frac{1}{4} & -\frac{1}{4} & -\frac{1}{4} \end{bmatrix}$$

$-\frac{2}{4}$(third row) + (first row)
$-\frac{1}{4}$(third row) + (second row)
$-\frac{1}{4}$(third row)

The matrix P which we seek now appears as the last three columns of our final matrix. To check, note that

$$PA = \begin{bmatrix} -\frac{2}{4} & \frac{2}{4} & -\frac{2}{4} \\ -\frac{5}{4} & \frac{7}{4} & -\frac{1}{4} \\ -\frac{1}{4} & -\frac{1}{4} & -\frac{1}{4} \end{bmatrix} \begin{bmatrix} 2 & -1 & -3 & 1 \\ 1 & 0 & -2 & 1 \\ -3 & 1 & 1 & 2 \end{bmatrix} = \begin{bmatrix} 1 & 0 & 0 & -1 \\ 0 & 1 & 0 & 0 \\ 0 & 0 & 1 & -1 \end{bmatrix}$$

THE INVERSE OF AN $n \times n$ MATRIX

If A is an $n \times n$ matrix, then the unique row echelon matrix equivalent to A may very well be I. Then the matrix P described in Theorem A would, in fact, be an inverse (we should, for the moment, say a left inverse) for A, and the procedure for finding P would be a procedure for finding the inverse of A.

Let us suppose for simplicity that A is an $n \times n$ matrix having the inverse $P = E_3 E_2 E_1$. Then

(2)
$$PA = E_3 E_2 E_1 A = I$$

Since E_3 is an elementary matrix, it has an inverse E_3^{-1}. Multiplication of (2) by E_3^{-1} gives

$$E_3^{-1}(E_3 E_2 E_1)A = E_3^{-1}I = E_3^{-1}$$

Associativity of multiplication gives

$$(E_3^{-1}E_3)E_2 E_1 A = E_3^{-1}$$

or
$$E_2 E_1 A = E_3^{-1}$$

This expression can now be multiplied by the inverse of the elementary matrix E_2.

$$E_2^{-1}(E_2 E_1 A) = E_2^{-1}E_3^{-1}$$

or
$$E_1 A = E_2^{-1}E_3^{-1}$$

A similar multiplication by E_1^{-1} gives us $A = E_1^{-1}E_2^{-1}E_3^{-1}$. From this it follows that

$$AP = (E_1^{-1}E_2^{-1}E_3^{-1})(E_3 E_2 E_1)$$

and several applications of the associative law lead to the conclusion that $AP = I$.

This same process of "peeling away" would work for any $P = E_k \dots E_2 E_1$. Thus, if an $n \times n$ matrix A is row equivalent to I so that the matrix P of Theorem A works as a left inverse, it will also work on the right; from $PA = I$ we are able to show that $AP = I$. The inverse $A^{-1} = P$, if it exists at all, works on either side.

Is it possible, given an $n \times n$ matrix A, that some matrix other than the one determined in Theorem A might work as an inverse? Suppose $BA = I$. Clearly the unique row echelon form of A is I, so the matrix P of Theorem A is in fact A^{-1}. Then $(BA)A^{-1} = IA^{-1}$, from which we conclude that $B = A^{-1}$.

> *If the square matrix A has a left inverse A^{-1}, then A^{-1} is in fact a two-sided inverse; $A^{-1}A = AA^{-1} = I$. Moreover, this inverse is unique.*　　　　*THEOREM B*

A matrix that has an inverse is said to be **invertible** or **nonsingular,** and of course a matrix that does not have an inverse is called **noninvertible** or **singular.**

Find the inverse of $\begin{bmatrix} 3 & -2 \\ -1 & 2 \end{bmatrix}$.　　　　*Example B*

We begin by copying the 2×2 matrix I to the right of the given matrix.

$$\begin{bmatrix} 3 & -2 & | & 1 & 0 \\ -1 & 2 & | & 0 & 1 \end{bmatrix}$$

$$\begin{bmatrix} -1 & 2 & | & 0 & 1 \\ 0 & 4 & | & 1 & 3 \end{bmatrix} \quad \begin{array}{l} \text{second row copied} \\ 3(\text{second row}) + (\text{first row}) \end{array}$$

$$\begin{bmatrix} 1 & -2 & | & 0 & -1 \\ 0 & 1 & | & \frac{1}{4} & \frac{3}{4} \end{bmatrix} \quad \begin{array}{l} -1(\text{first row}) \\ \frac{1}{4}(\text{second row}) \end{array}$$

$$\begin{bmatrix} 1 & 0 & | & \frac{2}{4} & \frac{2}{4} \\ 0 & 1 & | & \frac{1}{4} & \frac{3}{4} \end{bmatrix} \quad \begin{array}{l} 2(\text{second row}) + (\text{first row}) \\ \text{second row copied} \end{array}$$

The inverse now appears as the last two columns of our final matrix. To verify the answer, we multiply

$$\begin{bmatrix} \frac{2}{4} & \frac{2}{4} \\ \frac{1}{4} & \frac{3}{4} \end{bmatrix}\begin{bmatrix} 3 & -2 \\ -1 & 2 \end{bmatrix} = \begin{bmatrix} 1 & 0 \\ 0 & 1 \end{bmatrix} \quad \square$$

It is actually possible to write down a formula for the inverse of the
2×2 matrix $\begin{bmatrix} a & b \\ c & d \end{bmatrix}$. Close examination of the illustration opening this
section and of the result of Example A might enable you to guess the
formula (Problem 34). For larger matrices, however, there is no simple
formula and guesswork there will not help very much.

Example C   Find the inverse of

$$A = \begin{bmatrix} 2 & 0 & 1 & -1 \\ 1 & -1 & 0 & 2 \\ 0 & -1 & 2 & 1 \\ -2 & 1 & 3 & 0 \end{bmatrix}$$

We begin by copying the 4×4 matrix I to the right of A.

$$\begin{bmatrix} 2 & 0 & 1 & -1 & \vline & 1 & 0 & 0 & 0 \\ 1 & -1 & 0 & 2 & \vline & 0 & 1 & 0 & 0 \\ 0 & -1 & 2 & 1 & \vline & 0 & 0 & 1 & 0 \\ -2 & 1 & 3 & 0 & \vline & 0 & 0 & 0 & 1 \end{bmatrix}$$

$$\begin{bmatrix} 1 & -1 & 0 & 2 & \vline & 0 & 1 & 0 & 0 \\ 0 & -1 & 2 & 1 & \vline & 0 & 0 & 1 & 0 \\ 0 & 2 & 1 & -5 & \vline & 1 & -2 & 0 & 0 \\ 0 & -1 & 3 & 4 & \vline & 0 & 2 & 0 & 1 \end{bmatrix}$$
second row copied
third row copied
-2(second row) + (first row)
2(second row) + (fourth row)

$$\begin{bmatrix} 1 & 0 & -2 & 1 & \vline & 0 & 1 & -1 & 0 \\ 0 & 1 & -2 & -1 & \vline & 0 & 0 & -1 & 0 \\ 0 & 0 & 5 & -3 & \vline & 1 & -2 & 2 & 0 \\ 0 & 0 & 1 & 3 & \vline & 0 & 2 & -1 & 1 \end{bmatrix}$$
-1(second row) + (first row)
-1(second row)
2(second row) + (third row)
-1(second row) + (fourth row)

$$\begin{bmatrix} 1 & 0 & 0 & 7 & \vline & 0 & 5 & -3 & 2 \\ 0 & 1 & 0 & 5 & \vline & 0 & 4 & -3 & 2 \\ 0 & 0 & 1 & 3 & \vline & 0 & 2 & -1 & 1 \\ 0 & 0 & 0 & -18 & \vline & 1 & -12 & 7 & -5 \end{bmatrix}$$
2(fourth row) + (first row)
2(fourth row) + (second row)
fourth row copied
-5(fourth row) + (third row)

$$\begin{bmatrix} 1 & 0 & 0 & 0 & \vline & \frac{7}{18} & \frac{6}{18} & -\frac{5}{18} & \frac{1}{18} \\ 0 & 1 & 0 & 0 & \vline & \frac{5}{18} & \frac{12}{18} & -\frac{19}{18} & \frac{11}{18} \\ 0 & 0 & 1 & 0 & \vline & \frac{3}{18} & 0 & \frac{3}{18} & \frac{3}{18} \\ 0 & 0 & 0 & 1 & \vline & -\frac{1}{18} & \frac{12}{18} & -\frac{7}{18} & \frac{5}{18} \end{bmatrix}$$
$\frac{7}{18}$(fourth row) + (first row)
$\frac{5}{18}$(fourth row) + (second row)
$\frac{3}{18}$(fourth row) + (third row)
$-\frac{1}{18}$(fourth row)

GIVING YOU A LINE

We are sometimes helped by the simple observation that if M is an $m \times n$ matrix, then $M^t M$ is an $n \times n$ matrix. Consider, for example, the so-called **linear regression** problem in which we are asked to find the equation $y = ax + b$ of the line which best fits some given data (x_1, y_1), (x_2, y_2), ..., (x_m, y_m); see Figure 2-D. *Best fit* is understood in the sense of minimizing the squares of the vertical distances between the points and the line.

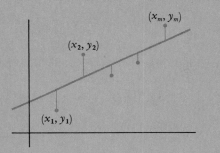

Figure 2-D

$$
\begin{array}{cc}
y \text{ values on line} & \text{differences} \\
y = ax_1 + b & y_1 - (ax_1 + b) \\
\vdots \quad \vdots \quad \vdots & \vdots \quad \vdots \quad \vdots \\
y = ax_m + b & y_m - (ax_m + b)
\end{array}
$$

A common technique is to use the methods of multivariable calculus to minimize the function

$$f(a, b) = [y_1 - (ax_1 + b)]^2 + \cdots + [y_m - (ax_m + b)]^2$$

A second approach, the one of interest to us, is to set

$$\mathbf{y} = \begin{bmatrix} y_1 \\ \vdots \\ y_m \end{bmatrix}, \qquad M = \begin{bmatrix} x_1 & 1 \\ \vdots & \vdots \\ x_m & 1 \end{bmatrix}, \qquad \boldsymbol{\ell} = \begin{bmatrix} a \\ b \end{bmatrix}$$

We note then that we are seeking $\boldsymbol{\ell}$ to satisfy

$$\mathbf{y} = M\boldsymbol{\ell} \qquad\qquad\qquad \text{(i)}$$

but that since there are more equations than variables in this system, we will probably not (unless all the points are colinear) find a solution. We thus come again to seeking $\boldsymbol{\ell}$ to minimize

$f(a, b)$ which may now be written

$$f(a, b) = | \mathbf{y} - M\boldsymbol{\ell}|^2 = (\mathbf{y} - M\boldsymbol{\ell})^t (\mathbf{y} - M\boldsymbol{\ell})$$
$$= \mathbf{y}^t \mathbf{y} - (M\boldsymbol{\ell})^t \mathbf{y} - \mathbf{y}^t M\boldsymbol{\ell} + (M\boldsymbol{\ell})^t M\boldsymbol{\ell}$$

Since $\qquad (M\boldsymbol{\ell})^t = \boldsymbol{\ell}^t M^t \quad$ and $\quad \mathbf{y}^t M = (M^t \mathbf{y})^t$

$$f(a, b) = \mathbf{y}^t \mathbf{y} - \boldsymbol{\ell}^t M^t \mathbf{y} - (M^t \mathbf{y})^t \boldsymbol{\ell} + \boldsymbol{\ell}^t M^t M\boldsymbol{\ell} \qquad \text{(ii)}$$

The appearance of $M^t \mathbf{y}$ together with our observation about the special properties of $M^t M$ motivates the next step. We cannot find $\boldsymbol{\ell}$ to satisfy (i), but we can find $\mathbf{k}$ to satisfy

$$M^t \mathbf{y} = M^t M\mathbf{k} \qquad \text{(iii)}$$

for unless the 2×2 matrix $M^t M$ has no inverse (a circumstance we'll not worry about since it can be shown to happen only when the given points are all in a vertical line),

$$\mathbf{k} = [M^t M]^{-1} M^t \mathbf{y} \qquad \text{(iv)}$$

With $\mathbf{k}$ thus determined, let us write the $\boldsymbol{\ell}$ that we seek in the form

$$\boldsymbol{\ell} = \mathbf{k} + \mathbf{r} \qquad \text{(v)}$$

Substitution of (iii) and (v) into (ii) gives, after simplification,

$$f(a, b) = \mathbf{y}^t \mathbf{y} - \mathbf{k}^t M^t M\mathbf{k} + \mathbf{r}^t M^t M\mathbf{r}$$

Since the last term is $(M\mathbf{r})^t M\mathbf{r} = |M\mathbf{r}|^2 \geq 0$, we will clearly minimize $f(a, b)$ with $\mathbf{r} = \mathbf{0}$, hence with $\boldsymbol{\ell} = \mathbf{k}$. Thus, (iv) gives us a formula for solving the problem.

The inverse that we seek is

$$\begin{bmatrix} \frac{7}{18} & \frac{6}{18} & -\frac{5}{18} & \frac{1}{18} \\ \frac{5}{18} & \frac{12}{18} & -\frac{19}{18} & \frac{11}{18} \\ \frac{3}{18} & 0 & \frac{3}{18} & \frac{3}{18} \\ -\frac{1}{18} & \frac{12}{18} & -\frac{7}{18} & \frac{5}{18} \end{bmatrix}$$

It is obvious that many of the entries could be reduced. We have left the common denominator since this makes multiplication of A^{-1} by any other matrix much easier. In particular, you should verify by multiplication that $A^{-1}A = AA^{-1} = I$. $\quad \square$

A final fact about inverses should be mentioned. If two $n \times n$ matrices A and B both have inverses, say A^{-1} and B^{-1}, then so does AB, and

$(AB)^{-1} = B^{-1}A^{-1}$. The proof is easy; just note that the inverse is always unique, and that $B^{-1}A^{-1}$ works because of Property 2.

SUMMARY

Let A be an arbitrary $m \times n$ matrix. Multiplication of A on the left by an elementary matrix E has the effect of performing a corresponding elementary operation on the rows of A. We learned in Section 1-4 that A is row equivalent to a unique matrix R in row echelon form. We now see that there is a product $P = E_k \ldots E_1$ of $m \times m$ elementary matrices such that $PA = R$; and we have learned a technique for finding P.

There is special interest in the case where A is a square matrix that is row equivalent to I. In this case, our ability to find P such that $PA = I$ amounts to being able to find the inverse A^{-1}. Although matrix multiplication is generally noncommutative, an invertible matrix does commute with its inverse; $A^{-1}A = AA^{-1} = I$.

Find inverses for the elementary matrices in Problems 1–6.

1. $\begin{bmatrix} 1 & 0 \\ 0 & 3 \end{bmatrix}$

2. $\begin{bmatrix} 1 & 0 \\ -2 & 1 \end{bmatrix}$

3. $\begin{bmatrix} 1 & 0 & 0 \\ 0 & 1 & 0 \\ -4 & 0 & 1 \end{bmatrix}$

4. $\begin{bmatrix} 1 & 0 & 0 \\ 0 & 1 & 5 \\ 0 & 0 & 1 \end{bmatrix}$

5. $\begin{bmatrix} 0 & 0 & 1 & 0 \\ 0 & 1 & 0 & 0 \\ 1 & 0 & 0 & 0 \\ 0 & 0 & 0 & 1 \end{bmatrix}$

6. $\begin{bmatrix} 1 & 0 & 0 & 0 \\ 0 & 0 & 0 & 1 \\ 0 & 0 & 1 & 0 \\ 0 & 1 & 0 & 0 \end{bmatrix}$

In Problems 7–12, find a matrix P such that PA is in row echelon form.

7. $A = \begin{bmatrix} 2 & 1 & 3 & 0 \\ -1 & 0 & 1 & 0 \\ 3 & -1 & 0 & 2 \end{bmatrix}$

8. $A = \begin{bmatrix} 4 & 1 & 1 & -1 \\ 2 & 0 & -4 & 0 \\ 6 & 1 & -5 & 1 \end{bmatrix}$

9. $A = \begin{bmatrix} 3 & 1 \\ 1 & 2 \\ 2 & -3 \end{bmatrix}$

10. $A = \begin{bmatrix} 2 & -3 \\ 4 & 3 \\ 1 & -2 \end{bmatrix}$

11. $A = \begin{bmatrix} 0 & 1 & -1 & -2 \\ 1 & 3 & 0 & -1 \\ -2 & -1 & -2 & 0 \\ 0 & 2 & 1 & 4 \end{bmatrix}$

12. $A = \begin{bmatrix} 4 & -3 & 1 & 2 \\ 5 & 1 & 4 & 5 \\ 3 & -1 & 2 & 1 \\ 2 & -1 & 0 & 3 \end{bmatrix}$

Find inverses, if they exist, for the 2 × 2 matrices given in Problems 13–18. Verify in each case that the inverse obtained is two-sided, that is, that it works on either the left or the right of the given matrix.

13. $\begin{bmatrix} -3 & 1 \\ -2 & 1 \end{bmatrix}$

14. $\begin{bmatrix} 5 & -2 \\ 2 & -1 \end{bmatrix}$

15. $\begin{bmatrix} 3 & -1 \\ 2 & 4 \end{bmatrix}$

16. $\begin{bmatrix} 2 & -3 \\ -1 & 5 \end{bmatrix}$

17. $\begin{bmatrix} 4 & 3 \\ 8 & 6 \end{bmatrix}$

18. $\begin{bmatrix} 9 & 3 \\ 6 & 2 \end{bmatrix}$

In Problems 19–24, find multiplicative inverses for the given matrices.

19. $\begin{bmatrix} 1 & 3 & -2 \\ -2 & 4 & 1 \\ 5 & 1 & -3 \end{bmatrix}$

20. $\begin{bmatrix} 1 & 2 & -1 \\ 2 & 3 & 2 \\ 4 & -2 & 3 \end{bmatrix}$

21. $\begin{bmatrix} 5 & 3 & 4 \\ -3 & 2 & 5 \\ 7 & 4 & 6 \end{bmatrix}$

22. $\begin{bmatrix} 3 & -2 & 4 \\ 5 & 3 & 3 \\ 2 & 5 & -2 \end{bmatrix}$

23. $\begin{bmatrix} 1 & -2 & 0 & 1 \\ 0 & 1 & 2 & -1 \\ 2 & -3 & 1 & 3 \\ -1 & 3 & -2 & 0 \end{bmatrix}$

24. $\begin{bmatrix} 1 & -1 & 3 & 0 \\ 2 & 3 & 0 & -1 \\ -1 & 2 & 1 & -2 \\ 0 & 1 & -1 & 3 \end{bmatrix}$

25. Prove that if $A^{-1} = B$, then $(A^2)^{-1} = B^2$.

26. Prove that if $A^{-1} = B$, then $(A^t)^{-1} = B^t$.

27. Prove that if A and B are $n \times n$ matrices for which AB is nonsingular, then both A and B are nonsingular.

28. Prove that if A and B are $n \times n$ matrices, A being nonsingular and B being singular, then AB and BA are both singular.

In Problems 29–32, suppose that an $n \times n$ matrix B and its inverse B^{-1} are known.

29. If C is formed by interchanging columns i and j of B, how can C^{-1} be obtained from B^{-1}?

30. If C is formed by multiplying the jth column of B by r, how can C^{-1} be obtained from B^{-1}?

31. If C is formed by adding r times the jth column of B to the ith column of B, how can C^{-1} be obtained from B^{-1}?

*32. If C is formed by replacing the jth column of B with a column vector $\mathbf{z}$ chosen so that C is invertible, how can C^{-1} be obtained from B^{-1}? *Hint:* This technique is often discussed in texts on linear programming because of its importance in the revised simplex method.

33. Show that the 2×2 matrix $A = \begin{bmatrix} a & b \\ c & d \end{bmatrix}$ has an inverse if and only if the number $D = ad - bc \neq 0$.

34. If the 2×2 matrix A of Problem 33 has an inverse, then A^{-1} can be found from a formula. Find the formula.

35. Obtain a straight line that best fits the points $(2, 8)$, $(5, 12)$, $(7, 13)$, and $(9, 16)$, using the method described in the vignette *Giving You a Line*. ▣

DESCRIBING THE SOLUTION SET 2-8

A linear system of m equations in n unknowns can be written in the form $A\mathbf{x} = \mathbf{b}$. With reference to such a system, two goals were mentioned in the introduction to this chapter. One goal was to develop an algebra for matrices so that in analogy with the elementary algebra solution of $ax = b$, we might solve the system $A\mathbf{x} = \mathbf{b}$ by writing $\mathbf{x} = A^{-1}\mathbf{b}$. The other goal was to develop terminology that would enable us to describe in precise terms the solution set of the system.

The first goal has been accomplished. Matrix multiplication has been defined and we know how to find a matrix P so that

$$PA = R \qquad (1)$$

is in row echelon form. This allows us to solve $A\mathbf{x} = \mathbf{b}$ by writing $P(A\mathbf{x}) = P\mathbf{b}$, or

$$R\mathbf{x} = P\mathbf{b} \qquad (2)$$

In particular, if A is a nonsingular $n \times n$ matrix, then $P = A^{-1}$, $R = I$, and (2) takes the desired form of

$$\mathbf{x} = A^{-1}\mathbf{b}$$

To reach the second goal, it will be useful to relate in some sense the size of the solution set to a characteristic property of the $m \times n$ coefficient matrix A. Toward this end, the subspace of R^n spanned by the row vectors $A_{1\cdot}, A_{2\cdot}, \cdots, A_{m\cdot}$ will be called the **row space** $\mathfrak{R}$ of matrix A, and the dimension of $\mathfrak{R}$ will be the **row rank** of A. In the same way, the subspace R^m spanned by the column vectors of A will be called the **column space** $\mathfrak{C}$. The dimension of $\mathfrak{C}$ is called the **column rank** of A.

Find the row and column rank of   *Example A*

$$A = \begin{bmatrix} -2 & 3 & 5 & 1 \\ -1 & 1 & 4 & -1 \\ 2 & -5 & 1 & -7 \end{bmatrix}$$

We begin by finding the unique equivalent row echelon form for A.

$$\begin{bmatrix} 1 & -1 & -4 & 1 \\ 0 & 1 & -3 & 3 \\ 0 & -3 & 9 & -9 \end{bmatrix} \quad \begin{array}{l} -1 \text{ (second row)} \\ -2 \text{ (second row)} + \text{(first row)} \\ 2 \text{ (second row)} + \text{(third row)} \end{array}$$

$$\begin{bmatrix} 1 & 0 & -7 & 4 \\ 0 & 1 & -3 & 3 \\ 0 & 0 & 0 & 0 \end{bmatrix} \quad \begin{array}{l} 1 \text{ (second row)} + \text{(first row)} \\ \text{second row copied} \\ 3 \text{ (second row)} + \text{(third row)} \end{array}$$

Thus, the row rank $r = 2$.

The column rank of A is determined by finding the row rank of A^t.

$$A^t = \begin{bmatrix} -2 & -1 & 2 \\ 3 & 1 & -5 \\ 5 & 4 & 1 \\ 1 & -1 & -7 \end{bmatrix}$$

Now the same technique can be used. Seek a row echelon matrix equivalent to A^t.

$$\begin{bmatrix} 1 & -1 & -7 \\ 0 & -3 & -12 \\ 0 & 4 & 16 \\ 0 & 9 & 36 \end{bmatrix} \quad \begin{array}{l} \text{fourth row copied} \\ 2 \text{ (fourth row)} + \text{(first row)} \\ -3 \text{ (fourth row)} + \text{(second row)} \\ -5 \text{ (fourth row)} + \text{(third row)} \end{array}$$

Multiplication of the second row by $-\frac{1}{3}$ gives us a 1 in the second column that can be used to obtain 0s elsewhere in the second column.

$$\begin{bmatrix} 1 & 0 & -3 \\ 0 & 1 & 4 \\ 0 & 0 & 0 \\ 0 & 0 & 0 \end{bmatrix}$$

The row rank of A^t, which is the column rank of A, is 2; $k = 2$.   □

We note that although the column space of the matrix in Example A was a subset of R^3 and the row space was a subspace of R^4, both subspaces had dimension 2. This is not a coincidence. To see why, we shall find the column rank for another matrix, this time paying more attention to the method than to obtaining an answer. We consider the matrix

$$A = \begin{bmatrix} 2 & 1 & -1 & 2 & -1 \\ 3 & -1 & -2 & 0 & 3 \\ -2 & 0 & 1 & -1 & 0 \\ 5 & -1 & -4 & 0 & 7 \end{bmatrix}$$

The column rank of A is, of course, the row rank of A^t. Since the row rank of A^t is the number of nonzero rows in the row echelon matrix that is row equivalent to A^t, we begin by finding in the usual way that

$$A^t = \begin{bmatrix} 2 & 3 & -2 & 5 \\ 1 & -1 & 0 & -1 \\ -1 & -2 & 1 & -4 \\ 2 & 0 & -1 & 0 \\ -1 & 3 & 0 & 7 \end{bmatrix} \sim \begin{bmatrix} 1 & 0 & 0 & 2 \\ 0 & 1 & 0 & 3 \\ 0 & 0 & 1 & 4 \\ 0 & 0 & 0 & 0 \\ 0 & 0 & 0 & 0 \end{bmatrix} = R$$

Since they are row equivalent, every row vector of A^t can be expressed as a linear combination of the row vectors of R. For instance, we can write

$$\begin{aligned} [2 \quad 3 \quad -2 \quad 5] \\ = s_{11}[1 \quad 0 \quad 0 \quad 2] + s_{21}[0 \quad 1 \quad 0 \quad 3] + s_{31}[0 \quad 0 \quad 1 \quad 4] \end{aligned}$$

Or, writing this again in terms of column vectors,

$$\begin{bmatrix} 2 \\ 3 \\ -2 \\ 5 \end{bmatrix} = s_{11}\begin{bmatrix} 1 \\ 0 \\ 0 \\ 2 \end{bmatrix} + s_{21}\begin{bmatrix} 0 \\ 1 \\ 0 \\ 3 \end{bmatrix} + s_{31}\begin{bmatrix} 0 \\ 0 \\ 1 \\ 4 \end{bmatrix}$$

It is in fact easy to see from the location of 0s and 1s in the vectors on the right that $S_{11} = 2, S_{21} = 3$, and $S_{31} = -2$. But we really don't need this much detail for our purposes. It is enough to note that we can write

$$\begin{bmatrix} 2 \\ 3 \\ -2 \\ 5 \end{bmatrix} = \begin{bmatrix} 1 & 0 & 0 \\ 0 & 1 & 0 \\ 0 & 0 & 1 \\ 2 & 3 & 4 \end{bmatrix}\begin{bmatrix} s_{11} \\ s_{21} \\ s_{31} \end{bmatrix}$$

Similar expressions can be obtained for the other column vectors of A, and the entire story can be summarized by writing

$$\begin{bmatrix} 2 & 1 & -1 & 2 & -1 \\ 3 & -1 & -2 & 0 & 3 \\ -2 & 0 & 1 & -1 & 0 \\ 5 & -1 & -4 & 0 & 7 \end{bmatrix} = \begin{bmatrix} 1 & 0 & 0 \\ 0 & 1 & 0 \\ 0 & 0 & 1 \\ 2 & 3 & 4 \end{bmatrix}\begin{bmatrix} s_{11} & s_{12} & s_{13} & s_{14} & s_{15} \\ s_{21} & s_{22} & s_{23} & s_{24} & s_{25} \\ s_{31} & s_{32} & s_{33} & s_{34} & s_{35} \end{bmatrix}$$

Again we observe that all the entries s_{ij} could be determined, but it is quite sufficient for our purposes to denote the matrix on the right by S, and to note that the number of rows of S will equal either the row rank of A^t or, equivalently, the column rank of A.

You may wish to check your understanding of the above product by verifying that ordinary matrix multiplication does give as the second column of the product

$$
\begin{bmatrix} 1 \\ -1 \\ 0 \\ 1 \end{bmatrix} = s_{12} \begin{bmatrix} 1 \\ 0 \\ 0 \\ 2 \end{bmatrix} + s_{22} \begin{bmatrix} 0 \\ 1 \\ 0 \\ 3 \end{bmatrix} + s_{32} \begin{bmatrix} 0 \\ 0 \\ 1 \\ 4 \end{bmatrix}
$$

If, however, we focus on obtaining the rows of the left-hand matrix then we get, using the fourth row as an example,

$$
\begin{aligned}
[5 \quad -1 \quad -4 \quad 0 \quad 7] = \; & 2[s_{11} \quad s_{12} \quad s_{13} \quad s_{14} \quad s_{15}] \\
& + 3[s_{21} \quad s_{22} \quad s_{23} \quad s_{24} \quad s_{25}] \\
& + 4[s_{31} \quad s_{32} \quad s_{33} \quad s_{34} \quad s_{35}]
\end{aligned}
$$

In the same way, every row vector of the given matrix A can be expressed as a linear combination of the row vectors of S. It follows that the dimension of the row space of A, that is, the row rank of A, cannot exceed the dimension of the row space of S, already observed to be the same as the column rank of A.

What we have just proved for A could in the same way be proved for every matrix. *The row rank of a matrix is less than or equal to its column rank.*

Let A be an arbitrary $m \times n$ matrix with row rank r and column rank c. The italicized remark tells us that $r \leq c$. But now consider the $n \times m$ matrix A^t which clearly has row rank c and column rank r. The italicized remark applied to A^t tells us that $c \leq r$. We have proved an important fact:

THEOREM A

  The row rank of a matrix equals its column rank.

This common number is called the **rank** of A. It is this number, intrinsically characteristic of A, that we can relate to the size of the solution set for the system $A\mathbf{x} = \mathbf{b}$.

HOMOGENEOUS SYSTEMS

We have already seen in Section 2-4 that if the coefficient matrix of a homogeneous system $A\mathbf{x} = \mathbf{0}$ is $m \times n$, then the solution set is a subspace of R^n. Suppose that the row rank of A is r, meaning that the row echelon matrix $PA = R$ has r nonzero rows, hence that the system $R\mathbf{x} = \mathbf{0}$ has r basic variables. These r basic variables can then be expressed in terms of the other $n - r$ variables. These ideas can be illustrated with another look at the homogeneous system studied in Section

2-4. That system,

$$\begin{bmatrix} 2 & 1 & -1 & 2 & -1 \\ 3 & -1 & -2 & 0 & 3 \\ -2 & 0 & 1 & -1 & 0 \\ 5 & -1 & -4 & 0 & 7 \end{bmatrix} \begin{bmatrix} x_1 \\ x_2 \\ x_3 \\ x_4 \\ x_5 \end{bmatrix} = \begin{bmatrix} 0 \\ 0 \\ 0 \\ 0 \end{bmatrix}$$

is equivalent to the system

$$\begin{bmatrix} 1 & 0 & 0 & 1 & -2 \\ 0 & 1 & 0 & 1 & -1 \\ 0 & 0 & 1 & 1 & -4 \\ 0 & 0 & 0 & 0 & 0 \end{bmatrix} \begin{bmatrix} x_1 \\ x_2 \\ x_3 \\ x_4 \\ x_5 \end{bmatrix} = \begin{bmatrix} 0 \\ 0 \\ 0 \\ 0 \end{bmatrix}$$

in which the coefficient matrix is in row echelon form. The row rank of A, therefore, is 3, and the three basic variables are easily expressed in terms of the $5 - 3$ other variables as

$$x_1 = -x_4 + 2x_5$$
$$x_2 = -x_4 + x_5$$
$$x_3 = -x_4 + 4x_5$$

The solution space may therefore be described by

$$(x_1, x_2, x_3, x_4, x_5) = (-x_4 + 2x_5, -x_4 + x_5, -x_4 + 4x_5, x_4, x_5)$$
$$= x_4(-1, -1, -1, 1, 0) + x_5(2, 1, 4, 0, 1)$$

Since this can only be the **0** vector if $x_4 = x_5 = 0$, the vectors $(-1, -1, -1, 1, 0)$ and $(2, 1, 4, 0, 1)$ which span the solution space are linearly independent, meaning that the dimension of the solution space is 2.

In the same way, it is easily seen that if a homogeneous system $A\mathbf{x} = \mathbf{0}$ has an $m \times n$ coefficient matrix A with row rank equal to r, then the solution subspace will have dimension $n - r$.

> *The homogeneous system $A\mathbf{x} = \mathbf{0}$ of m equations in n variables has a solution space of dimension $n - r$ where r is the rank of A.*

$\qquad\qquad$ *THEOREM B*

The entire story for a homogeneous system may therefore be summarized as follows:

1. The solution space, like any subspace, always contains **0**.
2. If $r = n$, then the solution space has dimension 0; the trivial solution is the only solution.

3. Since the row rank of A is r, then $m \geq r$. Therefore, if $n > m$, $n > r$ and the solution space has dimension $n - r > 0$.

If $n = m$, then the situation described in (2), rank of $A = n$, occurs if and only if matrix A is nonsingular. Notice that (3) is an improvement on Theorem B of Section 1-4 which asserted that when the number of variables exceeded the number of equations in a homogeneous system, there were sure to be an infinite number of nontrivial solutions.

NONHOMOGENEOUS SYSTEMS

Since the product of an $m \times n$ matrix A with a column vector on its right may be written as a linear combination of the column vectors of A, the system $A\mathbf{x} = \mathbf{b}$, that is,

$$\begin{bmatrix} a_{11} & \cdots & a_{1n} \\ a_{21} & \cdots & a_{2n} \\ \vdots & & \vdots \\ a_{m1} & \cdots & a_{mn} \end{bmatrix} \begin{bmatrix} x_1 \\ \vdots \\ x_n \end{bmatrix} = \begin{bmatrix} b_1 \\ \vdots \\ b_m \end{bmatrix}$$

can be written in the form

$$x_1 \begin{bmatrix} a_{11} \\ a_{21} \\ \vdots \\ a_{m1} \end{bmatrix} + \cdots + x_n \begin{bmatrix} a_{1n} \\ a_{2n} \\ \vdots \\ a_{mn} \end{bmatrix} = \begin{bmatrix} b_1 \\ b_2 \\ \vdots \\ b_m \end{bmatrix}$$

This makes it clear that there will be a solution to the system if and only if $\mathbf{b}$ can be written as a linear combination of the column vectors of A. Stated another way, there will be a solution if and only if $\mathbf{b}$ is in $\mathcal{C}$, the column space of A.

Obviously, if the rank of A is m, then $\mathcal{C}$ is all of R^m and there will be a solution; if the rank of $\mathcal{C}$ is less than m, there may not be a solution, depending entirely upon whether or not $\mathbf{b}$ is in $\mathcal{C}$.

Suppose that there is a vector $\mathbf{p}$ in the solution set $\mathcal{S}$ of the system $A\mathbf{x} = \mathbf{b}$; that is, suppose that $A\mathbf{p} = \mathbf{b}$. Suppose further that $\mathbf{s}$ is any vector in the solution space $\mathcal{N}$ of the homogeneous system $A\mathbf{x} = \mathbf{0}$; that is, suppose that $A\mathbf{s} = \mathbf{0}$. Now,

$$A(\mathbf{p} + \mathbf{s}) = A\mathbf{p} + A\mathbf{s} = \mathbf{b} + \mathbf{0} = \mathbf{b}$$

so we conclude that $\mathbf{p} + \mathbf{s}$ is also in $\mathcal{S}$. Consider now any other vector $\mathbf{r}$ in $\mathcal{S}$. Since $A\mathbf{r} = \mathbf{b}$, we see that

$$A(\mathbf{r} - \mathbf{p}) = A\mathbf{r} - A\mathbf{p} = \mathbf{b} - \mathbf{b} = \mathbf{0}$$

It follows that $\mathbf{r} - \mathbf{p}$ equals some vector $\mathbf{s}$ again in $\mathfrak{N}$, the solution space of $A\mathbf{x} = \mathbf{0}$. These results can be summarized as follows:

If the nonhomogeneous system $A\mathbf{x} = \mathbf{b}$ of m equations in n variables has at least one solution $\mathbf{p}$, then the solution set $\mathbb{S}$ is of the form

$$\mathbb{S} = \{\mathbf{p} + \mathbf{s}: \mathbf{s} \text{ is a solution to } A\mathbf{x} = \mathbf{0}\}$$

THEOREM C

The set $\mathbb{S}$ is said to be a **translate** of the solution subspace of $A\mathbf{x} = \mathbf{0}$. The equation $x_1 + x_2 = 1$ is a nonhomogeneous system of $m = 1$ equation in $n = 2$ variables. The solution set is a translate of (a line parallel to) the subspace (line through the origin) $x_1 + x_2 = 0$.

The solution set $\mathbb{S}$ for a nonhomogeneous system may therefore be described in one of three ways:

1. $\mathbb{S} = \varnothing$; the column vector $\mathbf{b}$ of constants is not in the span of the column space of the coefficient matrix.
2. $\mathbb{S}$ consists of a single point.
3. $\mathbb{S}$ is infinite; the dimension of the solution space of $A\mathbf{x} = \mathbf{0}$ is $n - r > 0$.

If $n = m$, then the solution described in (2) occurs if and only if A is nonsingular. In that case, the solution is $\mathbf{x} = A^{-1}\mathbf{b}$.

Example B

The solution set to the nonhomogeneous system

$$3x_1 + 4x_2 = 12$$

consists of all vectors of the form

$$(x_1, x_2) = (4 - \tfrac{4}{3}x_2, x_2) = (4, 0) + x_2(-\tfrac{4}{3}, 1)$$

The solution set

$$\mathbb{S} = \{(4, 0) + \mathbf{s}: \mathbf{s} \text{ is a solution to } 3x_1 + 4x_2 = 0\}$$

is a translate of the linear subspace $\mathbb{S}_1$ of R^2 spanned by $(-\tfrac{4}{3}, 1)$. See Figure 2-22. □

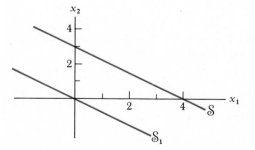

Figure 2-22

PROBLEM
SET 2-8

For each of the matrices in Problems 1–4, find a basis for the column space and a basis for the row space, thus verifying that the row rank equals the column rank.

1.
$$\begin{bmatrix} 3 & -1 & 2 \\ 2 & -2 & 3 \\ 0 & 1 & -1 \\ 4 & 2 & 0 \\ -1 & 2 & -2 \end{bmatrix}$$

2.
$$\begin{bmatrix} 5 & 3 & 4 \\ -3 & 1 & -2 \\ -1 & -5 & 1 \\ 4 & -1 & -3 \\ 1 & 0 & -2 \end{bmatrix}$$

3.
$$\begin{bmatrix} -1 & 2 & 5 & 3 \\ 4 & -2 & -1 & -5 \\ -4 & 4 & 5 & -1 \\ -3 & 1 & -2 & -1 \end{bmatrix}$$

4.
$$\begin{bmatrix} 3 & 1 & 2 & -4 \\ -5 & -2 & 1 & 2 \\ -5 & -3 & 2 & -3 \\ -2 & -1 & -1 & 1 \end{bmatrix}$$

For each of the systems given in Problems 5–8, find a basis for the solution subspace.

5.
$$\begin{aligned}
x_1 - x_2 - x_3 + x_4 &= 0 \\
2x_2 + x_3 - 3x_4 &= 0 \\
-x_1 - 2x_2 + x_3 - 2x_4 &= 0 \\
-2x_1 - 3x_2 + x_3 &= 0
\end{aligned}$$

6.
$$\begin{aligned}
2x_1 - x_2 + 3x_3 &= 0 \\
-2x_1 - 2x_2 - 4x_3 + x_4 &= 0 \\
3x_1 + x_2 + 2x_3 - x_4 &= 0 \\
-x_1 - 5x_2 + 2x_4 &= 0
\end{aligned}$$

7.
$$\begin{aligned}
x_1 - 2x_2 - 3x_3 + x_4 + x_5 &= 0 \\
-4x_1 - 5x_2 - 5x_3 + 5x_4 - x_5 &= 0 \\
x_1 + x_2 - x_4 + 2x_5 &= 0 \\
-2x_2 - 4x_3 + 2x_4 + 3x_5 &= 0 \\
-3x_1 + x_2 + 2x_3 - x_5 &= 0
\end{aligned}$$

8.
$$\begin{aligned}
x_1 - x_2 + x_3 - x_4 &= 0 \\
x_1 - 4x_2 + x_3 - x_4 &= 0 \\
2x_1 + x_2 + 2x_3 - 2x_4 &= 0 \\
x_1 - x_2 + x_3 - x_4 &= 0
\end{aligned}$$

Describe the complete solution set for each of the following systems of equations.

9.
$$\begin{aligned}
3x_1 - x_2 + x_3 - x_4 &= -6 \\
x_1 + x_2 - x_3 + 2x_4 &= 3 \\
2x_1 + 2x_2 + x_3 - 3x_4 &= -1
\end{aligned}$$

10.
$$\begin{aligned}
2x_1 - x_2 + x_3 + 5x_4 &= 1 \\
3x_1 + x_2 + x_3 - x_4 &= -2 \\
x_1 - x_2 + 2x_4 &= -1
\end{aligned}$$

11.
$$\begin{aligned}
x_1 + x_2 - 2x_3 - x_4 &= 0 \\
3x_1 - x_2 + x_3 + 2x_4 &= -3 \\
7x_1 + 3x_2 - 7x_3 - 2x_4 &= 5
\end{aligned}$$

12.
$$\begin{aligned}
x_1 - 3x_2 + x_3 - x_4 &= -8 \\
2x_1 - x_2 + 2x_3 + 3x_4 &= -1 \\
7x_1 - 11x_2 + 7x_3 + 3x_4 &= -10
\end{aligned}$$

Give reasons for the statements numbered 13–18.

13. The rank of a 4×7 matrix cannot exceed 4.

14. If A is a 4×6 matrix such that the system $A\mathbf{x} = \mathbf{0}$ has a nontrivial solution, then the rank of A is 3 or less.

15. If two $m \times n$ matrices are row equivalent, they have the same rank.

16. Given the matrix product $AB = C$, then rank $C \le$ rank A. *Hint:* Any column vector of C must be in the space spanned by the column vectors of A. Why?

17. Given the matrix product $AB = C$, then rank $C \le$ rank B. *Hint:* Any row vector of C must be in the space spanned by the row vectors of B.

18. If A and B are two $n \times n$ vectors, and if AB has rank n, then each matrix is nonsingular.

19. Prove that if the $m \times n$ matrix A has rank r, then A can be written as the sum of r matrices, each having a rank of 1.

TRANSPORTATION PROBLEMS

A department-store chain has 20 television sets in one warehouse and 35 identical television sets in another warehouse. They are to be shipped to three retail stores, 15 going to the first store, 15 to the second, and 25 to the third. The cost of shipping one unit from a warehouse to a store is indicated in the following table:

		1	2	3
Warehouses	1	2	3	2
	2	1	2	4
		1	2	3

Retail stores

How should the sets be shipped to minimize shipping costs?

This type of problem is called a **transportation problem.** In real situations it is typically encountered with many more sources and destinations.

Let x_{ij} be the number of units to be shipped from i to j. The problem then is to minimize

$$2x_{11} + 3x_{12} + 2x_{13} + x_{21} + 2x_{22} + 4x_{23}$$

subject to the constraints

$$x_{11} + x_{12} + x_{13} = 20$$
$$x_{21} + x_{22} + x_{23} = 35$$
$$x_{11} + x_{21} \qquad = 15$$
$$x_{12} + x_{22} \qquad = 15$$
$$x_{13} + x_{23} \qquad = 25$$

Answers to this problem must be integers if they are to be useful, and the method that has been devised for solving the problem depends on the fact that the augmented matrix of the system of constraints has rank 1 less than the number of equations. For the example at hand, the augmented matrix is

$$A = \begin{bmatrix} 1 & 1 & 1 & 0 & 0 & 0 & 20 \\ 0 & 0 & 0 & 1 & 1 & 1 & 35 \\ 1 & 0 & 0 & 1 & 0 & 0 & 15 \\ 0 & 1 & 0 & 0 & 1 & 0 & 15 \\ 0 & 0 & 1 & 0 & 0 & 1 & 25 \end{bmatrix}$$

To see that the rank is $5 - 1 = 4$, note first that the row rank is less than 5 because the row vectors are dependent;

$$\mathbf{A}_{1\cdot} + \mathbf{A}_{2\cdot} - \mathbf{A}_{3\cdot} - \mathbf{A}_{4\cdot} - \mathbf{A}_{5\cdot} = \mathbf{0}$$

On the other hand, the column rank is at least four because $\mathbf{A}_{\cdot 1}$, $\mathbf{A}_{\cdot 2}$, $\mathbf{A}_{\cdot 3}$, and $\mathbf{A}_{\cdot 4}$ are obviously linearly independent.

Can you show that in the case of m sources and n destinations, the rank of the augmented matrix will always be $m + n - 1$ so long as $a_1 + \cdots + a_m = b_1 + \cdots + b_n$?

CHAPTER 2
SELF-TEST

1. The vertices of a triangle are $A(1, -1, 2)$, $B(5, 0, 3)$, and $C(3, 1, 1)$. An altitude is dropped from C to a point D on the segment from A to B.
   (a) What are the coordinates of D?
   (b) What angle is formed at A?
   (c) Give the coordinates of a point E such that AB and CE are opposite sides of a parallelogram.

2. Five vectors in R^4 are given as follows:

$$\mathbf{r}_1 = (1, 0, -1, 0)$$
$$\mathbf{r}_2 = (-1, 0, 1, 1)$$
$$\mathbf{r}_3 = (0, 1, -1, -1)$$
$$\mathbf{r}_4 = (0, 1, -1, 0)$$
$$\mathbf{r}_5 = (1, 0, 0, -1)$$

   (a) Without any calculating, can you be certain of finding five real numbers x_i, not all zero, such that $x_1\mathbf{r}_1 + x_2\mathbf{r}_2 + x_3\mathbf{r}_3 + x_4\mathbf{r}_4 + x_5\mathbf{r}_5 = 0$? Why?
   (b) Does the set $B = \{\mathbf{r}_1, \mathbf{r}_2, \mathbf{r}_3, \mathbf{r}_4\}$ form a basis for R^4?
   (c) Is $\mathbf{s} = (1, -4, 3, 6)$ in the span of B?

3. What does it mean to say that a set $\mathcal{S}$ of vectors in R^n forms a subspace? Which of the following are subspaces of R^4?
   (a) The empty set

(b) All solutions to the system

$$\begin{aligned} x_1 - x_2 + x_3 - x_4 &= 1 \\ x_1 + x_2 + 2x_3 &= -2 \\ 2x_1 + x_3 + 3x_4 &= 2 \end{aligned}$$

(c) All vectors of the form $(1, -1, 2, 0) + a(2, 1, -1, 3)$.
(d) All vectors of the form $a_1(1, -1, 2, 0) + a_2(2, 1, -1, 3)$.
(e) All vectors $\mathbf{r} = (r_1, r_2, r_3, r_4)$ satisfying $r_1^2 + r_2^2 + r_3^2 + r_4^2 \leq 1$.
(f) All vectors $\mathbf{r}$ satisfying $r_1 + r_2 + r_3 + r_4 \leq 1$.
(g) All vectors $\mathbf{r}$ satisfying $r_1 + r_2 + r_3 + r_4 \leq 0$.
(h) All vectors $\mathbf{r}$ satisfying $r_1 + r_2 + r_3 + r_4 = 1$.
(i) All vectors $\mathbf{r}$ satisfying $r_1 + r_2 + r_3 + r_4 = 0$.

4. What is a basis of a subspace? Find a basis for the subspace spanned by the set B of Problem 2.

5. Suppose a set of nonzero vectors $\mathbf{s}_1$, $\mathbf{s}_2$, $\mathbf{s}_3$, and $\mathbf{s}_4$ in R^4 are mutually orthogonal, meaning that $\mathbf{s}_i \cdot \mathbf{s}_j = O$ whenever $i \neq j$. Prove that they form a basis of R^4. *Hint:* Suppose $a_1\mathbf{s}_1 + a_2\mathbf{s}_2 + a_3\mathbf{s}_3 + a_4\mathbf{s}_4 = \mathbf{0}$. Consider $\mathbf{s}_i \cdot (a_1\mathbf{s}_1 + a_2\mathbf{s}_2 + a_3\mathbf{s}_3 + a_4\mathbf{s}_4)$ for each i.

6. Suppose that $\mathbf{s}_1$ and $\mathbf{s}_2$ are orthogonal vectors in R^4. State and prove a natural version of the Theorem of Pythagoras.

7. Two students solve a system of homogeneous equations in four variables. One gives the set B_1 below as a basis for the solution space; the other gives B_2. Do they both describe the same subspace?

$$B_1 = \{(1, 1, 3, -1),\ (2, -1, -4, 1)\}$$
$$B_2 = \{(-1, 1, 1, -1),\ (-3, 2, 1, -2)\}$$

8. $A = \begin{bmatrix} 1 & 0 & 0 \\ 0 & 1 & 0 \\ -3 & 0 & 1 \end{bmatrix}$ and $B = \begin{bmatrix} 1 & -1 & 2 \\ 3 & 1 & -1 \\ -1 & 2 & -4 \end{bmatrix}$

(a) Find A^{-1} (by inspection).
(b) Find B^{-1} (by working).
(c) Find $(AB)^{-1}$ (without too much work).

9. Using matrix B from Problem 8, find a matrix A so that

$$AB = \begin{bmatrix} 1 & -1 & 0 \\ 0 & 2 & 1 \\ -1 & 0 & 1 \end{bmatrix}$$

10. $A = \begin{bmatrix} 1 & -1 & 2 & 0 & 3 \\ 0 & 2 & 1 & 3 & 1 \\ -1 & 1 & 5 & 1 & 0 \\ -1 & 0 & 1 & -1 & -2 \end{bmatrix}, \quad \mathbf{b} = \begin{bmatrix} -1 \\ -4 \\ 3 \\ 4 \end{bmatrix}$

(a) Find the rank of A.
(b) Describe the solution subspace to the system $A\mathbf{x} = \mathbf{0}$.
(c) Describe the solution set to the system $A\mathbf{x} = \mathbf{b}$.

3 DETERMINANTS

3-1 INTRODUCTION

Many of the matrix products included in Problem Set 2-5 were intended to suggest general properties of matrix multiplication. For instance, students who observe the products

$$\begin{bmatrix} 3 & -13 \\ -2 & 9 \end{bmatrix}\begin{bmatrix} 9 & 13 \\ 2 & 3 \end{bmatrix} = \begin{bmatrix} 1 & 0 \\ 0 & 1 \end{bmatrix} \quad \text{and} \quad \begin{bmatrix} -3 & 8 \\ -2 & 5 \end{bmatrix}\begin{bmatrix} 5 & -8 \\ 2 & -3 \end{bmatrix} = \begin{bmatrix} 1 & 0 \\ 0 & 1 \end{bmatrix}$$

often speculate that the inverse of

$$\begin{bmatrix} a & b \\ c & d \end{bmatrix}$$

might be obtained by interchanging a and d, and by changing the signs of b and c. Trying this a few times leads to momentary disappointment, however, when it is observed that

$$\begin{bmatrix} d & -b \\ -c & a \end{bmatrix}\begin{bmatrix} a & b \\ c & d \end{bmatrix} = \begin{bmatrix} ad - bc & 0 \\ 0 & ad - bc \end{bmatrix}$$

Zeros turn up where they are wanted, but we only get the 2×2 identity I if $ad - bc = 1$.

The difficulty is not hard to fix up so long as $D = ad - bc \neq 0$. Merely divide each member of the proposed inverse by D. It is then easy to verify that

$$\begin{bmatrix} \dfrac{d}{D} & -\dfrac{b}{D} \\ -\dfrac{c}{D} & \dfrac{a}{D} \end{bmatrix}\begin{bmatrix} a & b \\ c & d \end{bmatrix} = \begin{bmatrix} 1 & 0 \\ 0 & 1 \end{bmatrix}$$

And if $D = 0$, it is quite easy to show (see Problem 26 at the end of Section 3-2) that the 2×2 matrix A has no inverse.

The 2×2 matrix

$$A = \begin{bmatrix} a_{11} & a_{12} \\ a_{21} & a_{22} \end{bmatrix}$$

has an inverse if and only if $D = a_{11}a_{22} - a_{12}a_{21} \neq 0$. The number D is called the **determinant** of A, written $D = \det A$. The determinant is obviously related to A^{-1}, but its importance also reaches into other areas related to the matrix A. Take, for example, the area of the parallelogram P (Figure 3-1) that has as two adjacent sides the row vectors of matrix A,

$$\mathbf{A}_{1\cdot} = (a_{11}, a_{12}) \quad \text{and} \quad \mathbf{A}_{2\cdot} = (a_{21}, a_{22})$$

The area of the large rectangle in Figure 3-1 is $(a_{11} + a_{21})(a_{12} + a_{22})$; it is also the sum of the areas of the parallelogram P, two congruent rec-

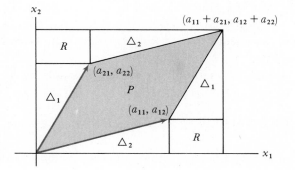

Figure 3-1

tangles R, two congruent triangles $\triangle_1$, and two congruent triangles $\triangle_2$. Thus,

$$(a_{11} + a_{21})(a_{12} + a_{22}) = \text{area } P + 2a_{12}a_{21} + 2(\tfrac{1}{2})a_{21}a_{22} + 2(\tfrac{1}{2})a_{11}a_{12}$$

When the algebraic dust has settled, we are left with

$$\text{area } P = a_{11}a_{22} - a_{12}a_{21} = \det A \tag{1}$$

Our goal in this chapter is to define the determinant of an $n \times n$ matrix; and we wish to do this in a way that will include the plane facts just described with $n = 2$, and will for $n = 3$ extend these concepts to 3-dimensional space. In particular, for

$$B = \begin{bmatrix} b_{11} & b_{12} & b_{13} \\ b_{21} & b_{22} & b_{23} \\ b_{31} & b_{32} & b_{33} \end{bmatrix}$$

we would like it to be true that

$$B^{-1} \text{ exists if and only if } \det B \neq 0$$
$$\text{volume } P = \det B$$

where P is the parallelepiped having as three adjacent edges the row vectors $\mathbf{B}_1.$, $\mathbf{B}_2.$, $\mathbf{B}_3.$ of matrix B.

3-2 | THE DEFINITION

It is clear that the elementary row operations play a key role in matrix algebra. We begin, therefore, by noting the effect of these operations on the determinant of a 2×2 matrix

$$A = \begin{bmatrix} a_{11} & a_{12} \\ a_{21} & a_{22} \end{bmatrix}$$

Let A_1 be obtained by interchanging the two rows of A. Then

$$\det A_1 = \det \begin{bmatrix} a_{21} & a_{22} \\ a_{11} & a_{12} \end{bmatrix} = a_{12}a_{21} - a_{11}a_{22} = -\det A$$

Let A_2 be obtained by multiplying the first row of A by the scalar r. Then

$$\det A_2 = \det \begin{bmatrix} ra_{11} & ra_{12} \\ a_{21} & a_{22} \end{bmatrix} = ra_{11}a_{22} - ra_{12}a_{21} = r \det A$$

Let A_3 be obtained by adding r times the first row of A to the second row of A.

$$\det A_3 = \det \begin{bmatrix} a_{11} & a_{12} \\ a_{21} + ra_{11} & a_{22} + ra_{12} \end{bmatrix}$$
$$= a_{11}(a_{22} + ra_{12}) - a_{12}(a_{21} + ra_{11}) = \det A$$

If to these observations we add the fact that the determinant of the 2×2 identity matrix I is 1, we have four properties of the determinant of a 2×2 matrix which we summarize as follows:

1. The interchange of two rows of a matrix A changes the sign of the determinant of A.
2. If a row of the matrix A is multiplied by a scalar r, then the value of the determinant is multiplied by r.
3. If to one row of matrix A we add r times another row, the value of the determinant is not changed.
4. The determinant of the identity matrix I is 1.

The delightful fact is that we can define the determinant of an $n \times n$ matrix

$$A = \begin{bmatrix} a_{11} & \cdots & a_{1j} & \cdots & a_{1n} \\ \vdots & & \vdots & & \vdots \\ a_{i1} & \cdots & a_{ij} & \cdots & a_{in} \\ \vdots & & \vdots & & \vdots \\ a_{n1} & \cdots & a_{nj} & \cdots & a_{nn} \end{bmatrix}$$

so that Properties 1, 2, 3, and 4 all remain valid. Another fact, this one not so delightful, is that the definition that must be used to preserve these four simple properties is anything but simple. We will be helped in stating the definition if we first introduce some terminology.

The determinant of the $(n-1) \times (n-1)$ submatrix obtained from matrix A by deleting the ith row and the jth column is called the **minor** of a_{ij}. Denote this minor by M_{ij}. Using the minors of entries in the first column, we define determinants inductively.

DEFINITION

The determinant of the $n \times n$ matrix A above is

$$\det A = a_{11}M_{11} - a_{21}M_{21} + \cdots + (-1)^{n+1}a_{n1}M_{n1}$$

Since we already have a definition for a 2×2 matrix, this definition enables us to obtain the determinant of any $n \times n$ matrix if $n \geq 2$. For completeness, we define the determinant of a 1×1 matrix by

$$\det [a_{11}] = a_{11}$$

Before looking at an example, there is one more notational device to mention. Brackets are commonly used, as we have used them, to enclose the entries of a matrix A. It is just as common to use vertical lines around the entries of A to indicate $\det A$. Thus, for $n = 2$,

$$\det \begin{bmatrix} a_{11} & a_{12} \\ a_{21} & a_{22} \end{bmatrix} = \begin{vmatrix} a_{11} & a_{12} \\ a_{21} & a_{22} \end{vmatrix} = a_{11}a_{22} - a_{12}a_{21}$$

Find $\det A$ where

Example A

$$A = \begin{bmatrix} 2 & 1 & 1 & 0 \\ 0 & -3 & 1 & 2 \\ 0 & -1 & 3 & -4 \\ 3 & 1 & -2 & 5 \end{bmatrix}$$

First, note from the definition just given that a sign is associated with the *position* of a a_{1i} regardless of the sign of a_{1i} itself. Thus, in the present case,

$$\det A = 2M_{11} - 0M_{21} + 0M_{31} - 3M_{41}$$

Using vertical lines to indicate determinants,

$$
\begin{vmatrix} 2 & 1 & 1 & 0 \\ 0 & -3 & 1 & 2 \\ 0 & -1 & 3 & -4 \\ 3 & 1 & -2 & 5 \end{vmatrix} = 2 \begin{vmatrix} -3 & 1 & 2 \\ -1 & 3 & -4 \\ 1 & -2 & 5 \end{vmatrix} - 0 \begin{vmatrix} 1 & 1 & 0 \\ -1 & 3 & -4 \\ 1 & -2 & 5 \end{vmatrix}
$$

$$
+ 0 \begin{vmatrix} 1 & 1 & 0 \\ -3 & 1 & 2 \\ 1 & -2 & 5 \end{vmatrix} - 3 \begin{vmatrix} 1 & 1 & 0 \\ -3 & 1 & 2 \\ -1 & 3 & -4 \end{vmatrix}
$$

To be sure, we have been kind to ourselves by including a goodly number of well-placed 0 entries, so we need only evaluate two of the four minors.

$$
M_{11} = \begin{vmatrix} -3 & 1 & 2 \\ -1 & 3 & -4 \\ 1 & -2 & 5 \end{vmatrix} = -3 \begin{vmatrix} 3 & -4 \\ -2 & 5 \end{vmatrix} - (-1) \begin{vmatrix} 1 & 2 \\ -2 & 5 \end{vmatrix} + 1 \begin{vmatrix} 1 & 2 \\ 3 & -4 \end{vmatrix}
$$

$$
= -3(15 - 8) + 1(5 + 4) + 1(-4 - 6) = -22
$$

$$
M_{41} = \begin{vmatrix} 1 & 1 & 0 \\ -3 & 1 & 2 \\ -1 & 3 & -4 \end{vmatrix} = 1 \begin{vmatrix} 1 & 2 \\ 3 & -4 \end{vmatrix} - (-3) \begin{vmatrix} 1 & 0 \\ 3 & -4 \end{vmatrix} + (-1) \begin{vmatrix} 1 & 0 \\ 1 & 2 \end{vmatrix}
$$

$$
= 1(-4 - 6) + 3(-4 - 0) - 1(2 - 0) = -24
$$

Thus, the determinant of the given 4×4 matrix is

$$
2(-22) - 3(-24) = 28 \quad \square
$$

The first column plays a distinguished role in our definition of a determinant but, as is true in so many introductions, that which at first seems important turns out to have no distinction beyond that of being first in line. A remarkable theorem tells us that we would get the same value for det A if we used the entries and their minors from any column or, for that matter, from any row. The definition that we have given, referred to as the expansion of det A by minors of the first column, is thus seen to be just one of many equivalent definitions that might have been used.

The promised theorem will be easier to state if the signs associated with positions in a matrix are attached to the minor of the entry in that position. We therefore define the **cofactor** A_{ij} of the matrix A by $A_{ij} = (-1)^{i+j} M_{ij}$. The signs attached to a position in a matrix are arranged in the same checkerboard style we have already observed in the first column.

$$\begin{bmatrix} + & - & + & \cdots & (-1)^{n+1} \\ - & + & - & \cdots & \\ + & - & + & \cdots & \\ \vdots & & & \ddots & \\ (-1)^{n+1} & & & & + \end{bmatrix}$$

We can now state the theorem that we have already called remarkable.

The determinant of a square matrix A may be found by choosing any row or column vector of A, multiplying each component of that vector by its cofactor, and adding these products.

The proof of this theorem is not difficult in the sense of involving new ideas, but it does involve a great deal of notation. We omit the proof, referring the interested reader to Hoffman, K., and Kunze, R. 1961. *Linear algebra*. Englewood Cliffs, N.J.: Prentice-Hall. p. 135. We content ourselves with illustrating it by evaluating again the determinant of the matrix in Example A, this time using cofactors of elements in the second row.

Find $\det A$ by using cofactors of the elements in the second row of

$$A = \begin{bmatrix} 2 & 1 & 1 & 0 \\ 0 & -3 & 1 & 2 \\ 0 & -1 & 3 & -4 \\ 3 & 1 & -2 & 5 \end{bmatrix}$$

By Theorem A, $\det A = 0A_{21} - 3A_{22} + A_{23} + 2A_{24}$. Now the sign attached to A_{22} is $+$, and its entries are obtained from A by deleting row 2 and column 2:

$$A_{22} = \begin{vmatrix} 2 & 1 & 0 \\ 0 & 3 & -4 \\ 3 & -2 & 5 \end{vmatrix} = 2\begin{vmatrix} 3 & -4 \\ -2 & 5 \end{vmatrix} - 1\begin{vmatrix} 0 & -4 \\ 3 & 5 \end{vmatrix}$$

$$= 2(15 - 8) - (0 + 12) = 2$$

The determinant A_{22} was evaluated using cofactors of the first row. We evaluate A_{23} the same way, remembering however that the sign attached to A_{23} is $-$.

$$A_{23} = -\begin{vmatrix} 2 & 1 & 0 \\ 0 & -1 & -4 \\ 3 & 1 & 5 \end{vmatrix} = -2\begin{vmatrix} -1 & -4 \\ 1 & 5 \end{vmatrix} + 1\begin{vmatrix} 0 & -4 \\ 3 & 5 \end{vmatrix}$$

$$= -2(-5 + 4) + (0 + 12) = 14$$

To take advantage of the 0, and for variety, A_{24} will be evaluated using cofactors of the second row.

$$A_{24} = \begin{vmatrix} 2 & 1 & 1 \\ 0 & -1 & 3 \\ 3 & 1 & -2 \end{vmatrix} = -1 \begin{vmatrix} 2 & 1 \\ 3 & -2 \end{vmatrix} - 3 \begin{vmatrix} 2 & 1 \\ 3 & 1 \end{vmatrix}$$

$$= -1(-4 - 3) - 3(2 - 3) = 10$$

Substitution gives

$$\det A = -3(2) + (14) + 2(10) = 28 \quad \square$$

Example C   Expand in two ways the determinant of the general 3×3 matrix,

$$A = \begin{bmatrix} a_{11} & a_{12} & a_{13} \\ a_{21} & a_{22} & a_{23} \\ a_{31} & a_{32} & a_{33} \end{bmatrix}$$

First, let us expand it using the second column. This gives

$$-a_{12} \begin{vmatrix} a_{21} & a_{23} \\ a_{31} & a_{33} \end{vmatrix} + a_{22} \begin{vmatrix} a_{11} & a_{13} \\ a_{31} & a_{33} \end{vmatrix} - a_{32} \begin{vmatrix} a_{11} & a_{13} \\ a_{21} & a_{23} \end{vmatrix}$$

Next, we use the third row.

$$a_{31} \begin{vmatrix} a_{12} & a_{13} \\ a_{22} & a_{23} \end{vmatrix} - a_{32} \begin{vmatrix} a_{11} & a_{13} \\ a_{21} & a_{23} \end{vmatrix} + a_{33} \begin{vmatrix} a_{11} & a_{12} \\ a_{21} & a_{22} \end{vmatrix}$$

If the determinants are all written out, it will be found that the same six terms will be obtained in either case.

(1)
$$a_{11}a_{22}a_{33} + a_{12}a_{23}a_{31} + a_{13}a_{21}a_{32}$$
$$- a_{13}a_{22}a_{31} - a_{11}a_{23}a_{32} - a_{12}a_{21}a_{33} \quad \square$$

ANOTHER DEFINITION

There is another way to describe the six terms written down in (1). Note that each term has been written so that the first subscripts of the three factors are always in 123 order. That is, each of the six terms is of the form

(2)
$$a_{1i_1} \quad a_{2i_2} \quad a_{3i_3}$$

where i_1, i_2, and i_3 are the numbers 1, 2, and 3, though not necessarily in that order. The term (2) may be described this way; it is a product of entries from the given determinant in which exactly one factor from each row and each column has been selected. The 123 order of the first subscripts assures us that we have an entry from row 1, row 2, and row 3, respectively. Now i_1 can be selected in any of three ways, but once it is selected, i_2 can only be selected in two ways (it must come from a column different from i_1). Finally, there is only one choice for i_3. Thus, there will be $3(2)(1)$ ways to get terms of the form (2). The product $3 \cdot 2 \cdot 1$ is called 3 factorial, written 3!

Similar reasoning easily shows that if we begin with an $n \times n$ determinant, we can form $n! = n(n-1) \ldots (2)(1)$ products of the form

$$a_{1i_1} a_{2i_2} \cdots a_{ni_n}$$

in which there is exactly one factor from each row and each column.

A second remarkable theorem tells us that the $n \times n$ determinant is equal to an expression involving sums and differences of these $n!$ products. The tricky part of stating this theorem is to explain how the sign of a particular term is determined. It is done by classifying the permutation of the second indices

$$i_1 i_2 \ldots i_n$$

as even or odd. (See the vignette, *Even and Odd Permutations,* on page 113.)

The value of an $n \times n$ determinant can be obtained as the sum of the $n!$ terms | THEOREM B

$$\pm a_{1i_1} a_{2i_2} \cdots a_{ni_n}$$

in which the sign of a term is chosen $+$ or $-$ depending upon whether $i_1 i_2 \ldots i_n$ is an even or odd permutation.

The proof of this theorem, like that of Theorem A, does not involve any new ideas, but it is a triumph of bookkeeping. Unlike Theorem A, this theorem is not useful as a practical tool for evaluating determinants. It is included here since our statement of Theorem B is frequently used as the definition of an $n \times n$ determinant. When this is done, then of course our definition must be derived as a theorem; difficulties can often be shifted around, seldom eliminated. Beginning with our definition, the proof of Theorem B is given in Hoffman, K., and Kunze, R. 1961. *Linear algebra.* Englewood Cliffs, N.J.: Prentice-Hall. p. 139–141.

Though Theorem B is not generally useful for evaluation of a determinant, it is of importance for theoretical purposes and it can be useful in special cases.

Show that if matrix A is in the **upper-triangular** form | *Example D*

$$\begin{bmatrix} a_{11} & a_{12} & a_{13} & \cdots & a_{1n} \\ 0 & a_{22} & a_{23} & \cdots & a_{2n} \\ 0 & 0 & a_{33} & \cdots & a_{3n} \\ \vdots & & & \ddots & \\ 0 & 0 & 0 & \cdots & a_{nn} \end{bmatrix}$$

in which all entries *below* the main diagonal are 0, then $\det A = a_{11}a_{22}\cdots a_{nn}$.

This is shown by noting that the term

$$a_{1i_1}a_{2i_2}\cdots a_{ni_n}$$

can only be nonzero if $i_1 = 1$. In this case, since $i_2 \neq i_1$, it can only be nonzero if $i_2 = 2$. The argument from here is all downhill, leading us to the conclusion that the only term of the form given in Theorem B that can possibly be nonzero is $a_{11}a_{22}\cdots a_{nn}$. ☐

Observe that the same argument would show that a **lower-triangular** matrix, one in which all terms *above* the main diagonal are 0, also has a determinant equal to the product of the diagonal elements. Either result shows that the determinant of the identity matrix I is 1.

Example E The determinant of an elementary matrix of the first kind is -1.

Consider the 4×4 elementary matrix

$$E_1 = \begin{bmatrix} 1 & 0 & 0 & 0 \\ 0 & 0 & 0 & 1 \\ 0 & 0 & 1 & 0 \\ 0 & 1 & 0 & 0 \end{bmatrix}$$

If evaluated using Theorem B, the only nonzero term would be $a_{11}a_{24}a_{33}a_{42}$ since any other term must contain at least one factor of 0. Now the permutation formed by the second subscripts is 1432, an odd permutation (in the technical sense). Hence, $\det E_1 = -1$. ☐

The same argument would apply to any $n \times n$ elementary matrix of the first kind. Only one nonzero term could be selected. By definition, one interchange of its second subscripts would restore them to natural order, meaning that they form an odd permutation. Thus, the term must be preceded by a minus sign.

PROBLEM SET 3-2

In Problems 1 and 2, classify the given permutations as even or odd.

1. (a) 4312 (b) 3241 2. (a) 1432 (b) 3241
   (c) 536421 (d) 534621 (c) 519463278 (d) 519483276

In Problems 3 and 4, find the determinant of the given elementary matrices.

3. (a) $\begin{bmatrix} 1 & 0 & 0 & 0 \\ 0 & 1 & 0 & 0 \\ -3 & 0 & 1 & 0 \\ 0 & 0 & 0 & 1 \end{bmatrix}$ (b) $\begin{bmatrix} 1 & 0 & 0 & 0 \\ 0 & 1 & 0 & 0 \\ 0 & 0 & 5 & 0 \\ 0 & 0 & 0 & 1 \end{bmatrix}$ (c) $\begin{bmatrix} 0 & 1 & 0 & 0 \\ 1 & 0 & 0 & 0 \\ 0 & 0 & 1 & 0 \\ 0 & 0 & 0 & 1 \end{bmatrix}$

(Problem set continues on p. 114)

EVEN AND ODD PERMUTATIONS

A **permutation** of the first seven positive integers is a listing, in any order whatsoever, of all seven integers. Thus, 3671254 is a permutation. Now suppose we set about restoring the sequence to its natural order in a series of steps, called **transpositions,** in which we are allowed to interchange just two integers on any given step. Working with our example, and indicating the transposition that interchanges 3 and 7 by (37), we show two procedures that will restore order.

Procedure I

(37) 3671254 = 7631254
(17) 7631254 = 1637254
(26) 1637254 = 1237654
(57) 1237654 = 1235674
(45) 1235674 = 1234675
(56) 1234675 = 1234576
(67) 1234576 = 1234567

Procedure II

(13) 3671254 = 1673254
(26) 1673254 = 1273654
(37) 1273654 = 1237654
(47) 1237654 = 1234657
(56) 1234657 = 1234567

Obviously procedure II was more efficient, taking five steps instead of the seven steps required in procedure I. We wish to focus attention, however, on the fact that both procedures took an odd number of steps. There are other sequences (like 7531624) that can be restored by an even number of transpositions. The point, however, is this:

> *Given a permutation of the numbers $12 \ldots n$, the number of transpositions required to restore the permutation to natural order will always be even, or it will always be odd.*

THEOREM

According to the theorem, if one procedure for restoring order to a permutation $i_1 i_2 \ldots i_n$ requires an even number of transpositions, then any other procedure—no matter how efficient or inefficient—will also require an even number of transpositions to restore order to $i_1 i_2 \ldots i_n$. This leads us to call such a permutation **even.** In the same way, a permutation is called **odd** if it requires an odd number of transpositions to restore order.

Given a permutation, the theorem guarantees that we can use any sequence of transpositions that occurs to us to restore order.

Counting the transpositions used will enable us to determine whether the permutation is even or odd.

The peculiarity of this theorem is that it is needed in the study of subjects (like determinant theory) that seem to have little to do with the study of permutations, and its proof calls for techniques that also seem far removed from the study of permutations. Its proof is often included in the study of symmetric polynomials in the theory of equations (see Birkhoff, G., and MacLane, S. 1953. *A survey of modern algebra.* rev. ed. New York: Macmillan. p. 145) or in group theory (see Spitznagel, E. L., Jr. 1968. Note on the alternating group. *American Math. Monthly* 75:68).

4. (a) $\begin{bmatrix} 1 & 0 & 0 & 0 \\ 0 & 1 & 0 & 0 \\ 0 & 0 & 1 & 0 \\ 0 & 5 & 0 & 1 \end{bmatrix}$ (b) $\begin{bmatrix} 1 & 0 & 0 & 0 \\ 0 & -3 & 0 & 0 \\ 0 & 0 & 1 & 0 \\ 0 & 0 & 0 & 1 \end{bmatrix}$ (c) $\begin{bmatrix} 1 & 0 & 0 & 0 \\ 0 & 0 & 0 & 1 \\ 0 & 0 & 1 & 0 \\ 0 & 1 & 0 & 0 \end{bmatrix}$

Evaluate the determinants in Problems 5–8 using the cofactors of elements in the second row. Check, using cofactors of elements in the last column.

5. $\begin{vmatrix} 1 & -2 & 2 \\ 3 & 1 & -1 \\ 1 & -1 & 3 \end{vmatrix}$

6. $\begin{vmatrix} -1 & -2 & 1 \\ 2 & 3 & -1 \\ 1 & 2 & -2 \end{vmatrix}$

7. $\begin{vmatrix} 2 & 1 & 1 & -1 \\ 1 & 0 & -2 & 0 \\ 3 & -1 & 1 & 2 \\ -1 & 2 & 3 & 0 \end{vmatrix}$

8. $\begin{vmatrix} 3 & -1 & 2 & 0 \\ 0 & 1 & 0 & -1 \\ 2 & 0 & 1 & 3 \\ -1 & -2 & 1 & 0 \end{vmatrix}$

The six terms of (1) of this section are sometimes remembered for 3×3 matrices by copying the first two columns over to the right of the matrix:

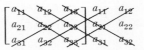

The three positive terms are the products found by multiplying factors on the diagonals sloping downward from left to right, while the negative terms are the products found along the upward sloping diagonals.

Caution: Do not try to make an "obvious" extension of this mnemonic device to 4×4 matrices. It won't work!

9. Use the method just described to evaluate the determinant of Problem 5.

10. Use the method just described to evaluate the determinant of Problem 6.

In Section 3-3, we will prove many properties of determinants. Try, as you evaluate the determinants given in Problems 11–24, to guess at the general properties being illustrated. Most of these problems refer to the following matrices:

$$A = \begin{bmatrix} 1 & -1 & 0 \\ 3 & 2 & -2 \\ -1 & 4 & 0 \end{bmatrix} \quad B = \begin{bmatrix} 1 & -1 & 0 \\ 3 & 2 & 1 \\ 0 & 1 & -2 \end{bmatrix}$$

$$C = \begin{bmatrix} 2 & 1 & -1 \\ 0 & 0 & 3 \\ -1 & 3 & 2 \end{bmatrix} \quad D = \begin{bmatrix} 2 & 3 & -1 \\ 0 & 1 & 2 \\ 1 & 0 & 1 \end{bmatrix}$$

11. Compare det A and

$$\begin{vmatrix} 0 & -1 & 1 \\ -2 & 2 & 3 \\ 0 & 4 & -1 \end{vmatrix}$$

12. Compare det C and

$$\begin{vmatrix} -1 & 3 & 2 \\ 0 & 0 & 3 \\ 2 & 1 & -1 \end{vmatrix}$$

13. Compare det B and

$$\begin{vmatrix} 1 & -1 & 0 \\ -6 & -4 & -2 \\ 0 & 1 & -2 \end{vmatrix}$$

14. Compare det D and

$$\begin{vmatrix} 2 & 3 & 3 \\ 0 & 1 & -6 \\ 1 & 0 & -3 \end{vmatrix}$$

15. Compare det A and det E where matrix

$$E = \begin{bmatrix} 1 & -1 & 2 \\ 3 & 2 & 4 \\ -1 & 4 & -2 \end{bmatrix}$$

has been obtained from A by adding 2 (column 1) to the last column.

16. Compare det C and det F where matrix

$$F = \begin{bmatrix} 2 & 1 & -1 \\ 3 & -9 & -3 \\ -1 & 3 & 2 \end{bmatrix}$$

has been obtained from C by adding -3 (row 3) to the second row.

17. Find det (B^t).

18. Find det (D^t).

19. Find det (AB).

20. Find det (CD).

21. Find det (A^{-1}).

22. Find det (C^{-1}).

23. Evaluate
$\begin{vmatrix} 1 & 1 & 1 & 1 \\ 1 & 2 & 4 & 8 \\ 1 & 3 & 9 & 27 \\ 1 & 4 & 16 & 64 \end{vmatrix}$

24. Evaluate
$\begin{vmatrix} 1 & 2 & 4 & 8 \\ 1 & 4 & 16 & 64 \\ 1 & 3 & 9 & 27 \\ 1 & -2 & 4 & -8 \end{vmatrix}$

25. Points $A(2, 3)$, $B(7, 8)$, and $C(10, -1)$ are vertices of a triangle in the plane (Figure 3-2). Find the area in two ways:

(a) area $= \frac{1}{2} |\overrightarrow{AC}| |\overrightarrow{DB}|$ where $\overrightarrow{DB} = \overrightarrow{AB} - \text{proj}_{\overrightarrow{AC}} \overrightarrow{AB}$.

(b) area $= |\frac{1}{2} \det A|$ where A is the matrix having $\overrightarrow{AB}$ and $\overrightarrow{AC}$ as two row vectors.

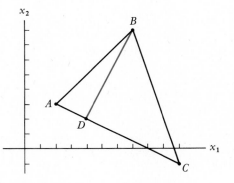

Figure 3-2

26. Prove that if $D = ad - bc = 0$, then the matrix

$$A = \begin{bmatrix} a & b \\ c & d \end{bmatrix}$$

has no inverse. *Hint:* Suppose that $LA = I$, and consider the product

$$(LA)\begin{bmatrix} d & -b \\ -c & a \end{bmatrix} = [L]\left(\begin{bmatrix} a & b \\ c & d \end{bmatrix}\begin{bmatrix} d & -b \\ -c & a \end{bmatrix}\right) = [L]\begin{bmatrix} 0 & 0 \\ 0 & 0 \end{bmatrix}$$

3-3 PROPERTIES OF DETERMINANTS

In Section 3-2 we listed four properties of the determinant of a 2×2 matrix. We claimed that if we were willing to pay the price of a rather complicated definition for the determinant of an $n \times n$ matrix, then these properties could be preserved. It seems appropriate to begin this section by showing that we got what we paid for.

Since $\det A$ is only defined when A is $n \times n$, that is, square, it is to be assumed throughout this chapter that the matrices of which we speak are square.

PROPERTY 1

The interchange of two rows of a matrix A changes the sign of the determinant of A.

Proof: We know the property holds for a 2×2 matrix. Consider a 3×3 matrix

$$A = \begin{bmatrix} a_{11} & a_{12} & a_{13} \\ a_{21} & a_{22} & a_{23} \\ a_{31} & a_{32} & a_{33} \end{bmatrix}$$

Suppose that rows 1 and 3 are interchanged to obtain the matrix

$$\mathcal{A} = \begin{bmatrix} a_{31} & a_{32} & a_{33} \\ a_{21} & a_{22} & a_{23} \\ a_{11} & a_{12} & a_{13} \end{bmatrix}$$

Expanding both $\det A$ and $\det \mathcal{A}$ by cofactors of the second row, we get

$$\det A = a_{21}A_{21} + a_{22}A_{22} + a_{23}A_{23}$$
$$\det \mathcal{A} = a_{21}\mathcal{A}_{21} + a_{22}\mathcal{A}_{22} + a_{23}\mathcal{A}_{23}$$

Now $\mathcal{A}_{21}$ is a 2×2 determinant that differs from the 2×2 determinant A_{21} in that two rows have been interchanged. In fact

$$A_{21} = \begin{vmatrix} a_{12} & a_{13} \\ a_{32} & a_{33} \end{vmatrix} \qquad \mathcal{A}_{21} = \begin{vmatrix} a_{32} & a_{33} \\ a_{12} & a_{13} \end{vmatrix}$$

Hence, $\mathcal{A}_{21} = -A_{21}$; similarly, $\mathcal{A}_{22} = -A_{22}$, $\mathcal{A}_{23} = -A_{23}$, and so $\det \mathcal{A} = -\det A$.

The same argument could be used no matter which two rows were interchanged. Simply evaluate by using cofactors of the row left unchanged.

Now what about a 4×4 matrix? Use the same argument, again evaluating by using cofactors of a row that was not changed. The cofactors in the two expansions, being determinants of 3×3 matrices that differ only in that two rows are interchanged, are known by the argument above to have values that differ only in sign. A formal proof for any value of n would use this same argument, proceeding by induction. $\square$

If a row of a matrix A is multiplied by a scalar r, then the value of the determinant is multiplied by r.

PROPERTY 2

Proof: Let the determinant of the new matrix be evaluated by cofactors of the row that has been multiplied by r. The value of the determinant is

$$ra_{k1}A_{k1} + ra_{k2}A_{k2} + \cdots + ra_{kn}A_{kn} = r \det A$$

Alternatively, appeal to Theorem B of the previous section. Every term of the form shown in Theorem B includes a factor $ra_{k i_k}$ taken from the kth row of the new matrix, so an r can be factored out of every term. $\square$

If to one row of matrix A we add r times another row, the value of the determinant is not changed.

Proof: A formal argument would again proceed by induction, just as in the proof of Property 1. To illustrate the procedure, suppose we have shown the property to hold for any 3×3 matrix, and suppose we wish to evaluate the determinant of

$$\mathcal{Q} = \begin{bmatrix} a_{11} & a_{12} & a_{13} & a_{14} \\ a_{21} & a_{22} & a_{23} & a_{24} \\ a_{31} + ra_{21} & a_{32} + ra_{22} & a_{33} + ra_{23} & a_{34} + ra_{24} \\ a_{41} & a_{42} & a_{43} & a_{44} \end{bmatrix}$$

obtained from A by adding $r \times$ (row 2) to (row 3). Expand det $\mathcal{Q}$ using cofactors of row 1 (row 4 would work just as well), getting

$$\det \mathcal{Q} = a_{11}\mathcal{Q}_{11} + a_{12}\mathcal{Q}_{12} + a_{13}\mathcal{Q}_{13} + a_{14}\mathcal{Q}_{14}$$

Now

$$\mathcal{Q}_{11} = \begin{vmatrix} a_{22} & a_{23} & a_{24} \\ a_{32} + ra_{22} & a_{33} + ra_{23} & a_{34} + ra_{24} \\ a_{42} & a_{43} & a_{44} \end{vmatrix}$$

which by hypothesis is equal to the determinant

$$\begin{vmatrix} a_{22} & a_{23} & a_{24} \\ a_{32} & a_{33} & a_{34} \\ a_{42} & a_{43} & a_{44} \end{vmatrix} = A_{11}$$

This is the cofactor of a_{11} in the given matrix A. Similarly, $\mathcal{Q}_{12} = A_{12}$, $\mathcal{Q}_{13} = A_{13}$, $\mathcal{Q}_{14} = A_{14}$, and so det $\mathcal{Q} = $ det A. $\square$

It is almost trivial to note that since an $n \times n$ identity matrix I is upper-triangular, its determinant is equal to the product of 1s on the diagonal.

The determinant of the $n \times n$ identity matrix is 1.

The determinant of an elementary matrix of the first kind has already been shown (Example E of the previous section) to be -1. An elementary matrix of the second kind, such as E_{II2} below, is both upper- and lower-triangular, so det $E = r$, the product of entries on the main diagonal. An elementary matrix of the third kind, such as E_{III3}, is either upper- or lower-triangular, so again its determinant is equal to the product of the diagonal entries; det $E_{III3} = 1$.

$$E_{II2} = \begin{bmatrix} 1 & 0 & 0 & 0 \\ 0 & r & 0 & 0 \\ 0 & 0 & 1 & 0 \\ 0 & 0 & 0 & 1 \end{bmatrix}, \qquad E_{III3} = \begin{bmatrix} 1 & 0 & 0 & 0 \\ 0 & 1 & 0 & 0 \\ 0 & 0 & 1 & 0 \\ 0 & r & 0 & 1 \end{bmatrix}.$$

These observations, together with Properties 1–4 above, show that if E is an elementary matrix, then

$$\det EA = (\det E)(\det A) \qquad (1)$$

The matrices which we have used in our examples of how to compute determinants have been kept small for good reason. If we start with even a 10×10 matrix and use the definition of Theorem A, Section 3-2, to expand it by minors of any row or column, we will at the next step have to evaluate ten determinants of 9×9 matrices. This method gets out of hand, and even out of computer range very quickly if we try to compute determinants of matrices of the size that turn up in actual applications. Fortunately, the properties of determinants that we have just developed provide an alternative.

We noted in our discussion of solving systems of equations that variations of Gaussian elimination with back substitution were the most efficient procedures to use in computers. A program to carry this out will, for an $n \times n$ matrix A, begin by using a sequence of elementary operations to produce a row equivalent upper-triangular matrix T. Since we now know the effect of each elementary operation on the value of a determinant, we can keep track of their cumulative effect until T is obtained. The determinant of T, as shown in Example D of the previous section, will be the product of the entries on the diagonal.

The method just described greatly improves efficiency, whether we program it for a computer or carry it out by hand. We will use it to evaluate once again the determinant first evaluated in Example A of the previous section.

Evaluate det A if   *Example A*

$$A = \begin{bmatrix} 2 & 1 & 1 & 0 \\ 0 & -3 & 1 & 2 \\ 0 & -1 & 3 & -4 \\ 3 & 1 & -2 & 5 \end{bmatrix}$$

$$\det A = 2 \begin{vmatrix} 1 & \frac{1}{2} & \frac{1}{2} & 0 \\ 0 & -3 & 1 & 2 \\ 0 & -1 & 3 & -4 \\ 3 & 1 & -2 & 5 \end{vmatrix} = 2 \begin{vmatrix} 1 & \frac{1}{2} & \frac{1}{2} & 0 \\ 0 & -3 & 1 & 2 \\ 0 & -1 & 3 & -4 \\ 0 & -\frac{1}{2} & -\frac{7}{2} & 5 \end{vmatrix}$$

The 2 comes as an application of Property 2; it may be thought of as having been "factored out" of the first row. The next step results from a type-III elementary operation, adding -3 (first row) to (third row), which does not change the value of the determinant. Continuing with the second column,

$$\det A = 2(-3) \begin{vmatrix} 1 & \frac{1}{2} & \frac{1}{2} & 0 \\ 0 & 1 & -\frac{1}{3} & -\frac{2}{3} \\ 0 & -1 & 3 & -4 \\ 0 & -\frac{1}{2} & -\frac{7}{2} & 5 \end{vmatrix} = -6 \begin{vmatrix} 1 & \frac{1}{2} & \frac{1}{2} & 0 \\ 0 & 1 & -\frac{1}{3} & -\frac{2}{3} \\ 0 & 0 & \frac{8}{3} & -\frac{14}{3} \\ 0 & 0 & -\frac{11}{3} & \frac{14}{3} \end{vmatrix}$$

Continuing in this methodical way to the third column,

$$\det A = -6(\tfrac{8}{3}) \begin{vmatrix} 1 & \frac{1}{2} & \frac{1}{2} & 0 \\ 0 & 1 & -\frac{1}{3} & -\frac{2}{3} \\ 0 & 0 & 1 & -\frac{7}{4} \\ 0 & 0 & -\frac{11}{3} & \frac{14}{3} \end{vmatrix} = -16 \begin{vmatrix} 1 & \frac{1}{2} & \frac{1}{2} & 0 \\ 0 & 1 & -\frac{1}{3} & -\frac{2}{3} \\ 0 & 0 & 1 & -\frac{7}{4} \\ 0 & 0 & 0 & -\frac{7}{4} \end{vmatrix}$$

Since we now have a matrix in upper-triangular form, its determinant is the product of $(1)(1)(1)(-\frac{7}{4})$. Hence, $\det A = -16(-\frac{7}{4}) = 28$. ☐

The procedure above, which has the methodicalness suited to a computer program, can be modified to avoid fractions when used as a hand calculation. Experience will make shortcuts evident.

PROPERTY 5

If matrix A has a row consisting entirely of zeros, then $\det A = 0$.

Proof: Appeal to Theorem B of Section 3-2. Every term has a factor of 0. ☐

PROPERTY 6

If two rows of a matrix are identical, then the determinant of the matrix is zero.

Proof: Let the value of the determinant be s. The interchange of the two identical rows should, according to Property 1, result in a new matrix with determinant $-s$. But clearly the determinant of the matrices must be the same; $s = -s$, $s = 0$. ☐

Suppose A has two identical *columns*. Then A^t has two identical rows, hence $\det A^t = 0$. We naturally wonder how $\det A^t$ is related to $\det A$. Let us initiate this inquiry with an example.

Find the determinants of $\qquad$ *Example B*

$$A = \begin{bmatrix} 2 & 3 & -1 \\ 1 & 2 & -2 \\ 3 & 1 & 4 \end{bmatrix} \quad \text{and} \quad A^t = \begin{bmatrix} 2 & 1 & 3 \\ 3 & 2 & 1 \\ -1 & -2 & 4 \end{bmatrix}$$

If we use the 1 sitting in the first column of A to eliminate other nonzero entries in that column, we get

$$\det A = \begin{vmatrix} 0 & -1 & 3 \\ 1 & 2 & -2 \\ 0 & -5 & 10 \end{vmatrix} = (-1) \begin{vmatrix} -1 & 3 \\ -5 & 10 \end{vmatrix}$$

$$= -1(-10 + 15) = -5$$

Similar use can be made of the -1 in the first column of A^t.

$$\det A^t = \begin{vmatrix} 0 & -3 & 11 \\ 0 & -4 & 13 \\ -1 & -2 & 4 \end{vmatrix} = (-1) \begin{vmatrix} -3 & 11 \\ -4 & 13 \end{vmatrix}$$

$$= (-1)(-39 + 44) = -5 \quad \square$$

This illustrates the next property of determinants:

$\det A = \det A^t$ $\qquad$ *PROPERTY 7*

Proof: This is another of the properties that can be proved by induction. Begin by noting that if

$$A = \begin{bmatrix} a_{11} & a_{12} \\ a_{21} & a_{22} \end{bmatrix}, \quad A^t = \begin{bmatrix} a_{11} & a_{21} \\ a_{12} & a_{22} \end{bmatrix}$$

then $\det A^t = a_{11}a_{22} - a_{12}a_{21} = \det A$. Next, evaluate the determinant of the transpose of a 3×3 matrix using the 2×2 minors of some row. The desired conclusion will follow from what you know about the determinant of the transpose of a 2×2 matrix. Then you can move up to the 4×4 case, and so on. $\quad \square$

The effect of Property 7 is to double the value of everything we have proved about determinants, since it enables us in all previous statements to replace the word *row* by *column*. The interchange of two columns, like the interchange of two rows, changes the sign of a determinant. To one column of a matrix, a multiple of another column may be added without changing the value of the determinant. A column of zeros, or two identical columns, in a matrix makes its determinant 0. The payoff is truly magnificent.

Example C

We shall illustrate these ideas by finding det A, det B, and det AB for

$$A = \begin{bmatrix} 2 & 4 & -3 \\ 3 & -2 & -1 \\ -2 & -1 & \boxed{1} \end{bmatrix}, \quad B = \begin{bmatrix} 2 & 1 & 0 \\ 1 & -1 & 1 \\ \boxed{-1} & 0 & 2 \end{bmatrix}, \quad AB = \begin{bmatrix} 11 & -2 & -2 \\ 5 & 5 & -4 \\ -6 & -1 & \boxed{1} \end{bmatrix}$$

Using the circled elements in each matrix to eliminate other entries in the same row,

$$\det A = \begin{vmatrix} -4 & 1 & -3 \\ 1 & -3 & -1 \\ 0 & 0 & 1 \end{vmatrix} = (1) \begin{vmatrix} -4 & 1 \\ 1 & -3 \end{vmatrix} = 1(12 - 1) = 11$$

$$\det B = \begin{vmatrix} 2 & 1 & 4 \\ 1 & -1 & 3 \\ -1 & 0 & 0 \end{vmatrix} = (-1) \begin{vmatrix} 1 & 4 \\ -1 & 3 \end{vmatrix} = (-1)(3 + 4) = -7$$

$$\det AB = \begin{vmatrix} -1 & -4 & -2 \\ -19 & 1 & -4 \\ 0 & 0 & 1 \end{vmatrix} = 1 \begin{vmatrix} -1 & -4 \\ -19 & 1 \end{vmatrix}$$

$$= (1)(-1 - 76) = -77 \quad \square$$

Our direction may be anticipated by noting in Example C that det $AB = -77 = 11(-7) = (\det A)(\det B)$. Before we can discuss the determinant of a product, however, we need to understand the relationship between the determinant of a matrix A and the existence of an inverse.

PROPERTY 8

An $n \times n$ matrix A is nonsingular if and only if det $A \neq 0$.

Proof: Given A, we know (Theorem A, Section 2-7) that there exists a product of elementary matrices $E_k \ldots E_1$ such that

$$E_k \ldots E_1 A = R$$

is in row echelon form. Since R is $n \times n$, either $R = I$ or else the last row (and perhaps other rows) of R consists entirely of zeros. In either case, repeated appeals to (1) give us

(2)

$$(\det E_k) \ldots (\det E_1)(\det A) = \det R$$

Consider first the case where $R = I$. Then A is nonsingular. And since det $R = \det I = 1$, there can be no factor of 0 on the left-hand side of (2). We conclude that det $A \neq 0$.

Now consider the case where $R \neq I$. Then, because of the last row of zeros in R, Property 5 implies that det $R = 0$. But this means that some factor on the left-hand side of (2) must be 0. Since no elementary matrix has a determinant of 0, it must be that det $A = 0$. $\quad \square$

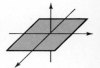

AN APPLICATION TO RANKS

Suppose we know that every minor of the matrix

$$A = \begin{bmatrix} a_{11} & a_{12} & a_{13} & a_{14} \\ a_{21} & a_{22} & a_{23} & a_{24} \\ a_{31} & a_{32} & a_{33} & a_{34} \\ a_{41} & a_{42} & a_{43} & a_{44} \end{bmatrix}$$

has value 0, but that

$$\begin{vmatrix} a_{12} & a_{14} \\ a_{32} & a_{34} \end{vmatrix} \neq 0$$

From this information, we may conclude that rank $A = 2$.

In the first place, we note that neither the first nor the third row of A is the **0** vector. Thus, if these vectors are linearly dependent, we must have

$$(a_{11}, a_{12}, a_{13}, a_{14}) = r(a_{31}, a_{32}, a_{33}, a_{34})$$

for some scalar r. But then it would follow that

$$\begin{vmatrix} a_{12} & a_{14} \\ a_{32} & a_{34} \end{vmatrix} = \begin{vmatrix} ra_{32} & ra_{34} \\ a_{32} & a_{34} \end{vmatrix} = 0$$

which contradicts the given information. Certainly then, rank $A \geq 2$.

Now suppose there are three linearly independent row vectors in A. Take them, for notational convenience, to be the first three. A fourth vector (b_1, b_2, b_3, b_4) may be adjoined to these first three row vectors of A to form a basis for R^4. The matrix

$$B = \begin{bmatrix} a_{11} & a_{12} & a_{13} & a_{14} \\ a_{21} & a_{22} & a_{23} & a_{24} \\ a_{31} & a_{32} & a_{33} & a_{34} \\ b_1 & b_2 & b_3 & b_4 \end{bmatrix}$$

is therefore row equivalent to the 4×4 identity matrix I, so $\det B \neq 0$. On the other hand, evaluation of $\det B$ using minors of the fourth row, all of them being minors of A, must give 0. This contradiction shows that there are not three linearly independent row vectors in A; rank $A < 3$.

We have proved that $2 \leq \text{rank } A < 3$, so rank $A = 2$. This result illustrates the relationship between determinants and the rank of a matrix:

THEOREM

> *The rank of an $n \times n$ matrix A is the largest integer k for which there is a $k \times k$ submatrix of A that has a nonzero determinant.*

PROPERTY 9

We are now prepared to prove the property hinted at in Example C.

$$\det (AB) = (\det A)(\det B)$$

Proof: Again we note that an arbitrary matrix A may be reduced to row echelon form R by multiplication on the left by a sequence of elementary matrices:

$$E_k \ldots E_1 A = R$$

Multiplication on the left by $E_1^{-1} \ldots E_k^{-1}$ gives us

$$A = E_1^{-1} \ldots E_k^{-1} R$$

which may be substituted in the product AB to give

$$AB = E_1^{-1} \ldots E_k^{-1} RB$$

Since the inverse of an elementary matrix is again an elementary matrix, we can use (1) to write

$$\det (AB) = (\det E_1^{-1}) \ldots (\det E_k^{-1}) \det (RB)$$

Now if A is nonsingular so that $R = I$, this becomes

$$\det AB = (\det E_1^{-1}) \ldots (\det E_k^{-1}) \det B$$
$$= \det (E_1^{-1} \ldots E_k^{-1}) \det B$$
$$= \det A \det B$$

The last equality is true because when $R = I$, $A = E_1^{-1} \ldots E_k^{-1}$. If A is singular so that $\det A = 0$, then the last row of RB, like the last row of R, must consist entirely of 0s. An appeal to Property 5 shows $\det RB = 0$, so (3) says $\det AB = 0$. Again, we have $\det (AB) = (\det A)(\det B)$.   □

PROBLEM SET 3-3

Evaluate the determinants in Problems 1–6 by any method that seems appropriate. Then check your answer (there are no answers in the back of the book for these problems) by evaluating the determinant a second way (expanding by cofactors of a different row or column).

1. $\begin{vmatrix} 2 & 5 & 1 \\ 3 & -2 & -1 \\ 7 & 4 & 3 \end{vmatrix}$

2. $\begin{vmatrix} -1 & 3 & 5 \\ 6 & 4 & 2 \\ -2 & 5 & 1 \end{vmatrix}$

3. $\begin{vmatrix} -1 & 0 & 2 & 0 \\ 2 & 3 & 1 & -1 \\ 1 & 0 & 4 & 3 \\ -2 & 1 & -1 & 2 \end{vmatrix}$

4. $\begin{vmatrix} 5 & 1 & -1 & 0 \\ -2 & 1 & -1 & 3 \\ 3 & 0 & 2 & -1 \\ 4 & 0 & 3 & -5 \end{vmatrix}$

5. $\begin{vmatrix} 5 & -2 & 2 & 3 \\ 3 & -3 & 7 & 2 \\ 4 & 3 & 5 & -2 \\ -2 & 2 & 4 & 5 \end{vmatrix}$

6. $\begin{vmatrix} 7 & -3 & 4 & 2 \\ 5 & 2 & 3 & -2 \\ 3 & 7 & 5 & 3 \\ -5 & 4 & -2 & 5 \end{vmatrix}$

Problems 7–10 refer to the matrices

$$A = \begin{bmatrix} 1 & -2 & 3 \\ 0 & 3 & 5 \\ 3 & 4 & -2 \end{bmatrix} \quad B = \begin{bmatrix} 5 & -1 & 0 \\ -3 & 2 & 4 \\ 2 & 5 & -3 \end{bmatrix}$$

7. Verify that $\det AB = \det A \det B$.

8. Verify that $\det BA = \det B \det A$.

9. Find A^{-1}, then $\det A^{-1}$.

10. Find B^{-1}, then $\det B^{-1}$.

11. Solve $\begin{vmatrix} 4 & -1 & 2 \\ 5 & x & 7 \\ x & -1 & 3 \end{vmatrix} = 0$

12. Solve $\begin{vmatrix} x & 3 & -4 \\ -2 & 7 & 3 \\ 22 & x & -19 \end{vmatrix} = 0$

13. Show that $\det AB = \det BA$.

14. Show that for any nonsingular matrix A, $\det A^{-1} = 1/\det A$.

15. Graph $\begin{vmatrix} x & y & 1 \\ -1 & 0 & 2 \\ 3 & 1 & 4 \end{vmatrix} = 0$

16. Graph $\begin{vmatrix} -1 & 2 & 0 \\ 1 & x & y \\ 2 & 5 & 3 \end{vmatrix} = 0$

17. Graph $\begin{vmatrix} x & y & 1 \\ 4 & -3 & 1 \\ -2 & 5 & 1 \end{vmatrix} = 0$

18. Graph $\begin{vmatrix} x & y & 1 \\ 5 & 3 & 1 \\ -2 & -4 & 1 \end{vmatrix} = 0$

19. Show without evaluating that

$$\begin{vmatrix} 1 & a & bc \\ 1 & b & ac \\ 1 & c & ab \end{vmatrix} = \begin{vmatrix} 1 & a & a^2 \\ 1 & b & b^2 \\ 1 & c & c^2 \end{vmatrix}$$

20. Show without evaluating that

$$\begin{vmatrix} a-b & b-c & c-a \\ 1 & 1 & 1 \\ a & b & c \end{vmatrix} = \begin{vmatrix} a & b & c \\ 1 & 1 & 1 \\ b & c & a \end{vmatrix}$$

21. Show that $\begin{vmatrix} 1 & 1 & 1 & \cdots & 1 \\ 1 & 2 & 2^2 & \cdots & 2^{n-1} \\ \vdots & \vdots & \vdots & & \vdots \\ 1 & n & n^2 & \cdots & n^{n-1} \end{vmatrix} = 1!\,2!\,3! \cdots (n-1)!$

Compare with Problem 23 of Problem Set 3-2.

22. Show that $\begin{vmatrix} 1 & a & a^2 \\ 1 & b & b^2 \\ 1 & c & c^2 \end{vmatrix} = (b-a)(c-a)(c-b)$

Compare with Problem 24 of Problem Set 3-2.

23. **Vandermonde's determinant** for $n = 4$. Show that

$$\begin{vmatrix} 1 & x_1 & x_1^2 & x_1^3 \\ 1 & x_2 & x_2^2 & x_2^3 \\ 1 & x_3 & x_3^2 & x_3^3 \\ 1 & x_4 & x_4^2 & x_4^3 \end{vmatrix} = (x_4 - x_3)(x_4 - x_2)(x_4 - x_1)(x_3 - x_2)(x_3 - x_1)(x_2 - x_1)$$

Hint: Consider the third degree equation in y defined by

$$\begin{vmatrix} 1 & x_1 & x_1^2 & x_1^3 \\ 1 & x_2 & x_2^2 & x_2^3 \\ 1 & x_3 & x_3^2 & x_3^3 \\ 1 & y & y^2 & y^3 \end{vmatrix} = 0$$

What are the roots? What is the coefficient of y^3?

24. Let $\mathbf{v}_1, \ldots, \mathbf{v}_n$ be n linearly independent vectors in R^n. Prove that

$$\begin{vmatrix} \mathbf{v}_1 \cdot \mathbf{v}_1 & \mathbf{v}_1 \cdot \mathbf{v}_2 & \cdots & \mathbf{v}_1 \cdot \mathbf{v}_n \\ \mathbf{v}_2 \cdot \mathbf{v}_1 & \mathbf{v}_2 \cdot \mathbf{v}_2 & \cdots & \mathbf{v}_2 \cdot \mathbf{v}_n \\ \vdots & \vdots & & \vdots \\ \mathbf{v}_n \cdot \mathbf{v}_1 & \mathbf{v}_n \cdot \mathbf{v}_2 & \cdots & \mathbf{v}_n \cdot \mathbf{v}_n \end{vmatrix} > 0$$

25. Using matrix A as defined for Problems 7–10 above, how many matrices B can you find such that $\det (A + B) = \det A + \det B$?

26. What can you say about det A in the following cases?
    (a) $A^2 = 0$ (the matrix consisting entirely of zeros)
    (b) $A^2 = I$ (the identity matrix)
    (c) $A\mathbf{x} = 0$ is a system of 4 equations in 4 unknowns having $(1, 1, 1, 1)$ as a solution.

27. Prove that the system $A\mathbf{x} = \mathbf{b}$ of n equations in n variables has a unique solution if and only if det $A \neq 0$.

28. Suppose that $A = P^{-1}BP$. What can you say about det P? About the relationship of det A and det B?

APPLICATIONS | 3-4

It has now been shown that our rather complicated definition of an $n \times n$ determinant has all the simple properties for which we had hoped, save one. We expressed the hope in the introduction to this chapter that if P is the parallelepiped in space having as adjacent edges the vectors $\mathbf{r} = (r_1, r_2, r_3)$, $\mathbf{s} = (s_1, s_2, s_3)$, and $\mathbf{t} = (t_1, t_2, t_3)$, then the volume of P should equal the determinant of

$$A = \begin{bmatrix} r_1 & r_2 & r_3 \\ s_1 & s_2 & s_3 \\ t_1 & t_2 & t_3 \end{bmatrix}$$

It is time to assess our success in this endeavor.

The volume of a parallelepiped is the area of the base multiplied by the height as measured on a perpendicular to the base. If three vectors $\mathbf{r}$, $\mathbf{s}$, and $\mathbf{t}$ are three adjacent edges of the parallelepiped, and if we take the base to be the parallelogram determined by $\mathbf{s}$ and $\mathbf{t}$ (Figure 3-3), then

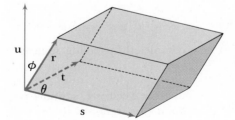

Figure 3-3

the area of the base is $|\mathbf{s}|\,|\mathbf{t}| \sin \theta$, where θ is the angle between $\mathbf{s}$ and $\mathbf{t}$. If $\mathbf{u}$ is normal (perpendicular) to the plane determined by $\mathbf{s}$ and $\mathbf{t}$, then the height of the parallelepiped is the length of the projection of $\mathbf{r}$ on $\mathbf{u}$; that is, it is $|\mathbf{r}| \cos \phi$ where ϕ is the angle between $\mathbf{u}$ and $\mathbf{r}$. Thus,

$$\text{volume} = \pm\, |\mathbf{r}|\,|\mathbf{s}|\,|\mathbf{t}| \sin \theta \cos \phi$$

Since $\cos \phi$ might be negative, depending on the orientation of $\mathbf{u}$ and $\mathbf{r}$, we agree here and in similar formulas in this section to choose the $\pm$ sign so as to make the volume positive.

THEOREM A   *The volume of the parallelepiped P having as adjacent edges the row vectors of the matrix*

$$A = \begin{bmatrix} r_1 & r_2 & r_3 \\ s_1 & s_2 & s_3 \\ t_1 & t_2 & t_3 \end{bmatrix}$$

is given by vol $(P) = \pm \det A$.

Proof: Expand $\det A$ by using minors of the first row:

$$\det A = r_1 \begin{vmatrix} s_2 & s_3 \\ t_2 & t_3 \end{vmatrix} - r_2 \begin{vmatrix} s_1 & s_3 \\ t_1 & t_3 \end{vmatrix} + r_3 \begin{vmatrix} s_1 & s_2 \\ t_1 & t_2 \end{vmatrix}$$

This may be viewed as the dot product of the vectors **r** and

$$\mathbf{u} = (s_2 t_3 - s_3 t_2, s_3 t_1 - s_1 t_3, s_1 t_2 - s_2 t_1)$$

This latter vector is exactly the vector noted in equations (2) and (3) of Section 2-3 to be perpendicular to the plane determined by **s** and **t**, and to satisfy

$$|\mathbf{u}|^2 = |\mathbf{s}|^2 |\mathbf{t}|^2 - (\mathbf{s} \cdot \mathbf{t})^2$$

Since, for any two vectors, $\mathbf{v} \cdot \mathbf{w} = |\mathbf{v}||\mathbf{w}| \cos \alpha$ where α is the angle between the vectors, we now have

$$\begin{aligned} (\det A)^2 = (\mathbf{r} \cdot \mathbf{u})^2 &= |\mathbf{r}|^2 |\mathbf{u}|^2 \cos^2 \phi \\ &= |\mathbf{r}|^2 \cos^2 \phi \, (|\mathbf{s}|^2 |\mathbf{t}|^2 - (\mathbf{s} \cdot \mathbf{t})^2) \\ &= |\mathbf{r}|^2 \cos^2 \phi \, (|\mathbf{s}|^2 |\mathbf{t}|^2 - |\mathbf{s}|^2 |\mathbf{t}|^2 \cos^2 \theta) \\ &= |\mathbf{r}|^2 |\mathbf{s}|^2 |\mathbf{t}|^2 \cos^2 \phi \, (1 - \cos^2 \theta) \\ &= |\mathbf{r}|^2 |\mathbf{s}|^2 |\mathbf{t}|^2 \cos^2 \phi \sin^2 \theta \end{aligned}$$

The proof is now completed by taking square roots of both sides.   □

Most people can see that three vectors **r**, **s**, and **t** positioned in space with a common initial point (Figure 3-3) will determine a parallelepiped P. It is simple then to describe the problem of trying to find the volume of P. Those who attempt to find a formula for this volume in terms of the coordinates of **r**, **s**, and **t** will soon discover that it is not simple to solve the problem. Yet we have found a very simple answer. Here we see the characteristics of mathematical elegance. A simple question approached in an obvious way leads to surprising difficulties, while a less obvious approach leads directly to a surprisingly concise answer.

The concept of mathematical elegance can be further illustrated by the problem of finding the volume of a tetrahedron T having vertices A, B, C, and D. If we think of ABC as the base of T, the desired volume can be obtained directly from

$$\text{vol } T = \tfrac{1}{3} \text{ (area of base } ABC)(\text{height})$$ *(1)*

The problem can be worked this way, but that is not elegant. The elegant way to do the problem is to think of AB, AC, and AD as adjacent edges of a parallelepiped P having $ACED$ as its base (Figure 3-4).

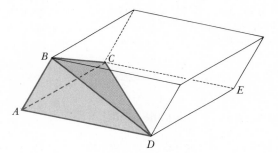

Figure 3-4

Then

$$\text{vol } T = \tfrac{1}{3} \, (\tfrac{1}{2} \text{ area of base } ACED)(\text{height})$$ *(2)*
$$\text{vol } T = \tfrac{1}{6} \, (\text{area of base } ACED)(\text{height})$$
$$\text{vol } T = \tfrac{1}{6} \, \text{vol } P$$

Since the volume of P can be computed using Theorem A, we have what might be called an elegant solution to our problem.

We are also in a position to get a neat formula for the area of a triangle in the plane if we know the coordinates of the three vertices. Again, we must allow for different orientations in the way we choose our points by introducing a $\pm$ to be chosen so that the area is positive.

The area of a triangle $\triangle$ in the plane having vertices $A(a_1, a_2)$,   *THEOREM B*
$B(b_1, b_2)$, and $C(c_1, c_2)$ is

$$\text{area } \triangle = \pm\tfrac{1}{2} \begin{vmatrix} a_1 & a_2 & 1 \\ b_1 & b_2 & 1 \\ c_1 & c_2 & 1 \end{vmatrix}$$ *(3)*

Proof: Draw the tetrahedron having vertices $\mathcal{A}(a_1, a_2, 1)$, $\mathcal{B}(b_1, b_2, 1)$, $\mathcal{C}(c_1, c_2, 1)$, and $\mathcal{O}(0, 0, 0)$. Considering $\mathcal{ABC}$ to be the base, it is clear that the area of the base is the area of $\triangle$, and that the height is 1 (Figure 3-5).

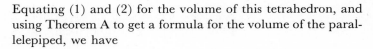

Figure 3-5

Equating (1) and (2) for the volume of this tetrahedron, and using Theorem A to get a formula for the volume of the parallelepiped, we have

$$\tfrac{1}{3}(\text{area } \triangle)(1) = \pm\tfrac{1}{6}\begin{vmatrix} a_1 & a_2 & 1 \\ b_1 & b_2 & 1 \\ c_1 & c_2 & 1 \end{vmatrix}$$

Multiplication by 3 gives the result. □

Example A

Find the area of the triangle with vertices $A(-2, 7)$, $B(5, 3)$, and $C(2, 1)$, as shown in Figure 3-6.

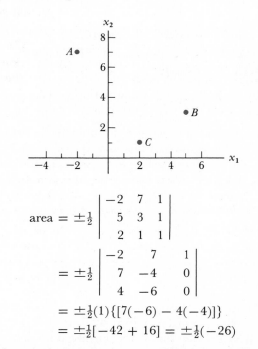

Figure 3-6

$$\text{area} = \pm\tfrac{1}{2}\begin{vmatrix} -2 & 7 & 1 \\ 5 & 3 & 1 \\ 2 & 1 & 1 \end{vmatrix}$$

$$= \pm\tfrac{1}{2}\begin{vmatrix} -2 & 7 & 1 \\ 7 & -4 & 0 \\ 4 & -6 & 0 \end{vmatrix}$$

$$= \pm\tfrac{1}{2}(1)\{[7(-6) - 4(-4)]\}$$

$$= \pm\tfrac{1}{2}[-42 + 16] = \pm\tfrac{1}{2}(-26)$$

We choose the minus sign to give a positive area of 13.

Since the dot product of $\overrightarrow{CA} = (-4, 6)$ and $\overrightarrow{CB} = (3, 2)$ is zero, we have a right triangle here that makes it easy to verify our answer using the familiar formula

$$\text{area} = \tfrac{1}{2}(\text{base})(\text{height}) = \tfrac{1}{2}|\overrightarrow{CA}|\,|\overrightarrow{BA}|$$
$$= \tfrac{1}{2}\sqrt{16 + 36}\,\sqrt{9 + 4} = \tfrac{1}{2}\sqrt{52(13)} = 13 \quad \square$$

If the three points A, B, and C are collinear, then of course the area, hence the determinant (3) is zero. This points the way to more applications. The set of all points $P(x_1, x_2)$ for which

$$\begin{vmatrix} x_1 & x_2 & 1 \\ a_1 & a_2 & 1 \\ b_1 & b_2 & 1 \end{vmatrix} = 0 \tag{4}$$

is, of course, the set of all points collinear with $A(a_1, a_2)$ and $B(b_1, b_2)$. That is, (4) is the equation of the line passing through A and B.

There is another way to see that (4) describes a line through A and B. Think of expanding (4) by cofactors of the first row. Clearly one will get an expression of the form

$$mx_1 + nx_2 + p = 0$$

which is the equation of a straight line. Now note that since $x_1 = a_1$, $x_2 = a_2$ satisfies (4) (because two rows are equal), the line passes through A. Similarly, it passes through $x_1 = b_1$, $x_2 = b_2$.

Once this idea is planted, it can grow in many directions. Think, for instance, of expanding by cofactors of the first row the determinant

$$\begin{vmatrix} x_1^2 + x_2^2 & x_1 & x_2 & 1 \\ a_1^2 + a_2^2 & a_1 & a_2 & 1 \\ b_1^2 + b_2^2 & b_1 & b_2 & 1 \\ c_1^2 + c_2^2 & c_1 & c_2 & 1 \end{vmatrix} = 0 \tag{5}$$

This equation describes a circle that passes through $A(a_1, a_2)$, $B(b_1, b_2)$, and $C(c_1, c_2)$. Problems 21–28 suggest more ways in which determinants can be used to write equations of standard curves.

We turn to another application. This one enables us, when a system of n equations in n unknowns has a unique solution, to use determinants to write a formula for the solution. Since the general method is clearly exposed by the case $n = 3$, we will understand $A\mathbf{x} = \mathbf{b}$ to describe the system

$$\begin{bmatrix} a_{11} & a_{12} & a_{13} \\ a_{21} & a_{22} & a_{23} \\ a_{31} & a_{32} & a_{33} \end{bmatrix} \begin{bmatrix} x_1 \\ x_2 \\ x_3 \end{bmatrix} = \begin{bmatrix} b_1 \\ b_2 \\ b_3 \end{bmatrix}$$

We rely now on the properties of determinants given in Section 3-3. Note first that the existence of a unique solution implies that det $A = |A| \neq 0$. Then from Property 2.

$$x_1|A| = \begin{vmatrix} a_{11}x_1 & a_{12} & a_{13} \\ a_{21}x_1 & a_{22} & a_{23} \\ a_{31}x_1 & a_{32} & a_{33} \end{vmatrix}$$

Two successive applications of Property 3 then give

$$x_1|A| = \begin{vmatrix} a_{11}x_1 + a_{12}x_2 & a_{12} & a_{13} \\ a_{21}x_1 + a_{22}x_2 & a_{22} & a_{23} \\ a_{31}x_1 + a_{32}x_2 & a_{32} & a_{33} \end{vmatrix} \qquad \text{(column 1)} + x_2 \text{ (column 2)}$$

$$x_1|A| = \begin{vmatrix} a_{11}x_1 + a_{12}x_2 + a_{13}x_3 & a_{12} & a_{13} \\ a_{21}x_1 + a_{22}x_2 + a_{23}x_3 & a_{22} & a_{23} \\ a_{31}x_1 + a_{32}x_2 + a_{33}x_3 & a_{32} & a_{33} \end{vmatrix} \qquad \text{(column 1)} + x_3 \text{ (column 3)}$$

The first column, being $A\mathbf{x}$, can now be replaced by $\mathbf{b}$. Division by the nonzero $|A|$ then gives

$$x_1 = \frac{\begin{vmatrix} b_1 & a_{12} & a_{13} \\ b_2 & a_{22} & a_{23} \\ b_3 & a_{32} & a_{33} \end{vmatrix}}{|A|}$$

Similar reasoning leads to

$$x_2 = \frac{\begin{vmatrix} a_{11} & b_1 & a_{13} \\ a_{21} & b_2 & a_{23} \\ a_{31} & b_3 & a_{33} \end{vmatrix}}{|A|} \qquad x_3 = \frac{\begin{vmatrix} a_{11} & a_{12} & b_1 \\ a_{21} & a_{22} & b_2 \\ a_{31} & a_{32} & b_3 \end{vmatrix}}{|A|}$$

Since there is no difficulty in carrying out the same reasoning for an $n \times n$ system, we can take as proved this result known as **Cramer's rule.**

THEOREM C

If the $n \times n$ system of equations $A\mathbf{x} = \mathbf{b}$ has a unique solutiuon, then

$$x_j = \frac{|\mathbf{A}_{\cdot 1} \cdots \mathbf{b} \cdots \mathbf{A}_{\cdot n}|}{|A|}$$

in which the numerator is the determinant of a matrix formed from A by replacing the jth column with $\mathbf{b}$.

Example B

Solve the system

$$3x_1 - 2x_2 = 19$$
$$4x_1 - 5x_2 = 30$$

Since the coefficient matrix has determinant

$$\begin{vmatrix} 3 & -2 \\ 4 & -5 \end{vmatrix} = -15 - (-8) = -7 \neq 0$$

we can use Cramer's rule.

$$x_1 = \frac{\begin{vmatrix} 19 & -2 \\ 30 & -5 \end{vmatrix}}{-7} = \frac{-95 + 60}{-7} = 5,$$

$$x_2 = \frac{\begin{vmatrix} 3 & 19 \\ 4 & 30 \end{vmatrix}}{-7} = \frac{90 - 76}{-7} = -2 \quad \square$$

Cramer's rule is not a practical way to solve a system of equations. The time used to evaluate $\det A$ might as well be invested in using the techniques of Chapter 1 to reduce the system to row echelon form. This will give some information about the system even when $\det A = 0$; and when $\det A \neq 0$, the solution is evident without evaluating n more determinants. The principal use of Cramer's rule is theoretic; there are times when it is useful to know that the formula for a solution can be written down.

We conclude this section with one more result that is more useful in theory than in computation. Suppose that corresponding to the matrix

$$A = \begin{bmatrix} a_{11} & a_{12} & a_{13} \\ a_{21} & a_{22} & a_{23} \\ a_{31} & a_{32} & a_{33} \end{bmatrix}$$ (6)

we form the sum

$$a_{11}A_{31} + a_{12}A_{32} + a_{13}A_{33}$$ (7)

$$= a_{11}\begin{vmatrix} a_{12} & a_{13} \\ a_{22} & a_{23} \end{vmatrix} - a_{12}\begin{vmatrix} a_{11} & a_{13} \\ a_{21} & a_{23} \end{vmatrix} + a_{13}\begin{vmatrix} a_{11} & a_{12} \\ a_{21} & a_{22} \end{vmatrix}$$

That's not a misprint. We do mean to consider the products of the entries in one row with the cofactors of the terms of some other row. Note that the same sum is obtained if we expand by cofactors of the third row the determinant of matrix

$$B = \begin{bmatrix} a_{11} & a_{12} & a_{13} \\ a_{21} & a_{22} & a_{23} \\ a_{11} & a_{12} & a_{13} \end{bmatrix}$$

Since $\det B = 0$ according to Property 6, it follows that (7) must be zero.

The same trick always works. If the elements of row i of a matrix A are multiplied by the cofactors of row $j \neq i$, the resulting expression will be identical to the determinant of a new matrix B obtained by replacing the jth row of A by the ith row. But then $\det B = 0$, and we have established what we shall call, following the sequence of Section 3-3, Property 10 of determinants.

PROPERTY 10

If the elements of one row of a matrix are multiplied by the cofactors of elements of another row, the sum of their products is 0.

Example C

Illustrate Property 10 by computing $a_{11}A_{31} + a_{12}A_{32} + a_{13}A_{33}$ for

$$A = \begin{bmatrix} 3 & 2 & 5 \\ -2 & 1 & 4 \\ -1 & -3 & 2 \end{bmatrix}$$

$$3 \begin{vmatrix} 2 & 5 \\ 1 & 4 \end{vmatrix} - 2 \begin{vmatrix} 3 & 5 \\ -2 & 4 \end{vmatrix} + 5 \begin{vmatrix} 3 & 2 \\ -2 & 1 \end{vmatrix} = 3(3) - 2(22) + 5(7) = 0 \quad \square$$

An immediate consequence of Property 10 is that it yields a formula for the inverse of a nonsingular matrix. We must first define the *adjoint* of a matrix A. Suppose that given A, we form a second matrix B in which the entries are the cofactors of the corresponding entries in A; $b_{ij} = A_{ij}$. The **adjoint** of A, written adj A, is the *transpose* of B:

$$\text{adj } A = [A_{ij}]^t$$

For the 3×3 matrix A in (6) above,

$$\text{adj } A = \begin{bmatrix} A_{11} & A_{21} & A_{31} \\ A_{12} & A_{22} & A_{32} \\ A_{13} & A_{23} & A_{33} \end{bmatrix}$$

THEOREM D

For a nonsingular matrix A,

$$A^{-1} = \frac{1}{\det A} \cdot \text{adj } A$$

Proof: The general proof is most easily comprehended by looking first at the product

$$\begin{bmatrix} a_{11} & a_{12} & a_{13} \\ a_{21} & a_{22} & a_{23} \\ a_{31} & a_{32} & a_{33} \end{bmatrix} \begin{bmatrix} A_{11} & A_{21} & A_{31} \\ A_{12} & A_{22} & A_{32} \\ A_{13} & A_{23} & A_{33} \end{bmatrix} = \begin{bmatrix} b_{11} & b_{12} & b_{13} \\ b_{21} & b_{22} & b_{23} \\ b_{31} & b_{32} & b_{33} \end{bmatrix}$$

$$b_{11} = a_{11}A_{11} + a_{12}A_{12} + a_{13}A_{13} = \det A$$

Similarly, $b_{22} = b_{33} = \det A$. Now from Property 10,

$$b_{12} = a_{11}A_{21} + a_{12}A_{22} + a_{13}A_{23} = 0$$

and in the same way, $b_{ij} = 0$ whenever $i \neq j$. This shows (for the 3×3 case) that

$$A \cdot \text{adj } A = (\det A)I$$

Clearly, the same equality is easily established for an $n \times n$ matrix (singular or not), and when A is nonsingular so that $\det A \neq 0$, the conclusion of the theorem follows when we divide by $\det A$. □

We'll say it again. Theorem D does not give us a practical way to compute A^{-1}, involving as it does the calculation of $n^2 + 1$ determinants. It is sometimes useful to know, however, that there is a formula for A^{-1}.

LEST WE FORGET

The interests of mathematicians, like those of other mortals, are subject to change. This is nicely illustrated in the well-documented history of determinants.

The rudiments of the theory of determinants can be seen in a 1693 paper by Gottfried Wilhelm von Leibnitz; and (Gabriel) Cramer's rule originated in his paper of 1750. Two important papers (1815 and 1829) by Augustin Cauchy developed much of the theory, but not until the work of Arthur Cayley and James Sylvester began in about 1840 did the subject achieve much visibility in the mathematical community. Yet it was already true in 1890 that Thomas Muir was complaining that known properties of determinants were being forgotten, rediscovered, and (apparently worst of all in his eyes) being attributed to mathematicians of a nationality different from that of the original discoverers. He did his best to rectify the situation, leaving us 1762 pages in a four volume work that traces developments from 1693 through 1900.

Even with such an effort, much has been forgotten. Consider, for example, the graph of $y^2 = px$ (Figure 3-A). Students of calculus may recall that the radius of the circle of curvature at a point (x_1, y_1) is the radius of the circle (the **osculating circle**) that most closely follows the curvature of the parabola at that point. Few, however, will have studied out of a book that included, as did a text by George Salmon in 1885, the information that the osculating circle at (x_1, y_1) is centered at $(3x_1 + \frac{1}{2}p, -4y_1^3/p^2)$. Still fewer will have wondered what would be the area of a triangle formed

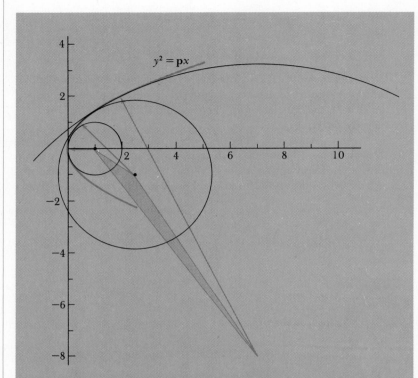

Figure 3-A

by joining the centers of three such osculating circles. But Salmon knew the answer to that too:

$$\text{area} = \frac{6}{p^3} \begin{vmatrix} 1 & 1 & 1 \\ x_1^2 & x_2^2 & x_3^2 \\ y_1^3 & y_2^3 & y_3^3 \end{vmatrix}$$

It is in tribute to Muir's monumental effort that we decided to mention at least one fact about determinants likely to be forgotten if not periodically called to the attention of the reading public by vigilant authors.

In Problems 1–4, find the volume of the parallelepiped for which the given vectors form three adjacent edges.

1. $\mathbf{A}_1 = \begin{bmatrix} 3 \\ 5 \\ 0 \end{bmatrix}$, $\mathbf{A}_2 = \begin{bmatrix} 6 \\ 1 \\ 0 \end{bmatrix}$, $\mathbf{A}_3 = \begin{bmatrix} 4 \\ 2 \\ 3 \end{bmatrix}$

2. $\mathbf{A}_1 = \begin{bmatrix} 2 \\ 4 \\ 0 \end{bmatrix}$, $\mathbf{A}_2 = \begin{bmatrix} 1 \\ 5 \\ 0 \end{bmatrix}$, $\mathbf{A}_3 = \begin{bmatrix} 1 \\ 4 \\ 5 \end{bmatrix}$

3. $\mathbf{A}_1 = \begin{bmatrix} -1 \\ 1 \\ 2 \end{bmatrix}, \qquad \mathbf{A}_2 = \begin{bmatrix} 3 \\ 4 \\ 7 \end{bmatrix}, \qquad \mathbf{A}_3 = \begin{bmatrix} -2 \\ 1 \\ -3 \end{bmatrix}$

4. $\mathbf{A}_1 = \begin{bmatrix} 2 \\ -1 \\ 3 \end{bmatrix}, \qquad \mathbf{A}_2 = \begin{bmatrix} 1 \\ 3 \\ 5 \end{bmatrix}, \qquad \mathbf{A}_3 = \begin{bmatrix} -1 \\ -5 \\ 4 \end{bmatrix}$

5. In Problem 1, the position vectors $\mathbf{A}_1$ and $\mathbf{A}_2$ lie in the $x_1 x_2$ plane.
   (a) Find the area of the parallelogram in the $x_1 x_2$ plane having $\mathbf{A}_1$ and $\mathbf{A}_2$ as adjacent sides.
   (b) Find the height of the parallelepiped by finding the length of the projection of $\mathbf{A}_3$ on the x_3 axis.
   (c) Use your answers to parts (a) and (b) to verify your previous calculation of the volume of the parallelepiped.

6. Follow the instructions for Problem 5, using vectors $\mathbf{A}_1$, $\mathbf{A}_2$, and $\mathbf{A}_3$ from Problem 2.

In Problems 7–8, you are given the vertices of a tetrahedron. Find the volume.

7. $A(2, -2, 1)$
   $B(3, -3, 1)$
   $C(1, -4, 2)$
   $D(2, -4, -2)$

8. $A(1, -1, 2)$
   $B(3, -5, -4)$
   $C(0, -9, -2)$
   $D(19, -12. 6)$

9. In Problem 7, if ABC is used as the base of the tetrahedron, then $\overrightarrow{BD}$ is perpendicular to the base.
   (a) Find the area of the base.
   (b) Use equation (1) to verify your previous calculation of the volume of the tetrahedron.

10. Follow the instructions for Problem 9 using points A, B, C, and D from Problem 8.

Use Cramer's rule to solve the systems of equations given in Problems 11–14.

11. $3x_1 - 4x_2 = 1$
    $5x_1 + 7x_2 = 29$

12. $4x_1 - 3x_2 = 8$
    $5x_1 + 4x_2 = 41$

13. $2x_1 - x_2 + x_3 = -1$
    $x_1 - 2x_2 + 3x_3 = -6$
    $3x_1 + x_2 - x_3 = 6$

14. $3x_1 + 2x_2 - x_3 = -6$
    $2x_1 - 3x_2 + 5x_3 = -7$
    $-x_1 + 5x_2 - 2x_3 = 11$

Find the adjoint of the given matrix in Problems 15–20. Where possible, go on to find the inverse matrix.

15. $\begin{bmatrix} 4 & 3 \\ -2 & 3 \end{bmatrix}$

16. $\begin{bmatrix} 5 & 2 \\ -5 & 3 \end{bmatrix}$

17. $\begin{bmatrix} 3 & 1 & -1 \\ 0 & 2 & 1 \\ 6 & -4 & -5 \end{bmatrix}$

18. $\begin{bmatrix} 1 & -1 & 0 \\ 2 & 1 & -2 \\ 1 & 2 & -1 \end{bmatrix}$

19. $\begin{bmatrix} 2 & 1 & -1 \\ 0 & 2 & 3 \\ -1 & -1 & 2 \end{bmatrix}$ 20. $\begin{bmatrix} 1 & 3 & -1 \\ -2 & 0 & -4 \\ 1 & 2 & 0 \end{bmatrix}$

Let $A(a_1, a_2)$, $B(b_1, b_2)$, and $C(c_1, c_2)$ be given points. It was shown in this section that one way to write the equation of a line through A and B, or of the circle through A, B, and C, is to expand by minors of the first row the determinants

$$(L) \quad \begin{vmatrix} x_1 & x_2 & 1 \\ a_1 & a_2 & 1 \\ b_1 & b_2 & 1 \end{vmatrix} = 0, \qquad (C) \quad \begin{vmatrix} x_1^2 + x_2^2 & x_1 & x_2 & 1 \\ a_1^2 + a_2^2 & a_1 & a_2 & 1 \\ b_1^2 + b_2^2 & b_1 & b_2 & 1 \\ c_1^2 + c_2^2 & c_1 & c_2 & 1 \end{vmatrix} = 0$$

These ideas are used in Problems 21–28.

21. Using (C), write the equation of a circle passing through $(-2, 7)$, $(5, 3)$, and $(2, 1)$. Verify your answer by using methods of analytic geometry to obtain the same equation.

22. Using (L), write the equation of a line passing through $(-2, 5)$ and $(4, -3)$. Verify your answer by using methods of analytic geometry to obtain the same equation.

23. Under what circumstances will the coefficient of $x_1^2 + x_2^2$ in the expansion of (C) be zero?

24. What will happen in the expansion of (C) if the points A, B, and C are colinear?

25. Write in a determinant form similar to (L) the equation of the plane determined by $A(a_1, a_2, a_3)$, $B(b_1, b_2, b_3)$, and $C(c_1, c_2, c_3)$.

26. Write in a determinant form similar to (C) the equation of a parabola having its axis of symmetry parallel to the x_1 axis, passing through (a_1, a_2), (b_1, b_2), and (c_1, c_2).

27. Use the formula obtained in Problem 25 above to write the equation of a plane passing through $(1, -2, 1)$, $(-3, 0, 4)$, and $(2, -1, 3)$. Write the same equation using other methods.

28. Use the formula obtained in Problem 26 above to write the equation of a parabola with vertex $(1, -1)$ and latus rectum endpoints of $(-1, 3)$ and

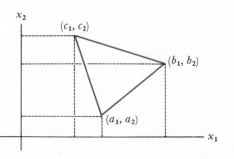

Figure 3-7

$(-1, -5)$. Verify your answer by using other methods to write the same equation.

29. Using Figure 3-7, prove directly that the area of the triangle is given by equation (3) of Theorem B.

*30. Prove that if a tetrahedron T has vertices $A(a_1, a_2, a_3)$, $B(b_1, b_2, b_3)$, $C(c_1, c_2, c_3)$, and $D(d_1, d_2, d_3)$, then

$$\text{vol } T = \pm\tfrac{1}{6} \begin{vmatrix} a_1 & a_2 & a_3 & 1 \\ b_1 & b_2 & b_3 & 1 \\ c_1 & c_2 & c_3 & 1 \\ d_1 & d_2 & d_3 & 1 \end{vmatrix}$$

Use the formula to verify your answers to Problems 7 and 8.

*31. Prove that the area A of a triangle having vertices of $A(a_1, a_2, a_3)$, $B(b_1, b_2, b_3)$, and $C(c_1, c_2, c_3)$ satisfies

$$A^2 = \tfrac{1}{4} \left\{ \begin{vmatrix} a_1 & b_1 & 1 \\ a_2 & b_2 & 1 \\ a_3 & b_3 & 1 \end{vmatrix}^2 + \begin{vmatrix} a_1 & c_1 & 1 \\ a_2 & c_2 & 1 \\ a_3 & c_3 & 1 \end{vmatrix}^2 + \begin{vmatrix} b_1 & c_1 & 1 \\ b_2 & c_2 & 1 \\ b_3 & c_3 & 1 \end{vmatrix}^2 \right\}$$

*32. A triangle is bounded by line segments lying on the three nonconcurrent lines

$$a_1 x_1 + a_2 x_2 + a_3 = 0$$
$$b_1 x_1 + b_2 x_2 + b_3 = 0$$
$$c_1 x_1 + c_2 x_2 + c_3 = 0$$

no two of which are parallel. Prove that the area of the triangle is given by

$$\text{area} = \frac{\pm 1}{2(b_1 c_2 - b_2 c_1)(a_2 c_1 - a_1 c_2)(a_1 b_2 - a_2 b_1)} \begin{vmatrix} a_1 & a_2 & a_3 \\ b_1 & b_2 & b_3 \\ c_1 & c_2 & c_3 \end{vmatrix}^2$$

1. Classify the permutation 7315426 as even or odd.

2. The determinant of an $n \times n$ matrix A is equal to the sum of all terms of the form $\pm a_{1i_1} a_{2i_2} \ldots a_{ni_n}$. Write all such nonzero terms to evaluate

$$\begin{bmatrix} -3 & 0 & 0 & 1 & -2 \\ 0 & 2 & 0 & 0 & 0 \\ 0 & 0 & 0 & 4 & 0 \\ 1 & 0 & -1 & 0 & 1 \\ 0 & -1 & 0 & 0 & 0 \end{bmatrix}$$

3. For a 3×3 matrix A, $\det A = a_{31}A_{31} + a_{32}A_{32} + a_{33}A_{33}$ where A_{ij} is the cofactor of a_{ij}. Use this fact to evaluate

$$\begin{vmatrix} 1 & -2 & 0 \\ 0 & 3 & 2 \\ 1 & -1 & 1 \end{vmatrix}$$

CHAPTER 3
SELF-TEST

4. The determinant of a 6 × 6 matrix A is 12. What is the determinant of EA where

$$E = \begin{bmatrix} 1 & 0 & 0 & 0 & 0 & 0 \\ 0 & 0 & 0 & 0 & 1 & 0 \\ 0 & 0 & \frac{1}{3} & 0 & 0 & 0 \\ -3 & 0 & 0 & 1 & 0 & 0 \\ 0 & 1 & 0 & 0 & 0 & 0 \\ 2 & 0 & 0 & 0 & 0 & 1 \end{bmatrix}$$

5. Evaluate $\begin{vmatrix} 2 & -1 & 4 & -5 \\ 3 & 2 & -2 & 2 \\ -2 & 5 & -10 & 7 \\ 4 & 3 & -3 & 2 \end{vmatrix}$

6. What is the adjoint of $A = \begin{bmatrix} 1 & -1 & 2 \\ 0 & 2 & 3 \\ -2 & 3 & 1 \end{bmatrix}$? How can you check your answer?

7. Find the area of a triangle in the plane having vertices $A(1, 3)$, $B(5, -2)$, and $C(-4, -5)$.

8. If A and B are $n \times n$ matrices, if $M = AA^t$ and $N = BB^t$, and if $\det M + \det N = 0$, then prove that both A and B are singular.

9. Prove that if there is a solution to the system

$$a_1 x_1 + a_2 x_2 + a_3 = 0$$
$$b_1 x_1 + b_2 x_2 + b_3 = 0$$
$$c_1 x_1 + c_2 x_2 + c_3 = 0$$

then $\begin{vmatrix} a_1 & a_2 & a_3 \\ b_1 & b_2 & b_3 \\ c_1 & c_2 & c_3 \end{vmatrix} = 0$

10. Suppose that the coefficient matrix of the system

$$a_{11} x_1 + a_{12} x_2 + a_{13} x_3 = 1$$
$$a_{21} x_1 + a_{22} x_2 + a_{23} x_3 = 0$$
$$a_{31} x_1 + a_{22} x_2 + a_{23} x_3 = 0$$

has a determinant of 1. Prove that $x_1 = 0$. *Hint:* Use Cramer's rule.

VECTORS AND VECTOR SPACES 4

INTRODUCTION | 4-1

We have seen that many of our concepts about vectors, though motivated by the pictures we draw for R^2 or R^3, can be extended to R^n for $n > 3$. Now we ask what essential features of n-tuples allowed us to stay on target even after our arrows were lost. They seem to be this. The n-tuples can be added together and they can be multiplied by real numbers in ways that obey familiar algebraic rules.

In this chapter we will not really define vectors at all, but shall simply insist that they can be added or multiplied by numbers in the now familiar way. We will see that this allows us to think of polynomials, matrices, differentiable functions, and sequences (to mention four examples) as vectors. What we learn about vectors thereby gives us information about a host of mathematical objects.

On the other hand, while an abstract vector meets tests of logic and convenience, it does not give us a picture of what is going on. Fortunately, we have the geometric concepts of R^2 and R^3 as visual support, and we shall continually refer back to these spaces for illustrations and guidance.

VECTOR SPACES | 4-2

Before listing the rules that define an abstract vector space, we wish to comment on our use of the word *scalar*. Up to this point, we have only talked about multiplying vectors by real numbers. Since there will be a few occasions near the end of this text where we will want to allow the multipliers to be complex numbers, we will use the more general term

scalar. Unless something is specifically said to the contrary, however, you should always assume that scalars are real numbers.

DEFINITION

Let V be a nonempty collection of elements that can be added together and can be multiplied by scalars. We call the elements **vectors,** and we say V is a **vector space** (or a **linear space**) if for arbitrary elements **u, v, w,** ... in V and arbitrary scalars r, s, t, ..., the following properties are satisfied:

1. The sum $\mathbf{u} + \mathbf{v}$ is in V.   (closure)
2. $(\mathbf{u} + \mathbf{v}) + \mathbf{w} = \mathbf{u} + (\mathbf{v} + \mathbf{w})$ (associative)
3. $\mathbf{u} + \mathbf{v} = \mathbf{v} + \mathbf{u}$ (commutative)
4. There is a **zero element 0** in V having the property that for any $\mathbf{u}$ in V, $\mathbf{u} + \mathbf{0} = \mathbf{0} + \mathbf{u} = \mathbf{u}$
5. For each $\mathbf{u}$ in V, there is another element, usually designated $(-\mathbf{u})$, such that $\mathbf{u} + (-\mathbf{u}) = (-\mathbf{u}) + \mathbf{u} = \mathbf{0}$.
6. The product of a scalar r and a vector $\mathbf{u}$, $r\mathbf{u}$, is in V.
7. $r(\mathbf{u} + \mathbf{v}) = r\mathbf{u} + r\mathbf{v}$
8. $(r + s)\mathbf{u} = r\mathbf{u} + s\mathbf{u}$
9. $r(s\mathbf{u}) = (rs)\mathbf{u}$
10. $1(\mathbf{u}) = \mathbf{u}$

Since these properties are nothing but an attempt to abstract the properties of vectors in R^n, it hardly seems necessary to say that R^n is one example of a vector space. Still, it is a very important vector space that should be kept in active storage. You will also be well served if you can keep the following examples in mind.

Example A

The set of all polynomials forms a vector space.

From among the infinite possibilities that exist, we shall choose for illustrative purposes the two polynomials

$$u(x) = x^3 - 3x + 7, \qquad v(x) = x^5 + 2x^3 - x^2 + 5x$$

The usual rules of algebra then give us

$$u(x) + v(x) = x^5 + 3x^3 - x^2 + 2x + 7$$
$$2u(x) = 2x^3 - 6x + 14$$
$$-u(x) = -x^3 + 3x - 7$$

The degree of a polynomial is the value of the largest exponent that appears in the polynomial. Polynomials $u(x)$ and $v(x)$ have degrees 3 and 5, respectively. A polynomial having degree 0, such as $w(x) = 4$, is a constant polynomial. For any x, the value of a constant polynomial is always the same. Note in particular the zero polynomial σ defined for all x by $\sigma(x) = 0$.

With this quick review of the algebra of polynomials, we are prepared to verify that the set of polynomials forms a vector space. The largest hurdle to get over, of course, is that of thinking of a polynomial as a vector. It then remains only to verify the ten required properties: The sum of any two polynomials $u(x)$ and $v(x)$ is again a polynomial (Property 1); addition is associative (Property 2) and commutative (Property 3); we do have the $\sigma(x)$ polynomial to act as a zero element (Property 4); for each $u(x)$ there is a $-u(x)$ such as was demonstrated above for our particular $u(x)$ (Property 5). Scalar multiplication is just as easily shown to satisfy Properties 6 through 10. □

The set of all $m \times n$ matrices forms a vector space. *Example B*

The problem here is not so much what to worry about as it is what not to worry about. There is no need, for instance, to worry about the fact that two $m \times n$ matrices can't be multiplied when $m \neq n$, because nothing in the definition of a vector space requires that we be able to multiply the vectors together. Neither should we think about the row or the column vectors within the matrix. There, in fact, is what we must concentrate upon. We must learn to think in the present context of the entire $m \times n$ matrix as *one* vector. The addition of two vectors refers to an addition of the form

$$\begin{bmatrix} a_{11} & \cdots & a_{1n} \\ \vdots & & \vdots \\ a_{m1} & \cdots & a_{mn} \end{bmatrix} + \begin{bmatrix} b_{11} & \cdots & b_{1n} \\ \vdots & & \vdots \\ b_{m1} & \cdots & b_{mn} \end{bmatrix} = \begin{bmatrix} a_{11} + b_{11} & \cdots & a_{1n} + b_{1n} \\ \vdots & & \vdots \\ a_{m1} + b_{m1} & \cdots & a_{mn} + b_{mn} \end{bmatrix}$$

and scalar multiplication refers to the multiplication

$$r\begin{bmatrix} a_{11} & \cdots & a_{1n} \\ \vdots & & \vdots \\ a_{m1} & \cdots & a_{mn} \end{bmatrix} = \begin{bmatrix} ra_{11} & \cdots & ra_{1n} \\ \vdots & & \vdots \\ ra_{m1} & \cdots & ra_{mn} \end{bmatrix}$$

Scalar multiplication of a matrix by r must be distinguished from the multiplication of det A by r. For a 2×2 matrix A,

$$rA = \begin{bmatrix} ra_{11} & ra_{12} \\ ra_{21} & ra_{22} \end{bmatrix}, \quad \text{whereas} \quad \begin{vmatrix} ra_{11} & ra_{12} \\ a_{21} & a_{22} \end{vmatrix} = r(\det A)$$

Using the matrix with every element equal to 0 for the $\mathcal{O}$ matrix, Properties 1 through 10 are easy to establish. □

The space of all real-valued functions that are defined and have a continuous derivative on an interval (a, b) form a vector space. *Example C*

A specific example of such a space must begin by specifying the interval. If we take $(a, b) = (-1, 1)$, then some vectors in the space are

$$f(x) = e^x, \qquad g(x) = x \sin x, \qquad h(x) = \frac{x}{x - 2}$$

Note that $h(x)$ would not qualify as a member of the space if we had taken $(a, b) = (1, 3)$. Why not? Also note that on the chosen interval $(-1, 1)$, $u(x) = |x|$ is not a vector in the space; it fails to have a derivative throughout the interval.

Once again, the biggest hurdle to get over is to think of a function as a vector. Addition and scalar multiplication are defined as usual. Thus, for the functions cited above

$$(f + g)(x) = e^x + x \sin x, \qquad rh(x) = \frac{rx}{x - 2}$$

The zero vector is the function $\sigma(x)$ that is constantly 0 for every value of x.

A principal question in such examples has to do with the very first property. If we begin with two members of the set under investigation and perform an addition, do we get another member of the set? In the present case, the elementary rules about differentiation speak to our problem: the derivative of a sum is the sum of the derivatives. We say that the set is **closed** under addition.

A question of closure also comes up with regard to scalar multiplication. If we multiply a member of a set by a scalar r, will the result still be in the set? Again, in this case, the elementary rules of differentiation come to our aid.

$$\frac{d}{dx} rf(x) = r \frac{d}{dx} f(x)$$

The rest of the properties are easily verified. □

The vector space just described turns up often enough in mathematics to have merited a name. It is called the vector space $C'(a, b)$. Similar spaces are similarly named. $C''[a, b]$ is the set of real-valued functions that have two continuous derivatives on $[a, b]$, and $C'''(a, b)$ is defined in the obvious way. Going in the other direction, $C(a, b]$ is the set of functions that are continuous on $(a, b]$. If in any of these notations the interval is not specified, it is assumed to be $(-\infty, \infty)$. Thus, C' would refer to the set of functions that have a continuous derivative for all x. $v(x) = x^{3/2}$ would be in C'; $w(x) = x^{1/2}$ would not be in C'.

Example D   The set of all infinite sequences forms a vector space.

Vectors in this space are of the form

$$\mathbf{u} = (u_1, u_2, u_3, \ldots), \qquad \mathbf{V} = (v_1, v_2, v_3, \ldots)$$

where each vector is a sequence of real numbers. Sums and scalar multiples may be defined in a natural way:

$$\mathbf{u} + \mathbf{v} = (u_1 + v_1, u_2 + v_2, \ldots)$$
$$r\mathbf{u} = (ru_1, ru_2, \ldots)$$

No problems are encountered in verifying the ten properties. □

You are perhaps ready to concede our up till now unspoken point that there are many examples of vector spaces. With this established, we shall turn to a closer examination of Properties 1 through 10. Like the ten commandments, they have far-reaching implications.

We have already used the term *closure* in connection with Properties 1 and 6. It is a simple idea, but an important one that is better understood by observing some sets that are not closed.

The set of odd integers is not closed under addition; the sum of two odd integers is not in the set. The set of 2×2 matrices having nonnegative entries would not be closed with respect to scalar multiplication, for multiplication by -1 would give a matrix no longer in the set. Sometimes the question of closure is not easy to settle. Consider the set of 2×2 matrices that have positive determinants. Is this set closed under addition? It may take you a few minutes to see that it is not.

Properties 1 through 10 enable us to prove other properties that will hold in any vector space. For instance, any vector space ever discovered will have only one zero vector. This is proved by supposing that some vector space V has two of them, say z_1 and z_2. Since z_1 is a zero vector, then for any vector u, according to Property 4,

$$u + z_1 = u$$

Similarly, since z_2 is also a zero vector, Property 4 tells us that for any vector w,

$$z_2 + w = w$$

The first equation, being true for all u, is true for $u = z_2$; and the second equation must be true for the particular choice of $w = z_1$. Taken together, we then have both $z_2 + z_1 = z_2$ and $z_2 + z_1 = z_1$, so $z_1 = z_2$. We have proved another property:

11. There is only one zero element 0 satisfying Property 4.

Other properties, none surprising, can also be proved. Their proofs are outlined in Problems 12–15 at the end of the section.

12. For a given u, there is only one element $(-u)$ that satisfies Property 5.
13. $0u = r0 = 0$
14. For every vector u, $(-1)u = (-u)$.
15. If $ru = 0$, then either $r = 0$ or $u = 0$.

We turn our attention to another important concept. A subcollection W of a vector space V is called a **subspace** of V if the elements of W themselves form a vector space using the operations of addition and scalar multiplication defined on V.

There are always two obvious subspaces: V is a subspace of V, and the **0** element taken by itself constitutes what we shall call the **zero subspace**. Any subspace other than these two special cases is called a **proper subspace** of V.

If a vector **u** is in a subspace W, then of course every scalar multiple $r\mathbf{u}$ is also in W (because W is a vector space). In particular, $0\mathbf{u} = \mathbf{0}$ must be in W, so any subspace of V shares a common zero vector with V.

Given a subcollection W of a vector space V, how do we decide whether or not it is a subspace? A quick reply might suggest checking to see if W satisfies all ten properties of a vector space, but this turns out to be unnecessary. Note that if **u**, **v**, and **w** are in W, then they are in V, so Properties 2, 3, 7, 8, 9, and 10 are satisfied immediately. Properties 1 and 6 are critical, obviously, since **u** and **v** being in a subcollection W does not imply that $\mathbf{u} + \mathbf{v}$ or $r\mathbf{u}$, for any scalar r, will also be in W. It turns out, however, that these are the only properties that need to be checked. For if a vector **u** is in W, and if Property 6 is known to be satisfied, then $0\mathbf{u} = \mathbf{0}$ and $(-1)\mathbf{u} = (-\mathbf{u})$ are also in W, satisfying Properties 4 and 5. These time-saving observations can be summarized this way:

THEOREM A

If W is a subcollection of elements of a vector space V, then W is a subspace of V if and only if $r\mathbf{u} + s\mathbf{v}$ is in W for every choice of scalars r and s and vectors $\mathbf{u}$ and $\mathbf{v}$ in W.

The characterization of subspaces given by Theorem A is, in fact, the one we used to define a subspace of R^n in Section 2-4. You may want to look again at Figure 2-21 to see some subspaces and non-subspaces of R^2. We shall conclude this section by discussing selected subspaces of the vector spaces described in Examples A, B, C, and D.

Example A'

The polynomials of degree 2 or less form a subspace (which we shall subsequently refer to as P_2) of the vector space of all polynomials.

We consider two such polynomials,

$$u(x) = a_2 x^2 + a_1 x + a_0, \qquad v(x) = b_2 x^2 + b_1 x + b_0$$

Now if r and s are any two scalars, then

$$ru(x) + sv(x) = (ra_2 + sb_2)x^2 + (ra_1 + sb_1)x + ra_0 + sb_0$$

is also a polynomial of degree 2 or less. An appeal to Theorem A shows that these polynomials form a subspace. □

Example B'

Let $\mathcal{C}$ be the set of all 2×2 matrices having the special form

$$\begin{bmatrix} a & -b \\ b & a \end{bmatrix}$$

ROOM FOR IMAGINATION

From the observation that the matrix I acts like 1, it is only a small step to notice that aI acts in matrix algebra the way the number a acts in real arithmetic. For 2×2 matrices,

$$aI + cI = \begin{bmatrix} a & 0 \\ 0 & a \end{bmatrix} + \begin{bmatrix} c & 0 \\ 0 & c \end{bmatrix}$$

$$= \begin{bmatrix} a + c & 0 \\ 0 & a + c \end{bmatrix} = (a + c)I$$

$$aI \cdot cI = \begin{bmatrix} a & 0 \\ 0 & a \end{bmatrix} \cdot \begin{bmatrix} c & 0 \\ 0 & c \end{bmatrix}$$

$$= \begin{bmatrix} ac & 0 \\ 0 & ac \end{bmatrix} = (ac)I$$

Next observe the product

$$\begin{bmatrix} 0 & -1 \\ 1 & 0 \end{bmatrix} \cdot \begin{bmatrix} 0 & -1 \\ 1 & 0 \end{bmatrix} = \begin{bmatrix} -1 & 0 \\ 0 & -1 \end{bmatrix} = (-1)I$$

We clearly have a matrix, quite simple and ordinary in appearance, which, when squared, gives a matrix corresponding to -1.

The set of all 2×2 matrices form a vector space. Consider the subspace—we'll call it C—spanned by the two "vectors"

$$I = \begin{bmatrix} 1 & 0 \\ 0 & 1 \end{bmatrix}, \qquad \text{Im} = \begin{bmatrix} 0 & -1 \\ 1 & 0 \end{bmatrix}$$

Any member of C, being a linear combination of these two vectors, is of the form

$$aI + b\,\text{Im} = \begin{bmatrix} a & -b \\ b & a \end{bmatrix}$$

Of course C is closed with respect to addition; any subspace is. The remarkable thing about C is that is also closed under multiplication.

$$(aI + b\,\text{Im})(cI + d\,\text{Im}) = \begin{bmatrix} a & -b \\ b & a \end{bmatrix}\begin{bmatrix} c & -d \\ d & c \end{bmatrix}$$

$$= \begin{bmatrix} ac - bd & -(ad + bc) \\ ad + bc & ac - bd \end{bmatrix}$$

$$= (ac - bd)I + (ad + bc)\,\text{Im}$$

If you have an i for this sort of thing, it can lead you into a complex subject.

in which entries in the upper left and lower right corners are equal, while those in the other two corners differ only in sign. The set $\mathcal{C}$ is a subspace of the vector space of 2×2 matrices.

By way of illustration, we note that $\mathcal{C}$ contains the matrices

$$\begin{bmatrix} -3 & 4 \\ -4 & -3 \end{bmatrix}, \quad \begin{bmatrix} 2 & -5 \\ 5 & 2 \end{bmatrix}, \quad \begin{bmatrix} 0 & 0 \\ 0 & 0 \end{bmatrix}$$

To show that $\mathcal{C}$ is a subspace, we note that for any two scalars r and s, and for two members of $\mathcal{C}$,

$$r\begin{bmatrix} a_1 & -b_1 \\ b_1 & a_1 \end{bmatrix} + s\begin{bmatrix} a_2 & -b_2 \\ b_2 & a_2 \end{bmatrix} = \begin{bmatrix} ra_1 + sa_2 & -(rb_1 + sb_2) \\ rb_1 + sb_2 & ra_1 + sa_2 \end{bmatrix}$$

The sum is again a matrix of the same form. It is in $\mathcal{C}$, so by Theorem A, $\mathcal{C}$ is a subspace. $\square$

Example C'

The set $\mathfrak{M}$ of all functions representable by a Maclaurin series convergent on $(-1, 1)$ is a subspace of the vector space $C'(-1, 1)$.

Two such functions can be expressed in the forms

$$f(x) = a_0 + a_1 x + a_2 x^2 + \cdots, \qquad g(x) = b_0 + b_1 x + b_2 x^2 + \cdots$$

Then the properties of power series allow us to conclude that for any scalars r and s,

$$rf(x) + sg(x) = (ra_0 + sb_0) + (ra_1 + sb_1)x + (ra_2 + sb_2)x^2 + \cdots$$

This series will have the same interval of convergence.

To be more specific about the subspace $\mathfrak{M}$, we note that

$$f(x) = \frac{1}{1 - x} = 1 + x + x^2 + \cdots$$

is in $\mathfrak{M}$, but

$$h(x) = \frac{1}{1 - 2x} = 1 + 2x + 4x^2 + \cdots$$

is not. Why not? $\square$

Example D'

The set of all arithmetic sequences forms a subspace of the vector space of all sequences.

Recall that arithmetic sequences are sequences which, like $(2, \frac{7}{2}, 5, \frac{13}{2}, 8, \ldots)$ are of the form

$$\mathbf{u} = (a_1, a_1 + d_1, a_1 + 2d_1, \ldots),$$
$$\mathbf{v} = (a_2, a_2 + d_2, a_2 + 2d_2, \ldots)$$

Thus, if r and s are scalars,

$$r\mathbf{u} + s\mathbf{v} = (ra_1, ra_1 + rd_1, ra_1 + 2rd_1, \ldots) + (sa_2, sa_2 + sd_2, sa_2 + 2sd_2, \ldots)$$
$$= (a_3, a_3 + d_3, a_3 + 2d_3, \ldots)$$

where $a_3 = ra_1 + sa_2$ and $d_3 = rd_1 + sd_2$. Clearly the sum is again an arithmetic sequence. □

ⓒ SOME SIGNIFICANT OBSERVATIONS

The vector space C'' consists of those functions that have, for all x, two continuous derivatives. Two such functions, with two of their derivatives, are

$$f(x) = e^{-x} \qquad g(x) = xe^{-x}$$
$$f'(x) = -e^{-x} \qquad g'(x) = -xe^{-x} + e^{-x}$$
$$f''(x) = e^{-x} \qquad g''(x) = xe^{-x} - 2e^{-x}$$

The functions f and g were not chosen arbitrarily. They share a common property. Whether we set $y = f(x)$ or $y = g(x)$, it will still be true that

$$y'' + 2y' + y = 0 \qquad\qquad (i)$$

It is easy to show, using Theorem A of this section, that the collection of all functions of the form

$$h(x) = r_1 e^{-x} + r_2 xe^{-x}$$

constitutes a subspace $\mathcal{S}$ of C''. And by direct substitution you can verify that any function of the form of $h(x)$ is also a solution of (i).

Equation (i), being an equation involving derivatives, is called a **differential equation.** The solution set S is the set of all functions u in C'' such that we get a solution to (i) by setting $y = u(x)$. If u and v are members of S, and s_1 and s_2 are arbitrary scalars, direct substitution in (i) shows that $y = s_1 u(x) + s_2 v(x)$ is also a solution. We conclude that S is a subspace of C''. From observations above about the subspace $\mathcal{S}$, it must be that $\mathcal{S}$ is contained in S. Does $\mathcal{S} = S$? The differential equation (i) is said to be a **second order** equation because the order of the highest derivative is two. Is it significant to observe that the functions $h(x)$ in $\mathcal{S}$ are all in some sense "built" from the two functions $f(x) = e^{-x}$ and $g(x) = xe^{-x}$? See Chapter 9.

1. Which of the following sets are closed under addition?
   (a) All polynomials $p(x)$ for which $p(1) = 0$.
   (b) All polynomials $p(x)$ for which $p(1) = 1$.
   (c) All polynomials $p(x)$ for which $p(1)$ is an integer.
   (d) All 2×2 matrices A for which $\det A = 0$.
   (e) All geometric sequences; that is, all sequences of the form $(a, ar, ar^2, ar^3, \ldots)$.
   (f) All points (x_1, x_2) in the plane that satisfy $x_1 + x_2 = 2$.
   (g) All points (x_1, x_2) in the plane that satisfy $x_1 + x_2 = 0$.

2. Which of the sets described in Problem 1 are closed under multiplication by real numbers?

3. Is the set of nonsingular matrices closed under addition? Under multiplication by real numbers?

4. Is the set of singular matrices closed under addition? Under multiplication by real numbers?

5. Which of the following sets are vector spaces?
   (a) All polynomials of degree three or less.
   (b) All polynomials of degree three.
   (c) All 3×3 matrices having their only nonzero elements on the **main diagonal** (that is, on the diagonal running from the upper left to the lower right corners).
   (d) All 3×3 matrices having nonzero elements on the main diagonal, zeros elsewhere.

6. Which of the following sets are vector spaces?
   (a) All polynomials $p(x)$ having 0 as the coefficient of x^3.
   (b) All polynomials $p(x)$ having 1 as the coefficient of x^3.
   (c) All 3×3 matrices for which the sum of the terms on the main diagonal (defined in Problem 5) is 0.
   (d) All 3×3 matrices for which the sum of the terms on the main diagonal is non-negative.

7. Do the polynomials $p(x)$ that satisfy $p(1) = p(2) = 0$ form a subspace of the vector space of all polynomials?

8. Do the polynomials $p(x)$ that satisfy $p(x) = p(-x)$ for all x form a subspace of the vector space of all polynomials?

9. Do the following sets form subspaces of R^3?
   (a) The set of vectors (x_1, x_2, x_3) where $x_2 = x_1 + x_3$.
   (b) The set of vectors $(1, x_2, 1)$ where x_2 is any number.
   (c) The set of vectors (x_1, x_2, x_3) for which the coordinates add up to 4.

10. Do the following sets form subspaces of R^3?
   (a) The set of vectors $(0, x_2, x_3)$ where x_2 and x_3 are arbitrary.
   (b) The set of vectors (x_1, x_2, x_3) where

$$\frac{3x_1 - 2}{6} = \frac{4x_2 - 3}{9} = \frac{x_3 - 1}{3}$$

   (c) The set of vectors (x_1, x_2, x_3) for which the coordinates add up to 0.

11. Prove that if **u** and **v** are members of a vector space V, then the set of all elements of the form $\mathbf{w} = r\mathbf{u} + s\mathbf{v}$ forms a subspace of V.

Using Properties 1 through 10 that define a vector space, prove the properties numbered 12 through 15.

12. For a given **u**, there is only one element $(-\mathbf{u})$ that satisfies Property 5. *Hint:* Suppose there is a second element **w** such that $\mathbf{w} + \mathbf{u} = \mathbf{0}$. Show that $\mathbf{w} = (-\mathbf{u})$.

13. To show that $0\mathbf{u} = \mathbf{0}$, note that $0\mathbf{u} = (0 + 0)\mathbf{u} = 0\mathbf{u} + 0\mathbf{u}$. Now add $(-0\mathbf{u})$ to both sides.

14. To show that $(-1)\mathbf{u} = (-\mathbf{u})$, note that $(-1)\mathbf{u} + \mathbf{u} = [(-1) + 1]\mathbf{u} = 0\mathbf{u} = \mathbf{0}$ and appeal to Property 12.

15. To show that $r\mathbf{u} = \mathbf{0}$ implies either $r = 0$ or $\mathbf{u} = \mathbf{0}$, suppose $r \neq 0$. Multiply by $1/r$.

16. Let L and M be subspaces of a vector space V. Are the following sets also subspaces of V?
    (a) $L \cup M = \{$the set of all vectors belonging to L or M or both$\}$
    (b) $L \cap M = \{$the set of all vectors belonging to both L and $M\}$
    (c) $L + M = \{$the set of all vectors **w** that can be written as a sum of a vector **x** in L and **y** in $M\}$

17. Let $L = \{(x_1, x_2, x_3): x_1 + x_2 + x_3 = 0\}$ and let $M = \{(x_1, x_2, x_3): x_1 = x_2 = x_3\}$. Both are subspaces of R^3. Give geometric descriptions of L and M; then similarly describe the sets $L \cup M$, $L \cap M$, and $L + M$ defined in Problem 16.

18. The set of all complex numbers $z = a + bi$ may be thought of as the set of vectors in the plane (Figure 4-1). If scalars r are taken to be complex numbers, rz is the usual product of two complex numbers.

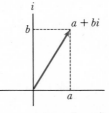

Figure 4-1

(a) Plot $z = 2 + i$ and rz where $r = \dfrac{1}{2} + \dfrac{i\sqrt{3}}{2}$.

(b) Plot $z = -1 + i$ and rz where $r = \dfrac{1}{2} + \dfrac{i\sqrt{3}}{2}$.

(c) Describe the geometric effect of the two scalar multiplications in parts (a) and (b).

(d) Prove that if $r = \dfrac{1}{2} + \dfrac{i\sqrt{3}}{2}$ and $z = a + bi$ is arbitrary, the geometric representation of rz is obtained by rotating the geometric representation of z through an angle of $\pi/3$.

(e) What is the effect of multiplying $z = a + bi$ by $r = \cos\theta + i\sin\theta$?

*19. Let V be the set of all 3×3 matrices, and suppose that the "sum" of two such matrices M and N is defined to be M "$+$" $N = MN$ (the usual product). Does V with this definition of addition form a vector space? If not, tell which properties fail.

V *20. Verify the claim (see the vignette *Some Significant Observations*) that if $u(x)$ and $v(x)$ are both solutions of the differential equation $y'' + 2y' + y = 0$, then so is $s_1u(x) + s_2v(x)$ for any real numbers s_1 and s_2. Could the same claim be made for the differential equation $y'' + 2y' + y = x^2 + 1$? (1)

4-3 | THE SPAN OF A SET

The essential features of vector addition and scalar multiplication first encountered in R^n were used in the last section to define the notion of a vector space. This section continues the program of extending into a more general setting some of the concepts found useful in our study of R^n.

To begin with, we shall say that

(1)
$$\mathbf{w} = r_1\mathbf{v}_1 + \cdots + r_k\mathbf{v}_k$$

is a **linear combination** of the vectors $\mathbf{v}_1, \ldots, \mathbf{v}_k$. And given a set of vectors S, the **span** of S, often written span S, is the set of all vectors $\mathbf{w}$ that can, for some choice of scalars $r_1, \ldots, r_k$ and some choice of vectors $\mathbf{v}_1, \ldots, \mathbf{v}_k$ in S, be written in the form (1). Note in particular that the definition allows the possibility that the set S may contain an infinite number of vectors, but if $\mathbf{w}$ is in the span of S, then there must be some k vectors $\mathbf{x}_1, \ldots, \mathbf{x}_k$ in S and some k real numbers $r_1, \ldots, r_k$ such that $\mathbf{w} = r_1\mathbf{x}_1 + \cdots + r_k\mathbf{x}_k$.

As in R^n, so in any vector space V, the span of a set S is a subspace $\mathcal{S}$ of V (Figure 4-2). The proof follows directly from Theorem A of the last

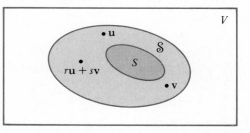

Figure 4-2

section in which we saw that a subset $\mathcal{S}$ is a subspace if, whenever $\mathbf{u}$ and $\mathbf{v}$ are in $\mathcal{S}$, the linear combination $r\mathbf{u} + s\mathbf{v}$ is also in $\mathcal{S}$.

THEOREM A

Let S be any collection of vectors in a vector space V. Then span S is a subspace of V.

Proof: Suppose that $\mathbf{u}$ and $\mathbf{v}$ are in the span of S, meaning of course that there are vectors $\mathbf{u}_1, \ldots, \mathbf{u}_m$ and $\mathbf{v}_1, \ldots, \mathbf{v}_k$ in S such that

$$\mathbf{u} = r_1 \mathbf{u}_1 + \cdots + r_m \mathbf{u}_m \qquad \mathbf{v} = t_1 \mathbf{v}_1 + \cdots + t_k \mathbf{v}_k$$

Then for any scalars r and s,

$$r\mathbf{u} + s\mathbf{v} = r(r_1 \mathbf{u}_1 + \cdots + r_m \mathbf{u}_m) + s(t_1 \mathbf{v}_1 + \cdots + t_k \mathbf{v}_k)$$

showing that $r\mathbf{u} + s\mathbf{v}$ is also in span S. It follows that span S is a subspace. $\square$

In Section 4-2, we gave four examples of vector spaces, and for each vector space we showed some subset to be a subspace. It is instructive to see how Theorem A can be used (or abused) in trying to show in another way that these subsets are subspaces.

The polynomials of degree 2 or less form a subspace P_2 of the vector space of all polynomials.   *Example A*

Let S be the set of three polynomials

$$p_2(x) = x^2, \qquad p_1(x) = x, \qquad p_0(x) = 1$$

According to Theorem A, the span of this set must be a subspace in the space of all polynomials. But the span of S is the set of all polynomials of the form

$$a_2 p_2(x) + a_1 p_1(x) + a_0 p_0(x) = a_2 x^2 + a_1 x + a_0$$

which is precisely the set P_2. $\square$

The vector space of all 2×2 matrices has as a subspace the set $\mathcal{C}$ of all   *Example B*
2×2 matrices that are of the form

$$\begin{bmatrix} a & -b \\ b & a \end{bmatrix}$$

The set $\mathcal{C}$ is easily seen to be the span of the set S consisting of the two matrices

$$I = \begin{bmatrix} 1 & 0 \\ 0 & 1 \end{bmatrix} \quad \text{and} \quad R = \begin{bmatrix} 0 & -1 \\ 1 & 0 \end{bmatrix}$$

since any matrix in $\mathcal{C}$ can be written in the form $aI + bR$. $\square$

The set $\mathfrak{M}$ of all functions representable by a Maclaurin series convergent on $(-1, 1)$ is a subspace of the vector space $C'(-1, 1)$.   *Example C*

C

It might be tempting in this case to let S be the infinite collection of polynomials

$$p_0(x) = 1, \quad p_1(x) = x, \quad p_2(x) = x^2, \quad p_3(x) = x^3, \ldots$$

It is surely true that any function f representable as a Maclaurin series can be written

$$f(x) = \sum_{n=0}^{\infty} a_n x^n = \sum_{n=0}^{\infty} a_n p_n(x)$$

The hitch is this. We have not shown f to be in the span of S because we have not written f as a linear combination of a *finite* number of polynomials in S. Sad to say, Theorem A will not do us any good in this example. □

Example D The set of all arithmetic sequences forms a subspace of the vector space of all sequences.

Let S consist of just two sequences

$$\mathbf{s}_1 = (1, 1, 1, 1, \ldots) \quad \text{and} \quad \mathbf{s}_2 = (0, 1, 2, 3, \ldots)$$

The span of these two vectors consists of all sequences of the form

$$a\mathbf{s}_1 + d\mathbf{s}_2 = (a, a, a, a, \ldots) + (0, d, 2d, 3d, \ldots)$$
$$= (a, a + d, a + 2d, a + 3d, \ldots)$$

These, being the set of all arithmetic sequences, are seen by Theorem A to form a subspace. □

Let us return to the vector space of all 2×2 matrices, considering the span of

$$S_1 = \left\{ A = \begin{bmatrix} 2 & 1 \\ 1 & 0 \end{bmatrix}, B = \begin{bmatrix} 0 & 1 \\ 1 & 2 \end{bmatrix}, C = \begin{bmatrix} 1 & 1 \\ 1 & 1 \end{bmatrix} \right\}$$

The subspace spanned by S_1 obviously contains the matrix

$$M = A - B + C = \begin{bmatrix} 3 & 1 \\ 1 & -1 \end{bmatrix}$$

Our interest here is due to the fact that M can also be written in the form

$$M = \tfrac{3}{2}A - \tfrac{1}{2}B = \begin{bmatrix} 3 & 1 \\ 1 & -1 \end{bmatrix}$$

Subtracting the last expression for M from the first gives

(2)
$$\mathcal{O} = -\tfrac{1}{2}A - \tfrac{1}{2}B + C$$

from which it follows that $C = \frac{1}{2}(A + B)$. Obviously then, any matrix N in the span of S_1 can be written

$$N = r_1A + r_2B + r_3C = r_1A + r_2B + \frac{r_3}{2}(A + B)$$

$$= \left(r_1 + \frac{r_3}{2}\right)A + \left(r_2 + \frac{r_3}{2}\right)B$$

so any vector in the span of S_1 is also in the span of

$$S_2 = \left\{ A = \begin{bmatrix} 2 & 1 \\ 1 & 0 \end{bmatrix}, B = \begin{bmatrix} 0 & 1 \\ 1 & 2 \end{bmatrix} \right\}$$

Since it is obvious that any vector in the span of S_2 is surely in the span of S_1, we see that the sets S_1 and S_2 span the same subspace.

We have just seen that two sets, S_1 and S_2, containing different numbers of vectors, can span the same subspace. This being the case, there is certainly something more efficient about using the spanning set having the fewest vectors. Our next definitions, again carried forward from R^n to an arbitrary vector space V, allow us to be more specific in saying what it is that makes a spanning set efficient.

A set of nonzero vectors $v_1, \ldots, v_k$ is **linearly dependent** if there are scalars $r_1, \ldots, r_k$, not all zero, such that

$$r_1v_1 + \cdots + r_kv_k = 0$$

The set is **linearly independent** if this equation can only be satisfied by setting $r_1 = \cdots = r_k = 0$.

Because (2) holds, the set S_1 is linearly dependent. Moreover, the method by which (2) was obtained can obviously be used to establish the fact that if a vector can be expressed in more than one way as a linear combination of vectors in a set S, then S is a linearly dependent set. The importance of this fact makes it worth restating.

If S is a linearly independent set and w is in the span of S, then the representation of w as a linear combination of vectors in S is unique.   THEOREM B

Now let us ask whether S_2 is or is not linearly independent. Appealing directly to the definition, we ask whether we can find r_1 and r_2, not both 0, such that

$$r_1\begin{bmatrix} 2 & 1 \\ 1 & 0 \end{bmatrix} + r_2\begin{bmatrix} 0 & 1 \\ 1 & 2 \end{bmatrix} = \begin{bmatrix} 2r_1 & r_1 + r_2 \\ r_1 + r_2 & 2r_2 \end{bmatrix} = \begin{bmatrix} 0 & 0 \\ 0 & 0 \end{bmatrix}$$

Equating corresponding entries shows immediately that $r_1 = r_2 = 0$. Sets S_1 and S_2 both span the same vector space; the difference is that S_2

is linearly independent, while S_1 is not. A linearly independent set that spans a vector space V is said to form a **basis** for V.

All of the spanning sets mentioned in the examples above are actually bases for the spaces that they span. To verify this, it is necessary in each case to establish that the set is linearly independent. Our remaining examples show how this is done.

Example A'

The vector space of all polynomials of degree 2 or less has a basis of $S = \{ p_2(x) = x^2, \, p_1(x) = x, \, p_0(x) = 1 \}$.

We have already seen that S spans the space. To verify that S is a linearly independent set, suppose that

$$a_2 p_2(x) + a_1 p_1(x) + a_0 p_0(x) = a_2 x^2 + a_1 x + a_0 = \sigma(x)$$

Now the right-hand side of the equation is, remember, an element in the vector space; that is, it is a polynomial. It is the polynomial that is 0 for *every* x. Thus, we seek real numbers a_0, a_1, and a_3 so that

$$a_2 x^2 + a_1 x + a_0 = 0$$

for every x. This can only happen if $a_0 = a_1 = a_2 = 0$. The polynomials in S must, therefore, be linearly independent. □

Example B'

The vector space of 2×2 matrices of the form

$$\begin{bmatrix} a & -b \\ a & a \end{bmatrix}$$

has as a basis the set

$$S = \left\{ I = \begin{bmatrix} 1 & 0 \\ 0 & 1 \end{bmatrix}, \, R = \begin{bmatrix} 0 & -1 \\ 1 & 0 \end{bmatrix} \right\}$$

We have shown that the space is spanned by S, and the linear independence of S is established by asking what values of r_1 and r_2 can be used to obtain

$$r_1 \begin{bmatrix} 1 & 0 \\ 0 & 1 \end{bmatrix} + r_2 \begin{bmatrix} 0 & -1 \\ 1 & 0 \end{bmatrix} = \begin{bmatrix} 0 & 0 \\ 0 & 0 \end{bmatrix}$$

Clearly, only $r_1 = r_2 = 0$ will do the job. □

Example C'
C

In the vector space $\mathfrak{M}$ of all functions representable by a Maclaurin series convergent on $(-1, 1)$, the infinite set

$$S = \{ p_0(x) = 1, \, p_1(x) = x, \, p_2(x) = x^2, \ldots \}$$

is linearly independent.

We have already dismissed the possibility of S being a basis, for we

BARYCENTRIC COORDINATES

Since any vector in R^2 can be written as a linear combination of two basis vectors, it may seem superfluous to ever consider representing 2-dimensional vectors as linear combinations of three fixed vectors in R^2. Yet there are times when this turns out to be very convenient.

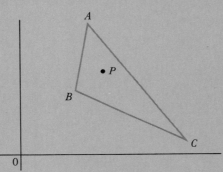

Figure 4-A

Given a triangle ABC (Figure 4-A) the representation of a vector $\overrightarrow{OP}$ from the origin to an arbitrary point P in the form

$$\overrightarrow{OP} = t_1\overrightarrow{OA} + t_2\overrightarrow{OB} + t_3\overrightarrow{OC}$$

will generally not be unique. We can make it unique, however, if we require that t_1, t_2, and t_3 be positive real numbers such that $t_1 + t_2 + t_3 = 1$. In this case, the point P will necessarily fall in the interior or on the boundary of $\triangle ABC$, and components of (t_1, t_2, t_3) are called the **barycentric coordinates** of point P. The barycentric coordinates of the vertices A, B, and C are $(1, 0, 0)$, $(0, 1, 0)$, and $(0, 0, 1)$.

Barycentric coordinates are useful in some areas of engineering and in certain branches of geometry. Some interesting properties can be pointed out without digressing into more advanced topics. For instance:

1. If weights w_1, w_2, and w_3 are located at the respective vertices A, B, and C of an otherwise uniform triangular plate then, with $W = w_1 + w_2 + w_3$, the barycentric coordinates of the center of mass (the "balance point", the point at which such a plate should be bolted down to minimize vibration) will be $(w_1/W, w_2/W, w_3/W)$.

2. In particular, if $w_1 = w_2 = w_3 = 1$ so that $W = 3$, then the center of mass (in this case called the **centroid**) is known to lie at the intersection of the medians. The barycentric coordinates are $(\frac{1}{3}, \frac{1}{3}, \frac{1}{3})$.

3. If the area of $\triangle ABC$ is a, and if line segments are drawn from the vertices to a point P with barycentric coordinates (t_1, t_2, t_3), then the three subtriangles I, II, and III (Figure 4-B) will have the respective areas $t_1 a$, $t_2 a$, and $t_3 a$.

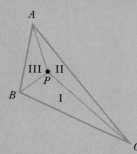

Figure 4-B

(See Coxeter, H. S. M. 1961. *Introduction to geometry*. New York: Wiley. pp. 216–21.)

showed in Example C that S does not span $\mathfrak{M}$. It is still instructive to establish the linear independence of $\mathfrak{M}$, however, for it will give us an example of a linearly independent set with an infinite number of elements. Recalling that linear combinations always involve just a *finite* number of terms, we ask whether any linear combination

$$r_1 p_{i_1}(x) + r_2 p_{i_2}(x) + \cdots + r_k p_{i_k}(x) = \sigma(x)$$

can equal the zero polynomial for all x. As before, it is easy to see that this can only happen if $r_1 = r_2 = \cdots = r_k = 0$.   □

Example D′

The vector space of all arithmetic sequences has as a basis the set

$$S = \{s_1 = (1, 1, 1, \ldots), s_2 = (0, 1, 2, 3, \ldots)\}$$

For any real numbers r_1 and r_2,

$$r_1 s_1 + r_2 s_2 = (r_1, r_1 + r_2, r_1 + 2r_2, \ldots)$$

If this sum is the vector $\mathbf{0} = (0, 0, 0, \ldots)$, then we see by equating the first two entries that $r_1 = 0$ and $r_1 + r_2 = 0$. Hence, $r_1 = r_2 = 0$. The sequences $\mathbf{s}_1$ and $\mathbf{s}_2$ are linearly independent. And we knew that they spanned the space, so S is a basis.   □

In Problems 1–10, decide whether **w** *is in the span of the given set of vectors* $\mathbf{v}_1, \ldots, \mathbf{v}_k$.

1. $\mathbf{v}_1 = \begin{bmatrix} -1 & 3 \\ 2 & 0 \end{bmatrix}$, $\mathbf{v}_2 = \begin{bmatrix} 0 & -4 \\ -2 & +2 \end{bmatrix}$, $\mathbf{v}_3 = \begin{bmatrix} 5 & 1 \\ 4 & -2 \end{bmatrix}$, $\mathbf{w} = \begin{bmatrix} 1 & -2 \\ 0 & 1 \end{bmatrix}$

2. $\mathbf{v}_1 = \begin{bmatrix} 1 & -1 \\ 0 & -1 \end{bmatrix}$, $\mathbf{v}_2 = \begin{bmatrix} -1 & 3 \\ 1 & 0 \end{bmatrix}$, $\mathbf{v}_3 = \begin{bmatrix} 1 & 3 \\ 2 & -3 \end{bmatrix}$, $\mathbf{w} = \begin{bmatrix} 2 & 4 \\ 3 & -5 \end{bmatrix}$

3. $\mathbf{v}_1(x) = x^2 + x + 1$
   $\mathbf{v}_2(x) = x^2 - x + 1$
   $\mathbf{v}_3(x) = x^2 + 1$
   $\mathbf{w}(x) = x^2 + 3x + 1$

4. $\mathbf{v}_1(x) = \frac{1}{2}x(x - 1)$
   $\mathbf{v}_2(x) = -x^2 + 1$
   $\mathbf{v}_3(x) = \frac{1}{2}x(x + 1)$
   $\mathbf{w}(x) = 2x^2 - x - 3$

5. $\mathbf{v}_1(x) = \sin x$
   $\mathbf{v}_2(x) = \cos x$
   $\mathbf{v}_3(x) = \sin^2 x$
   $\mathbf{v}_4(x) = \cos^2 x$
   $\mathbf{w}(x) = \cos 2x$

6. $\mathbf{v}_1(x) = \sin x$
   $\mathbf{v}_2(x) = \cos x$
   $\mathbf{v}_3(x) = \sin^2 x$
   $\mathbf{v}_4(x) = \cos^2 x$
   $\mathbf{w}(x) = \sin 2x$

7. $\mathbf{v}_0(x) = \ln 1$, $\mathbf{v}_1(x) = \ln x$, $\mathbf{v}_2(x) = \ln x^2$, $\ldots$, $\mathbf{w}(x) = \ln(x^2 + 2x)$

8. $\mathbf{v}_0(x) = e^{\circ}$, $\mathbf{v}_1(x) = e^x$, $\mathbf{v}_2(x) = e^{2x}$, $\ldots$, $\mathbf{w}(x) = e^{x + x^2}$

9. $\mathbf{v}_1 = (14, 16, 18, 20, 22, \ldots)$
   $\mathbf{v}_2 = (1, 4, 7, 10, \ldots)$
   $\mathbf{w} = (9, 16, 23, 30, 37, \ldots)$

10. $\mathbf{v}_1 = (4, -2, 6, -2, 8, -2, \ldots)$
    $\mathbf{v}_2 = (7, -1, 13, -1, 19, -1, \ldots)$
    $\mathbf{v}_3 = (1, 1, 3, 1, 5, 1, \ldots)$
    $\mathbf{w} = (-3, 1, -5, 1, -7, 1, \ldots)$

11–20. *Decide, for the vectors listed in Problems 1–10, whether* $\mathbf{v}_1, \ldots, \mathbf{v}_k$ *are linearly independent.*

In Problems 21–24, decide whether the given set of vectors in R^4 *is linearly independent.*

21. $\mathbf{x}_1 = (1, -1, 0, -2)$
    $\mathbf{x}_2 = (-1, 0, 2, 0)$
    $\mathbf{x}_3 = (-1, 2, -1, 3)$
    $\mathbf{x}_4 = (-1, 1, 1, 1)$

22. $\mathbf{x}_1 = (1, -1, 0, 2)$
    $\mathbf{x}_2 = (0, 1, 2, 3)$
    $\mathbf{x}_3 = (-1, 0, 1, 0)$
    $\mathbf{x}_4 = (2, 1, 0, 3)$

23. $\mathbf{x}_1 = (3, -2, 5, 4)$
    $\mathbf{x}_2 = (1, -3, 3, 2)$
    $\mathbf{x}_3 = (3, 5, 1, 2)$

24. $\mathbf{x}_1 = (4, 1, -5, 2)$
    $\mathbf{x}_2 = (1, 3, -6, -1)$
    $\mathbf{x}_3 = (5, -7, 6, 7)$

In Problems 25–28, decide whether the sets of vectors $\mathbf{v}_1, \mathbf{v}_2, \mathbf{v}_3$ *form a basis for* R^3, *and whether* **w** *is in their span.*

25. $\mathbf{v}_1 = (5, 11, -9)$
    $\mathbf{v}_2 = (7, 4, -8)$
    $\mathbf{v}_3 = (17, -31, -3)$
    $\mathbf{w} = (3, 18, -10)$

26. $\mathbf{v}_1 = (7, -3, 5)$
    $\mathbf{v}_2 = (4, 1, 2)$
    $\mathbf{v}_3 = (1, -14, 5)$
    $\mathbf{w} = (-2, 9, -4)$

27. $\mathbf{v}_1 = (1, -1, 2)$ 28. $\mathbf{v}_1 = (-1, 0, 3)$
    $\mathbf{v}_2 = (-2, 0, 2)$ $\mathbf{v}_2 = (2, -2, -3)$
    $\mathbf{v}_3 = (1, 2, -7)$ $\mathbf{v}_3 = (-1, 4, -3)$
    $\mathbf{w} = (1, 2, 4)$ $\mathbf{w} = (2, -4, 5)$

29. Decide whether the following statements are true or false.
    (a) Any subset of a linearly independent set is linearly independent.
    (b) If three vectors in R^3 do not span R^3, then at least two of them must be parallel.
    (c) If a subset of a collection S of vectors is linearly dependent, then S must be linearly dependent.
    (d) Let $\mathbf{v}_1$, $\mathbf{v}_2$, $\mathbf{v}_3$, and $\mathbf{v}_4$ be vectors in a vector space V. If $\mathbf{v}_4$ is not in the subspace spanned by $\mathbf{v}_1$, $\mathbf{v}_2$, and $\mathbf{v}_3$, then $\{\mathbf{v}_1, \mathbf{v}_2, \mathbf{v}_3, \mathbf{v}_4\}$ is a linearly independent set.

30. Suppose $\mathbf{v}_1, \ldots, \mathbf{v}_n$ are linearly independent vectors in R^n, and that M is the $n \times n$ matrix having the $\mathbf{v}_i$ as column vectors. Does it follow that $\det M \neq 0$?

V 31. Refer to Figure 4-A. Prove that for any P interior to $\triangle ABC$, vector $\overrightarrow{OP}$ can be represented in the form $\overrightarrow{OP} = t_1\overrightarrow{OA} + t_2\overrightarrow{OB} + t_3\overrightarrow{OC}$ where t_1, t_2, and t_3 are positive real numbers such that $t_1 + t_2 + t_3 = 1$, and that there is only one such representation of $\overrightarrow{OP}$.

V 32. In the triangle with vertices $A(2, 1)$, $B(5, 4)$, and $C(11, -8)$, $\overrightarrow{OP} = \frac{1}{3}\overrightarrow{OA} + \frac{1}{2}\overrightarrow{OB} + \frac{1}{6}\overrightarrow{OC}$ determines a point P such that $\triangle ABP$ is a right triangle. Verify the property of barycentric coordinates that relates the area of $\triangle ABP$ to that of $\triangle ABC$.

4-4 BASES AND THE DIMENSION OF A SPACE

We said in the last section that a basis for a vector space V is a linearly independent set of vectors that spans V. Stated another way, the set B is a basis for the vector space V if

1. The set B is linearly independent.
2. For every vector $\mathbf{w}$ in V, we can find a finite number of vectors $\mathbf{v}_1, \ldots, \mathbf{v}_k$ in B and a finite number of scalars $r_1, \ldots, r_k$ such that

$$\mathbf{w} = r_1\mathbf{v}_1 + \cdots + r_k\mathbf{v}_k$$

According to Theorem A of the last section, there is only one way to represent $\mathbf{w}$ as a linear combination of vectors in B.

Several facts about bases are best emphasized by way of examples.

Each of the sets

$$B_1 = \{p_2(x) = x^2, p_1(x) = x, p_0(x) = 1\}$$
$$B_2 = \{q_1(x) = \tfrac{1}{2}x(x - 1), q_2(x) = -x^2 + 1, q_3(x) = \tfrac{1}{2}x(x + 1)\}$$

forms a basis for the vector space P_2 of polynomials of degree 2 or less.

The set B_1 was shown in Example A′ of the last section to be a basis. The linear independence of B_2 can be established directly from the definition. The usefulness of knowing about B_2 as a basis can be shown by the following demonstration that it spans P_2. We begin by noting that the span of B_2 includes all polynomials of the form

$$p(x) = a_1\left(\frac{x}{2}\right)(x - 1) + a_2(-x^2 + 1) + a_3\left(\frac{x}{2}\right)(x + 1) \qquad (1)$$

This is obviously a polynomial of degree 2 or less. Direct substitution shows that

$$p(-1) = a_1, \qquad p(0) = a_2, \qquad p(1) = a_3$$

Now to be specific, let us consider an arbitrarily chosen second degree polynomial $q(x) = 2x^2 - 3x + 7$. Substitution shows that $q(-1) = 12$, $q(0) = 7$, and $q(1) = 6$. Thus, if in (1) we set $a_1 = 12$, $a_2 = 7$, and $a_3 = 6$, then $p(x)$ and $q(x)$ will both pass through $(-1, 12)$, $(0, 7)$, and $(1, 6)$. But if two polynomials of degree 2 or less agree at three points, they agree everywhere (see Problem 22 at the end of this section).

This same procedure can be used to obtain any polynomial $q(x)$ of degree 2 or less from (1), showing that polynomials of the form (1) do span the vector space P_2. $\square$

The beginner who looks at the bases B_1 and B_2 above is naturally attracted to B_1 as the preferable basis to work with. But the ease with which the basis B_2 allows us to obtain a quadratic function with specified values at $x = -1$, $x = 0$, and $x = 1$ makes B_2 a particularly useful basis in applications (see the vignette *Simpson's rule*).

Now comes the lesson from this example: Not only does a vector space have many different sets that can serve as a basis, but for different purposes we will want to use different bases for the same vector space.

The infinite set of polynomials

$$S = \{p_0(x) = 1, p_1(x) = x, p_2(x) = x^2, \ldots\}$$

forms a basis for the vector space of all polynomials.

The linear independence of the polynomials in S is easily established by the usual considerations, and was in fact demonstrated in Example C of Section 4-3. It is obvious that these polynomials span the space, since an arbitrary polynomial $a_n x^n + \cdots + a_1 x + a_0$ can be written as the linear combination $a_n p_n(x) + \cdots + a_1 p_1(x) + a_0 p_0(x)$. □

The lesson of Example C of Section 4-3 is that a basis may have an infinite number of members. We wish to stress, however, that though the basis may be infinite, we must be able to express any vector in the space as a linear combination of just a finite number of elements selected from the basis. (This is the reason that the set S of Example B was rejected in Example C of Section 4-3 as a basis for the vector space $\mathfrak{M}$ of functions representable by a Maclaurin series.)

If a basis is not infinite, then it is finite. Now comes the question: Since a vector space V has many different bases, is it possible that one basis may be finite, another infinite? Our next theorem says it can't happen because if a finite basis has n elements, then any set (finite or infinite) having more than n elements must be a linearly dependent set, hence certainly not a basis.

THEOREM A

If a vector space V has a basis of n vectors, and if $S = \{\mathbf{w}_1, \ldots, \mathbf{w}_m\}$ is any set of m vectors in V, $m > n$, then set S is linearly dependent.

Proof: Let $B = \{\mathbf{u}_1, \ldots, \mathbf{u}_n\}$ be a basis of n vectors. Then each $\mathbf{w}_j$ may be expressed as a linear combination of $\mathbf{u}_1, \ldots, \mathbf{u}_n$.

$$\mathbf{w}_1 = a_{11}\mathbf{u}_1 + a_{21}\mathbf{u}_2 + \cdots + a_{n1}\mathbf{u}_n$$
$$\mathbf{w}_2 = a_{12}\mathbf{u}_1 + a_{22}\mathbf{u}_2 + \cdots + a_{n2}\mathbf{u}_n$$
$$\vdots \qquad \vdots \qquad \vdots \qquad \vdots$$
$$\mathbf{w}_m = a_{1m}\mathbf{u}_1 + a_{2m}\mathbf{u}_2 + \cdots + a_{nm}\mathbf{u}_n$$

We wish to show that we can find $r_1, \ldots, r_m$, not all zero, such that

(2)
$$r_1\mathbf{w}_1 + r_2\mathbf{w}_2 + \cdots r_m\mathbf{w}_m = \mathbf{0}$$

This says that

$$r_1(a_{11}\mathbf{u}_1 + a_{21}\mathbf{u}_2 + \cdots + a_{n1}\mathbf{u}_n)$$
$$+ r_2(a_{12}\mathbf{u}_1 + a_{22}\mathbf{u}_2 + \cdots + a_{n2}\mathbf{u}_n)$$
$$\vdots \qquad \vdots \qquad \vdots \qquad \vdots$$
$$+ r_m(a_{1m}\mathbf{u}_1 + a_{2m}\mathbf{u}_2 + \cdots + a_{nm}\mathbf{u}_n) = \mathbf{0}$$

This last expression is, however, a linear combination of the vectors $\mathbf{u}_1, \ldots, \mathbf{u}_n$ that adds up to $\mathbf{0}$. Since $\mathbf{u}_1, \ldots, \mathbf{u}_n$ are linearly independent, this means that the coefficient of each $\mathbf{u}_i$ must be 0. Hence,

$$a_{11}r_1 + a_{12}r_2 + \cdots + a_{1m}r_m = 0 \qquad \text{(coefficient of } \mathbf{u}_1)$$
$$a_{21}r_1 + a_{22}r_2 + \cdots + a_{2m}r_m = 0 \qquad \text{(coefficient of } \mathbf{u}_2)$$
$$\vdots \qquad \vdots \qquad \vdots$$
$$a_{n1}r_1 + a_{n2}r_2 + \cdots + a_{nm}r_m = 0 \qquad \text{(coefficient of } \mathbf{u}_n)$$

C SIMPSON'S RULE

Suppose we wish to find the area under $y = 2^{-x^2}$ between $x = -1$ and $x = 1$. The area is equal to $\int_{-1}^{1} 2^{-x^2}\,dx$, but we cannot evaluate this integral using the fundamental theorem of calculus because 2^{-x^2} has no elementary function as an antiderivative. Noting from the graph of $y = 2^{-x^2}$ (Figure 4-C) that the curve passes

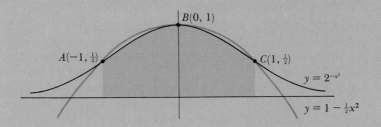

Figure 4-C

through $A(-1, \frac{1}{2})$, $B(0, 1)$, and $C(1, \frac{1}{2})$, it appears that the desired area might be approximated by the area under a parabola passing through A, B, and C. Moreover, we know from Example A that the parabola that passes through $(-1, a_1)$, $(0, a_2)$, and $(1, a_3)$ is

$$p(x) = a_1 \tfrac{1}{2} x (x - 1) + a_2(-x^2 + 1) + a_3 \tfrac{1}{2} x (x + 1)$$

The desired parabola in this case is, therefore,

$$p(x) = \tfrac{1}{2}[\tfrac{1}{2} x(x - 1)] + 1[-x^2 + 1] + \tfrac{1}{2}[\tfrac{1}{2} x(x + 1)]$$
$$= 1 - \tfrac{1}{2} x^2$$

Our integral is now approximated by

$$\int_{-1}^{1} 2^{-x^2}\,dx \approx \int_{-1}^{1} \left(1 - \frac{1}{2} x^2\right) dx = \frac{5}{3}$$

The polynomials

$$q_1(x) = \tfrac{1}{2} x(x - 1), \qquad q_2(x) = -x^2 + 1, \qquad q_3(x) = \tfrac{1}{2} x(x + 1)$$

are called the **Lagrange interpolating polynomials.** They can easily be modified to provide a basis convenient to use in passing a parabola through the points (x_1, a_1), $(\tfrac{1}{2}(x_1 + x_2), a_2)$, and (x_2, a_3) in place of $(-1, a_1)$, $(0, a_2)$, and $(1, a_3)$; see Problem 23. Modified in this way, they provide the basis for Simpson's rule, sometimes introduced in calculus as a numerical method for evaluating integrals.

Now this homogeneous system of n equations in the m unknowns $r_1, \ldots, r_m$ has more unknowns than equations, so (Theorem B of Section 1-4) there is an infinite number of nontrivial solutions. We can, therefore, find $r_1, \ldots, r_m$, not all zero, so that (2) holds. □

We can, in fact, do better than to say that if one basis is finite, any other basis must be finite.

THEOREM B

If V has one basis of n elements, then every basis has n elements.

Proof: Suppose that $B = \{\mathbf{u}_1, \ldots, \mathbf{u}_n\}$ and $C = \{\mathbf{v}_1, \ldots, \mathbf{v}_m\}$ are both bases. Since the definition of C as a basis requires that $\mathbf{v}_1, \ldots, \mathbf{v}_m$ be linearly independent, it follows from Theorem A that $m \leq n$. Similarly, since $\mathbf{u}_1, \ldots, \mathbf{u}_n$ must be linearly independent and C is a basis, we have $n \leq m$. Hence, $n = m$. □

If a vector space V has a basis of n elements, we now know that any other basis will also have n elements. Such a space is said to be a **finite-dimensional vector space of dimension n.** If no finite basis exists, we say V is an **infinite-dimensional** vector space. To wrap everything up, the trivial vector space consisting of just the zero element **0** is said to have dimension 0.

Since the basis of a vector space is an important, we might even say basic, concept in all that is to follow, a review of this section is in order. Let B be a basis of the vector space V. Then

- Every member of V can be expressed as a linear combination of vectors in B, and this expression is unique.
- There will always be sets other than B that form a basis.
- The set B may consist of a finite or an infinite set of members.
- If B is infinite, every basis will be infinite.
- If B is finite, having n members, every basis will have n members. V is then said to be an n-dimensional space.

*PROBLEM
SET 4-4*

Decide whether the vector spaces indicated in Problems 1–12 are infinite- or finite-dimensional. If the dimension is finite, give its value.

1. All 4-tuples having the second and fourth entries equal.

2. All 4-tuples having the second and fourth entries equal to 0.

3. All 4-tuples (x_1, x_2, x_3, x_4) in which $x_1 + x_2 + x_3 = 0$.

4. All 4-tuples (x_1, x_2, x_3, x_4) in which $x_1 + x_4 = 0$.

5. All 3×3 matrices.

6. All 3×3 matrices in which the only nonzero entries occur on the main diagonal.

7. All 3×3 matrices in which the only nonzero entries occur in the corners.

8. All 3×2 matrices.

9. All polynomials $p(x)$ satisfying $p(1) = 0$.

10. All polynomials whose graph passes through the origin.

11. All arithmetic sequences; see Examples D and D' of Section 4-3.

12. All sequences $a_1, a_2, a_3, \ldots$ where each a_i is a real number.

Give a reason for answering Problems 13–18 with yes or no.

13. Can a proper subspace of a given infinite-dimensional vector space be infinite-dimensional?

14. Can a proper subspace of a given infinite-dimensional vector space be finite-dimensional?

15. Can we choose r_1, r_2, r_3, and r_4 so that the polynomials

$$p_1(x) = (x - r_2)(x - r_3)$$
$$p_2(x) = (x - r_3)(x - r_4)$$
$$p_3(x) = (x - r_4)(x - r_1)$$
$$p_4(x) = (x - r_1)(x - r_2)$$

are linearly independent?

16. Can we choose r_1, r_2, r_3, and r_4 so that the matrices

$$A_1 = \begin{bmatrix} r_1 & 0 \\ 0 & 0 \end{bmatrix}, \qquad A_2 = \begin{bmatrix} 0 & r_2 \\ 0 & 0 \end{bmatrix},$$

$$A_3 = \begin{bmatrix} 0 & 0 \\ r_3 & 0 \end{bmatrix}, \qquad A_4 = \begin{bmatrix} 0 & 0 \\ 0 & r_4 \end{bmatrix}$$

are linearly independent?

17. Can we find a set of three polynomials $\{p_1(x), p_2(x), p_3(x)\}$ that will form a basis for the vector space of all polynomials of degree three or less?

18. Let it be given that the graphs of each of the equations $y = p_1(x)$, $y = p_2(x)$, and $y = p_3(x)$ are parabolas. Can we be sure that there are scalars r_1, r_2, and r_3 such that

$$\tfrac{2}{3}x^2 - \tfrac{1}{2}x + \tfrac{3}{4} = r_1 p_1(x) + r_2 p_2(x) + r_3 p_3(x)$$

is true?

19. In the space C' of all functions having everywhere continuous first derivatives, what is the dimension of the space spanned by the following functions?

$$f_1(x) = \sin^2 x, \qquad f_2(x) = \cos^2 x,$$
$$f_3(x) = \cos 2x, \qquad f_4(x) = \sin 2x$$

C

C

20. In the space C', what is the dimension of the space spanned by the following functions?

$$g_1(x) = \sin x, \qquad g_2(x) = \cos x,$$
$$g_3(x) = \sin 2x, \qquad g_4(x) = \cos 2x$$

21. Show that the Lagrange Interpolating Polynomials

$$q_1(x) = \tfrac{1}{2}x(x - 1), \qquad q_2(x) = -x^2 + 1, \qquad q_3(x) = \tfrac{1}{2}x(x + 1)$$

are linearly independent.

22. Suppose that $p(x) = a_2x^2 + a_1x + a_0$ and $q(x) = b_2x^2 + b_1x + b_0$ agree at the three distinct points x_1, x_2, and x_3. That is, suppose that $p(x_i) = q(x_i)$ for $i = 1, 2, 3$. Prove that $p(x)$ and $q(x)$ are the same polynomial; that is, $a_i = b_i$ for $i = 1, 2, 3$. *Hint:* Form $r(x) = p(x) - q(x)$, a second degree polynomial. How many distinct roots does $r(x)$ have? How can this be?

23. Find three linearly independent second degree polynomials $r_1(x)$, $r_2(x)$, and $r_3(x)$ such that

$$r(x) = a_1r_1(x) + a_2r_2(x) + a_3r_3(x)$$

passes through the points (x_1, a_1), $(\tfrac{1}{2}(x_1 + x_2), a_2)$, and (x_2, a_3).

V

*24. Use the Lagrange Interpolating Polynomials $q_1(x)$, $q_2(x)$, and $q_3(x)$ (see the vignette *Simpson's rule*) to estimate

$$\int_{-1}^{1} \frac{dx}{1 + x^2}$$

Compare your answer with $\text{Arctan}(1) - \text{Arctan}(-1)$.

4-5   COORDINATES AND BASES

Attention is restricted in this section to an n-dimensional vector space V having a basis $B = \{\mathbf{u}_1, \ldots, \mathbf{u}_n\}$. In this context, we know that any vector $\mathbf{w}$ in V can be written uniquely in the form

$$\mathbf{w} = x_1\mathbf{u}_1 + \cdots + x_n\mathbf{u}_n$$

Uniqueness of representation means that if $\mathbf{w}$ is given, the n-tuple $(x_1, \ldots, x_n)$ is determined; and conversely, n given numbers determine $\mathbf{w}$. The components of $(x_1, \ldots, x_n)$ are called the **coordinates of w with respect to the basis B.** Since the coordinates so clearly depend on B (not just the vectors in B, but also on the order in which those vectors are listed), we write $(x_1, \ldots, x_n)_B$ if there is any possibility of confusion.

The possibility of confusion does exist because, as has been emphasized, the same vector space has many different bases, and different bases are useful for different purposes. You may still wonder, however, why anyone would risk confusion by working with two bases in the same problem. Why not choose the best basis for a particular problem, whatever that basis might be, and then stick with it? The answer is that some problems are made easier precisely because we can shift from one basis to another during the process of solution. We will digress to consider a classic problem of this kind which is sometimes used as an organizing theme for a linear algebra course.

ROTATION OF AXES

Most students will get stalled if asked to graph

$$4x_1x_2 - 3x_2^2 = 8 \qquad\qquad (1)$$

but many of these same students have learned techniques that make it relatively easy for them to sketch the graph (Figure 4-3) of

$$\frac{x_1^2}{8} - \frac{x_2^2}{2} = 1 \qquad\qquad (2)$$

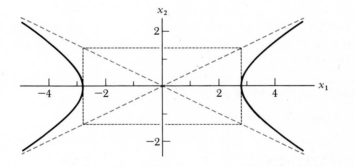

Figure 4-3

We shall see, however, that a simple change of basis will transform equation (1) into equation (2).

The usual association between equation (1) and a graph in the plane can be described in the following manner. Along mutually perpendicular axes in the plane, draw from the origin two arrows of unit length representing $\mathbf{e}_1 = (1, 0)$ and $\mathbf{e}_2 = (0, 1)$ (Figure 4-4). Then a pair such as $(\frac{5}{2}, 2)$ that satisfies (1) is represented in the plane as the terminal point P of $\overrightarrow{OP} = \frac{5}{2}\mathbf{e}_1 + 2\mathbf{e}_2$.

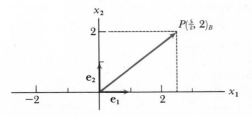

Figure 4-4

In terms of the notation introduced above, the pair has become $(\frac{5}{2}, 2)_B$, the coordinates of P with respect to the standard basis $B = \{\mathbf{e}_1, \mathbf{e}_2\}$. The set of all pairs satisfying (1) in this way correspond to a set of points P in the plane called the graph of the given equation.

Now, think of rotating the arrows representing $\mathbf{e}_1$ and $\mathbf{e}_2$ both through an angle θ. The resulting arrows (Figure 4-5) represent the vectors

$$\mathbf{v}_1 = (\cos \theta, \sin \theta)_B, \qquad \mathbf{v}_2 = (-\sin \theta, \cos \theta)_B$$

Figure 4-5

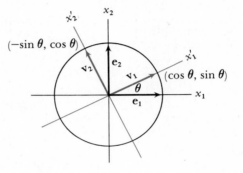

We now have two bases for R^2,

$$B = \{\mathbf{e}_1, \mathbf{e}_2\} \quad \text{and} \quad B' = \{\mathbf{v}_1, \mathbf{v}_2\}$$

An arbitrary vector $\mathbf{w}$ in R^2 has the respective coordinates $(x_1, x_2)_B$ and $(x_1', x_2')_{B'}$. This means

$$\mathbf{w} = x_1 \mathbf{e}_1 + x_2 \mathbf{e}_2 = x_1' \mathbf{v}_1 + x_2' \mathbf{v}_2$$

Rewritten,

$$x_1(1, 0) + x_2(0, 1) = x_1'(\cos \theta, \sin \theta) + x_2'(-\sin \theta, \cos \theta)$$

The rules for scalar multiplication, addition, and equality of vectors give

(3)
$$x_1 = x_1' \cos \theta - x_2' \sin \theta$$
$$x_2 = x_1' \sin \theta + x_2' \cos \theta$$

Using these expressions for x_1 and x_2, substitution into (1) gives

$$4(x_1' \cos \theta - x_2' \sin \theta)(x_1' \sin \theta + x_2' \cos \theta) - 3(x_1' \sin \theta + x_2' \cos \theta)^2 = 8$$

Multiplication and collection of terms gives

(4)
$$(x_1')^2(4 \sin \theta \cos \theta - 3 \sin^2 \theta)$$
$$+ x_1' x_2'(4 \cos^2 \theta - 4 \sin^2 \theta - 6 \sin \theta \cos \theta)$$
$$+ (x_2')^2(-4 \sin \theta \cos \theta - 3 \cos^2 \theta) = 8$$

The idea of this substitution is to choose θ so that the coefficient of $x_1' x_2'$ turns out to be 0; that is, choose θ so that

$$4 \cos^2 \theta - 4 \sin^2 \theta - 6 \sin \theta \cos \theta = 0$$

Our purposes here do not require that we take the trouble to solve this trigonometric equation. Indeed, one of the triumphs of linear algebra is that we can later transform (1) to (2) with no trigonometry at all. Suffice to say, since we started this problem, that θ should be chosen so that $\sin \theta = 1/\sqrt{5}$ and $\cos \theta = 2/\sqrt{5}$. Using these values in (4) gives $(x_1')^2 - 4(x_2')^2 = 8$ which (except for the primes) is equation (2). The primes remind us that the graph (Figure 4-6) is to be drawn with respect to the $x_1' x_2'$ axes, the rotated axes which are determined by

$$\mathbf{v}_1 = (\cos \theta, \sin \theta) = \left(\frac{2}{\sqrt{5}}, \frac{1}{\sqrt{5}} \right)$$

$$\mathbf{v}_2 = (-\sin \theta, \cos \theta) = \left(-\frac{1}{\sqrt{5}}, \frac{2}{\sqrt{5}} \right)$$

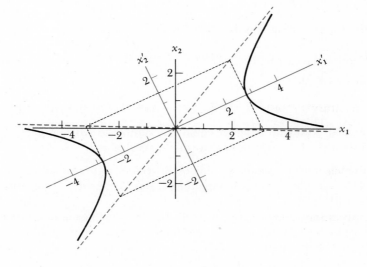

Figure 4-6

It is to be emphasized that the graph that has been drawn in Figure 4-6 is the graph of (1). The coordinate system $x_1' x_2'$ has served its purpose of enabling us to draw the graph, and may now be dropped (Figure 4-7). The pair (x_1, x_2) lies on the graph if and only if (x_1, x_2) satisfies (1).

A FORMULA FOR A CHANGE OF BASIS

Assured that the need to change from one basis to another does arise, let us see how it is done. Begin by assuming that V is a vector space, that w

Figure 4-7

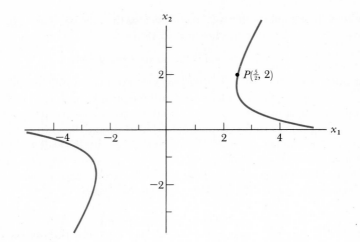

is an arbitrary vector in V, and that we have two bases

$$B = \{\mathbf{u}, \ldots, \mathbf{u}_n\}, \qquad B' = \{\mathbf{v}_1, \ldots, \mathbf{v}_n\}$$

with corresponding coordinates of $\mathbf{w}$ given by

$$(x_1, \ldots, x_n)_B, \qquad (x'_1, \ldots, x'_n)_{B'}$$

This means that

(5)
$$\mathbf{w} = x_1\mathbf{u}_1 + \cdots + x_n\mathbf{u}_n = x'_1\mathbf{v}_1 + \cdots + x'_n\mathbf{v}_n$$

Our goal is to find a set of equations similar to (3) that relates the coordinates of $\mathbf{w}$ with respect to the bases B and B'.

For notational convenience, think of all vectors written as columns, and let U be the $n \times n$ matrix having the column vectors $\mathbf{u}_1, \ldots, \mathbf{u}_n$; that is, $U = [\mathbf{u}_1 \quad \cdots \quad \mathbf{u}_n]$. Similarly, form $V = [\mathbf{v}_1 \quad \cdots \quad \mathbf{v}_n]$. Having previously observed (Property 5 of Section 2-6) that a matrix multiplied on the right by a column vector results in a linear combination of the columns of the matrix, we can rewrite (5) in the form

(6)
$$\mathbf{w} = U\begin{bmatrix} x_1 \\ \vdots \\ x_n \end{bmatrix} = V\begin{bmatrix} x'_1 \\ \vdots \\ x'_n \end{bmatrix}$$

Since $B = \{\mathbf{u}_1, \cdots, \mathbf{u}_n\}$ is a basis, we can find real numbers $p_{11}, \ldots, p_{n1}$ such that

$$\mathbf{v}_1 = p_{11}\mathbf{u}_1 + \cdots + p_{n1}\mathbf{u}_n = U\begin{bmatrix} p_{11} \\ \vdots \\ p_{n1} \end{bmatrix}$$

Similar expressions can be written for all the vectors of B'.

$$\mathbf{v}_1 = U \begin{bmatrix} p_{11} \\ \vdots \\ p_{n1} \end{bmatrix}, \dots, \mathbf{v}_j = U \begin{bmatrix} p_{1j} \\ \vdots \\ p_{nj} \end{bmatrix}, \dots, \mathbf{v}_n = U \begin{bmatrix} p_{1n} \\ \vdots \\ p_{nn} \end{bmatrix}$$

Thus,

$$V = [\mathbf{v}_1 \quad \cdots \quad \mathbf{v}_n] = \begin{bmatrix} U \begin{bmatrix} p_{11} \\ \vdots \\ p_{n1} \end{bmatrix} \quad \cdots \quad U \begin{bmatrix} p_{1n} \\ \vdots \\ p_{nn} \end{bmatrix} \end{bmatrix}$$

or

$$V = U \begin{bmatrix} p_{11} & \cdots & p_{1n} \\ \vdots & & \vdots \\ p_{n1} & \cdots & p_{nn} \end{bmatrix} = UP$$

The matrix P is called the **transition matrix.** It is also the matrix that relates the coordinates of $\mathbf{w}$ with respect to the two bases, for since $V = UP$, equation (6) becomes

$$\mathbf{w} = U \begin{bmatrix} x_1 \\ \vdots \\ x_n \end{bmatrix} = V \begin{bmatrix} x'_1 \\ \vdots \\ x'_n \end{bmatrix} = UP \begin{bmatrix} x'_1 \\ \vdots \\ x'_n \end{bmatrix}$$

We know that the representation of $\mathbf{w}$ with respect to the column vectors of U is unique, so it must be that

$$\begin{bmatrix} x_1 \\ \vdots \\ x_n \end{bmatrix} = P \begin{bmatrix} x'_1 \\ \vdots \\ x'_n \end{bmatrix} \qquad (7)$$

Our results may be summarized as follows:

THEOREM A

Suppose that a vector space V has two bases

$$B = \{\mathbf{u}_1, \cdots, \mathbf{u}_n\} \quad and \quad B' = \{\mathbf{v}_1, \cdots, \mathbf{v}_n\}$$

Use the vectors of B and B' as column vectors to form matrices U and V, respectively. Then there is a matrix P such that $V = UP$, and the coordinates $(x_1, \dots, x_n)_B$ and $(x'_1, \dots, x'_n)_{B'}$ of an arbitrary point are related by (7).

Note that the entries in the jth column of P are scalars $p_{1j}, \dots, p_{nj}$ that enable us to write the jth vector $\mathbf{v}_j$ as a linear combination of the vectors in B; that is,

$$\mathbf{v}_j = p_{1j}\mathbf{u}_1 + \cdots + p_{nj}\mathbf{u}_n$$

Two bases for R^2 are

Example A

$$B = \{\mathbf{u}_1 = (-3, -2), \mathbf{u}_2 = (-1, 2)\}$$
$$B' = \{\mathbf{v}_1 = (-1, -6), \mathbf{v}_2 = (2, 4)\}$$

Given that $\mathbf{w} = (3, 2)_{B'}$, find the coordinates of $\mathbf{w}$ with respect to the basis B.

Our first task is to determine p_{11}, p_{21}, p_{12}, p_{22} so that

$$(-1, -6) = p_{11}(-3, -2) + p_{21}(-1, 2)$$
$$(2, 4) = p_{12}(-3, -2) + p_{22}(-1, 2)$$

This requires that we solve the system

$$
\begin{aligned}
-3p_{11} - p_{21} &= -1 \\
-2p_{11} + 2p_{21} &= -6 \\
-3p_{12} - p_{22} &= 2 \\
-2p_{12} + 2p_{22} &= 4
\end{aligned}
$$

While this is a system of four equations in four unknowns, there are so many zero coefficients that, without too much effort, we determine

$$p_{11} = 1, \qquad p_{21} = -2, \qquad p_{12} = -1, \qquad p_{22} = 1$$

According to Theorem A, the coordinates $(x_1, x_2)_B$ and $(x'_1, x'_2)_{B'}$ of an arbitrary point in R^2 are related by

$$\begin{bmatrix} x_1 \\ x_2 \end{bmatrix} = \begin{bmatrix} 1 & -1 \\ -2 & 1 \end{bmatrix}\begin{bmatrix} x'_1 \\ x'_2 \end{bmatrix}$$

In particular, for $\mathbf{w} = (3, 2)_{B'}$,

$$\begin{bmatrix} x_1 \\ x_2 \end{bmatrix} = \begin{bmatrix} 1 & -1 \\ -2 & 1 \end{bmatrix}\begin{bmatrix} 3 \\ 2 \end{bmatrix} = \begin{bmatrix} 1 \\ -4 \end{bmatrix}$$

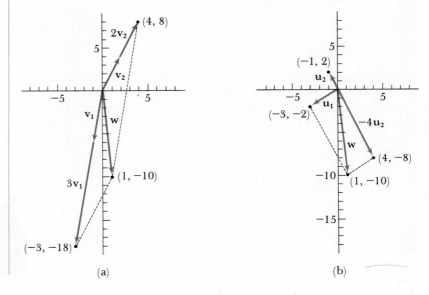

Figure 4-8

(a)   (b)

In Figure 4-8a, **w** is shown as the sum of $3\mathbf{v}_1 + 2\mathbf{v}_2$; in Figure 4-8b, **w** is shown as the sum of $1\mathbf{u}_1 - 4\mathbf{u}_2$. It will be noted that

$$\mathbf{w} = 3(-1, -6) + 2(2, 4) = 1(-3, -2) - 4(-1, 2)$$
$$= (1, -10) \quad \square$$

THE CORRECT BASIS CAN MAKE IT CRYSTAL CLEAR

A fundamental idea of chemistry is that each molecule of a pure compound is a duplicate of all other such molecules. Crystals, being composed of systematic repetitions of molecules, are naturally classified by studying the nature of these systematic repetitions, much as wallpaper patterns are classified into 17 groups (they really are: see Budden, F. J. 1971. *The fascination of groups.* New York: Cambridge Univ. Press.) according to the system used to duplicate the basic motif (Figure 4-D).

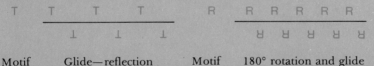

| Motif | Glide—reflection | Motif | 180° rotation and glide |

Figure 4-D

Classification has nothing to do with the shape of either the basic motif or the shape of the resulting pattern; it has everything to do with the rule for creating the repetition. The systems used to describe the repetition may be likened to the instructions we give to a visitor looking for a certain store in our neighborhood. Ignoring the fact that blocks on north-south streets may be longer than those on east-west streets, or even that the streets to be followed may not be perpendicular, we advise the visitor to go three blocks up this street, turn right, and go two more blocks. That is, we give directions using the natural grid provided by the streets (Figure 4-E).

Figure 4-E

It turns out that repetitive patterns in crystals may always be conveniently described using one of six different grids in space. Crystallographers have named these grids, five of which are described in terms of three basis vectors **a, b,** and **c** and the angles between them (Figure 4-F).

These natural grids, called **crystal systems,** nicely emphasize our point. There are times when it is most convenient to describe even physical phenomena in three dimensions using something other than the standard basis!

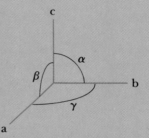

Reference system

Figure 4-F

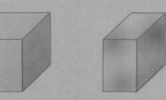

Cubic
$|a| = |b| = |c|$
$\alpha = \beta = \gamma = 90°$

Tetragonal
$|a| = |b| \neq |c|$
$\alpha = \beta = \gamma = 90°$

Orthorhombic
$|a| \neq |b| \neq |c|$
$\alpha = \beta = \gamma = 90°$

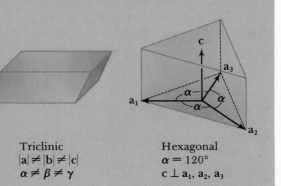

Monoclinic
$|a| \neq |b| \neq |c|$
$\alpha = \beta = 90°$

Triclinic
$|a| \neq |b| \neq |c|$
$\alpha \neq \beta \neq \gamma$

Hexagonal
$\alpha = 120°$
$c \perp a_1, a_2, a_3$

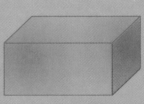

Even though the hexagonal grid is 3-dimensional, it turns out to be most convenient to locate points with respect to four (clearly not linearly independent) vectors.

(Reference: Buerger, M. J. 1970. *Contemporary crystallography.* New York: McGraw-Hill.)

Problems 1–4 refer to the two bases for R^2 described by

$$B = \{\mathbf{u}_1 = (2, -2), \mathbf{u}_2 = (-1, 3)\}$$
$$B' = \{\mathbf{v}_1 = (-1, -2), \mathbf{v}_2 = (1, -3)\}$$

PROBLEM
SET 4-5

The coordinates of $\mathbf{w}$ *are given with respect to the basis* B'. *Find the transition matrix for the change from* B *to* B' *and use it to find the coordinates of* $\mathbf{w}$ *with respect to* B.

1. $\mathbf{w} = (-4, 3)_{B'}$ 2. $\mathbf{w} = (-\frac{12}{5}, \frac{23}{5})_{B'}$

3. $\mathbf{w} = (2, -5)_{B'}$ 4. $\mathbf{w} = (3, -2)_{B'}$

Problems 5–8 refer to the two bases for R^3 described by

$$B = \{\mathbf{u}_1 = (1, 0, -1), \mathbf{u}_2 = (0, 1, 1), \mathbf{u}_3 = (-1, 1, 0)\}$$
$$B' = \{\mathbf{v}_1 = (-2, 2, -2), \mathbf{v}_2 = (3, -2, -3), \mathbf{v}_3 = (2, -1, -1)\}$$

The coordinates of $\mathbf{w}$ *are given with respect to the basis* B'. *Find the transition matrix for the change from* B *to* B' *and use it to find the coordinates of* $\mathbf{w}$ *with respect to* B.

5. $\mathbf{w} = (2, 1, 1)_{B'}$ 6. $\mathbf{w} = (1, 0, -1)_{B'}$

7. $\mathbf{w} = (1, -1, 2)_{B'}$ 8. $\mathbf{w} = (3, 1, -2)_{B'}$

Problems 9 and 10 refer to two bases we have used for P_2, the space of polynomials of degree 2 or less.

$$B_1 = \{p_0(x) = 1, p_1(x) = x, p_2(x) = x^2\}$$
$$B_2 = \{q_1(x) = \tfrac{1}{2}x(x - 1), q_2(x) = -x^2 + 1, q_3(x) = \tfrac{1}{2}x(x + 1)\}$$

9. (a) The polynomial $r(x) = 4q_1(x) - 3q_2(x) - 6q_3(x)$ has coordinates $(4, -3, -6)_{B_2}$. Find the coordinates with respect to B_1.
   (b) Find the transition matrix P so that, given the coordinates $(x_1', x_2', x_3')_{B_2}$ of a polynomial $r(x)$, the coordinates $(x_1, x_2, x_3)_{B_1}$ can be found from

$$\begin{bmatrix} x_1 \\ x_2 \\ x_3 \end{bmatrix} = P \begin{bmatrix} x_1' \\ x_2' \\ x_3' \end{bmatrix}$$

   (c) Use matrix P to verify your answer to (a).

10. (a) The polynomial $r(x) = 2p_0(x) - 5p_1(x) + 3p_2(x)$ has coordinates $(2, -5, 3)_{B_1}$. Find the coordinates with respect to B_2.
   (b) Find the transition matrix Q so that, given the coordinates $(x_1, x_2, x_3)_{B_1}$ of a polynomial $r(x)$, the coordinates $(x'_1, x'_2, x'_3)_{B_2}$ can be found from

$$\begin{bmatrix} x'_1 \\ x'_2 \\ x'_3 \end{bmatrix} = Q \begin{bmatrix} x_1 \\ x_2 \\ x_3 \end{bmatrix}$$

   (c) Use matrix Q to verify your answer to (a).

11. The standard basis for R^2 is $S = \{e_1 = (1, 0), e_2 = (0, 1)\}$. The grid of parallel lines superimposed on the plane in Figure 4-9 uses $B = \{v_1 =$

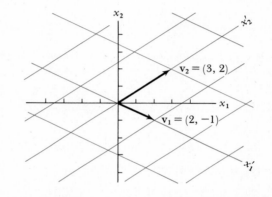

Figure 4-9

$(2, -1)_S, v_2 = (3, 2)_S\}$ as a basis. Find a transition matrix P such that

$$\begin{bmatrix} x'_1 \\ x'_2 \end{bmatrix}_B = P \begin{bmatrix} x_1 \\ x_2 \end{bmatrix}_S$$

Locate $w = (-4, 8)_S$ in Figure 4-9, showing also the graphical representation of $w = x'_1 v_1 + x'_2 v_2$.

12. Follow the instructions for Problem 11, using the grid of parallel lines in Figure 4-10 which has $B = \{v_1 = (3, -1)_S, v_2 = (-1, 2)_S\}$ as a basis.

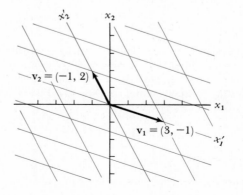

Figure 4-10

13. How are the transition matrices P and Q of Problems 9 and 10 related?

14. Prove that the transition matrix is always nonsingular.

15. Suppose the triclinic crystal grid system is to be used where $\alpha = 60°$, $\beta = 60°$, $\tau = 45°$, $|\mathbf{a}| = |\mathbf{b}| = 2$, and $|\mathbf{c}| = 3$ (refer to Figure 4-F). With respect to the standard coordinate system, what are the coordinates of the basis vectors $\mathbf{v}_1$, $\mathbf{v}_2$, and $\mathbf{v}_3$ of this crystal grid system?

INNER PRODUCT SPACES | 4-6

The one concept defined for vectors in R^n which has not yet been generalized to arbitrary vector spaces is that of the dot product. If you associate the dot product with the concept of angle (two vectors are orthogonal if and only if their dot product is zero) or with distance ($|\mathbf{x}|^2 = \mathbf{x} \cdot \mathbf{x}$), it probably will not surprise you to learn that we cannot generalize the concept to arbitrary vector spaces. It probably will surprise you, in fact, to find that the dot product can be generalized as often as it can. We will see, for instance, that it will be possible to talk meaningfully about orthogonal polynomials.

For vectors $\mathbf{u}$ and $\mathbf{v}$ in an arbitrary vector space V, the dot product, if it exists, is designated not by $\mathbf{u} \cdot \mathbf{v}$, but by $\langle \mathbf{u}, \mathbf{v} \rangle$; and it is called an **inner product** rather than a dot product. An inner product on V, of which the dot product on R^n is an example, is a real number $\langle \mathbf{u}, \mathbf{v} \rangle$ defined for any $\mathbf{u}$, $\mathbf{v}$, and $\mathbf{w}$ in V in such a way as to satisfy the following properties:

1. $\langle \mathbf{u}, \mathbf{v} \rangle = \langle \mathbf{v}, \mathbf{u} \rangle$
2. $\langle \mathbf{u} + \mathbf{v}, \mathbf{w} \rangle = \langle \mathbf{u}, \mathbf{w} \rangle + \langle \mathbf{v}, \mathbf{w} \rangle$
3. $\langle r\mathbf{u}, \mathbf{v} \rangle = \langle \mathbf{u}, r\mathbf{v} \rangle = r\langle \mathbf{u}, \mathbf{v} \rangle$ for any scalar r.
4. $\langle \mathbf{u}, \mathbf{u} \rangle \geq 0$, equality holding if and only if $\mathbf{u} = \mathbf{0}$.

A vector space V having an inner product defined between its elements is called an **inner product space.** It is not surprising that R^n is an inner product space, but some surprises follow:

The vector space P_2 of polynomials of degree 2 or less is an inner product space having an inner product defined by

Example A

$$\langle a_2x^2 + a_1x + a_0, b_2x^2 + b_1x + b_0 \rangle = a_2b_2 + a_1b_1 + a_0b_0$$

Verification of the four properties proceeds in a straightforward fashion. Property 4, for instance, follows from the fact that

$$\langle a_2x^2 + a_1x + a_0, a_2x^2 + a_1x + a_0 \rangle = a_2^2 + a_1^2 + a_0^2 \geq 0$$

with equality possible only if $a_2 = a_1 = a_0 = 0$. □

Example B
C

The vector space $C[a, b]$ of all real-valued functions continuous on $[a, b]$ is an inner product space having an inner product defined by

$$\langle f, g \rangle = \int_a^b f(x)g(x)\, dx$$

Properties 1 through 4 are, in this case, statements about the properties of the integral. □

It should be noted that the space P_2 of Example A is a subspace of the vector space $C[a, b]$ of Example B, and that the inner product of Example B could therefore be used to obtain an inner product on P_2 different from the one used in Example A. Obviously then, there may be more than one way to define an inner product on the same vector space.

As is usual when properties are written down to characterize a concept, the four properties of an inner product which have been listed above have consequences that have not been listed. We now list two further properties that can be proved using Properties 1 through 4:

5. $\langle \mathbf{0}, \mathbf{v} \rangle = \langle \mathbf{v}, \mathbf{0} \rangle = 0$
6. $\langle \mathbf{w}, \mathbf{u} + \mathbf{v} \rangle = \langle \mathbf{w}, \mathbf{u} \rangle + \langle \mathbf{w}, \mathbf{v} \rangle$

The proof of Property 5 is left as an exercise (Problem 13). The proof of Property 6 follows quickly from Properties 1, 2, and 1 again.

$$\langle \mathbf{w}, \mathbf{u} + \mathbf{v} \rangle = \langle \mathbf{u} + \mathbf{v}, \mathbf{w} \rangle$$
$$= \langle \mathbf{u}, \mathbf{w} \rangle + \langle \mathbf{v}, \mathbf{w} \rangle$$
$$= \langle \mathbf{w}, \mathbf{u} \rangle + \langle \mathbf{w}, \mathbf{v} \rangle$$

In any inner product space, we immediately have a concept of length and a concept of distance. We define

$$\text{length of } \mathbf{u} = \|\mathbf{u}\| = \langle \mathbf{u}, \mathbf{u} \rangle^{1/2}$$
$$\text{distance from } \mathbf{u} \text{ to } \mathbf{v} = \|\mathbf{u} - \mathbf{v}\| = \langle \mathbf{u} - \mathbf{v}, \mathbf{u} - \mathbf{v} \rangle^{1/2}$$

C

It follows that Example B has given us a way to define the distance between two functions continuous on $[a, b]$.

$$\|f - g\| = \langle f - g, f - g \rangle^{1/2} = \left\{ \int_a^b [f(x) - g(x)]^2\, dx \right\}^{1/2}$$

Geometrically it means that we are using the square of the area between the graphs of $y = f(x)$ and $y = g(x)$ as a measure of "distance" between them (Figure 4-11).

There are certain ideas we naturally associate with length. The zero vector should have length 0; any other vector should have positive length. If a vector $\mathbf{v}$ is multiplied by a scalar r, its length should be $|r|$

times the original length; that is, $\|r\mathbf{v}\| = |r|\, \|\mathbf{v}\|$. All of these properties follow immediately from the definition, $\|\mathbf{v}\| = \langle \mathbf{v}, \mathbf{v} \rangle^{1/2}$

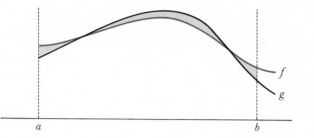

Figure 4-11

There is a third property commonly associated with length which says that for any two vectors $\mathbf{u}$ and $\mathbf{v}$,

$$\|\mathbf{u} + \mathbf{v}\| \le \|\mathbf{u}\| + \|\mathbf{v}\|$$

Called the **triangle inequality** (Figure 4-12), its proof depends on an important inequality that comes up in many different areas of mathe-

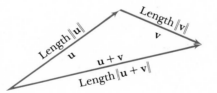

Figure 4-12

matics. It is variously attributed to Cauchy, Bunyakovsky, or Schwartz (CBS).

> *The* CBS *inequality: If* $\mathbf{u}$ *and* $\mathbf{v}$ *are vectors in an inner product space V,* *then*

$$\langle \mathbf{u}, \mathbf{v} \rangle \le \|\mathbf{u}\|\, \|\mathbf{v}\|$$

THEOREM A

In R^2 and R^3, where we commonly write $\mathbf{u} \bullet \mathbf{v}$ instead of $\langle \mathbf{u}, \mathbf{v} \rangle$, the CBS inequality is obvious. It follows from the fact that the angle θ between $\mathbf{u}$ and $\mathbf{v}$ can be found from

$$\mathbf{u} \bullet \mathbf{v} = |\mathbf{u}|\, |\mathbf{v}| \cos \theta$$

and from the fact that $|\cos \theta| \le 1$. The proof in a general inner product space is quite tricky, though it is not difficult to follow if someone shows you the tricks (see Problem 17).

Armed with the CBS inequality, we can easily establish the triangle inequality using Properties 1 through 4.

$$\|\mathbf{u} + \mathbf{v}\|^2 = \langle \mathbf{u} + \mathbf{v}, \mathbf{u} + \mathbf{v} \rangle \qquad \text{(Def.)}$$
$$= \langle \mathbf{u}, \mathbf{u} + \mathbf{v} \rangle + \langle \mathbf{v}, \mathbf{u} + \mathbf{v} \rangle \qquad (2)$$
$$= \langle \mathbf{u} + \mathbf{v}, \mathbf{u} \rangle + \langle \mathbf{u} + \mathbf{v}, \mathbf{v} \rangle \qquad (1)$$
$$= \langle \mathbf{u}, \mathbf{u} \rangle + \langle \mathbf{v}, \mathbf{u} \rangle + \langle \mathbf{u}, \mathbf{v} \rangle + \langle \mathbf{v}, \mathbf{v} \rangle \qquad (2)$$
$$= \langle \mathbf{u}, \mathbf{u} \rangle + \langle \mathbf{u}, \mathbf{v} \rangle + \langle \mathbf{u}, \mathbf{v} \rangle + \langle \mathbf{v}, \mathbf{v} \rangle \qquad (3)$$
$$\|\mathbf{u} + \mathbf{v}\|^2 = \|\mathbf{u}\|^2 + 2\langle \mathbf{u}, \mathbf{v} \rangle + \|\mathbf{v}\|^2 \qquad (1)$$

Now is the time to appeal to the CBS inequality to obtain

$$\langle \mathbf{u}, \mathbf{v} \rangle \leq \|\mathbf{u}\|\|\mathbf{v}\|$$

Thus,

$$\|\mathbf{u} + \mathbf{v}\|^2 \leq \|\mathbf{u}\|^2 + 2\|\mathbf{u}\|\|\mathbf{v}\| + \|\mathbf{v}\|^2 = (\|\mathbf{u}\| + \|\mathbf{v}\|)^2$$

Taking square roots gives

$$\|\mathbf{u} + \mathbf{v}\| \leq \|\mathbf{u}\| + \|\mathbf{v}\|$$

Carrying forward the terminology of R^n, let us agree that two vectors $\mathbf{u}$ and $\mathbf{v}$ in an inner product will be called **orthogonal** if $\langle \mathbf{u}, \mathbf{v} \rangle = 0$. Having just proved for any two vectors $\mathbf{u}$ and $\mathbf{v}$ that

$$\|\mathbf{u} + \mathbf{v}\|^2 = \|\mathbf{u}\|^2 + 2\langle \mathbf{u}, \mathbf{v} \rangle + \|\mathbf{v}\|^2$$

it follows that for two orthogonal vectors,

$$\|\mathbf{u} + \mathbf{v}\|^2 = \|\mathbf{u}\|^2 + \|\mathbf{v}\|^2$$

This is the inner product space version of the **Theorem of Pythagoras** (Figure 4-13).

Figure 4-13

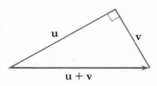

PROBLEM
SET 4-6

Problems 1 and 2 refer to the vector space of all ordered triples of real numbers. You are to decide which of the suggested operations $\mathbf{x} * \mathbf{y}$ on $\mathbf{x} = (x_1, x_2, x_3)$ and $\mathbf{y} = (y_1, y_2, y_3)$ define an inner product on the space.

1. (a) $\mathbf{x} * \mathbf{y} = x_1 y_1 + x_2 y_2$
   (b) $\mathbf{x} * \mathbf{y} = x_1 y_1 + 2x_2 y_2 + 3x_3 y_3$
   (c) $\mathbf{x} * \mathbf{y} = (x_1 y_1)^2 + (x_2 y_2)^2 + (x_3 y_3)^2$

2. (a) $\mathbf{x} * \mathbf{y} = (x_1 - y_1) + (x_2 - y_2) + (x_3 - y_3)$
   (b) $\mathbf{x} * \mathbf{y} = x_1 y_1 - x_2 y_2 + x_3 y_3$
   (c) $\mathbf{x} * \mathbf{y} = r_1 x_1 y_1 + r_2 x_2 y_2 + r_3 x_3 y_3$
      where r_1, r_2, and r_3 are positive numbers.

(Problem set continues on p. 182)

BOXED IN WITH THE
CBS INEQUALITY

Here is a problem typical of those found in a calculus text after a discussion of Lagrange multipliers and their use in minimizing a function of several variables, given some constraint on the variables. We shall show how the problem can be solved using only algebra, together with the CBS inequality.

The problem is to minimize the length of the diagonal of a rectangular box having edges of x_1, x_2, and x_3 (Figure 4-G). The constraint, quite appropriately in this case called a side-condition, is that the total length of all the sides, $L = 4(x_1 + x_2 + x_3)$, is to remain constant.

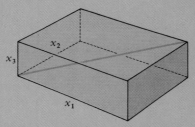

Figure 4-G

Begin by noting that if $\mathbf{x} = (x_1, x_2, x_3)$ and $\mathbf{u} = (1, 1, 1)$, then $L/4 = \mathbf{x} \cdot \mathbf{u} = x_1 + x_2 + x_3$ is constant. But by the CBS inequality,

$$\frac{L}{4} = x_1 + x_2 + x_3$$

$$= \mathbf{x} \cdot \mathbf{u} \leq \|\mathbf{x}\| \|\mathbf{u}\|$$

$$= \sqrt{x_1^2 + x_2^2 + x_3^2} \, \sqrt{3}$$

This says that no matter how the box is shaped,

$$\frac{L}{4\sqrt{3}} \leq \sqrt{x_1^2 + x_2^2 + x_3^2}$$

The length of the diagonal can never be shorter than the fixed number $L/4\sqrt{3}$. Now, can it ever be that short? Well, when do we have equality in the CBS inequality? We have equality (see Problem 20) if and only if the two vectors are linearly dependent, that is if and only if $(x_1, x_2, x_3) = r(1, 1, 1)$. This means that the required rectangle is a cube.

Problems 3 and 4 refer to the vector space P_2 of all polynomials of degree 2 or less. You are to decide which of the suggested operations $p(x) * q(x)$ on $p(x) = a_0 + a_1 x + a_2 x^2$ and $q(x) = b_0 + b_1 x + b_2 x^2$ define an inner product on P_2.

3. (a) $p(x) * q(x) = a_0 b_0 + a_1 b_1 + a_2 b_2$
   (b) $p(x) * q(x) = p(0)q(0)$

4. (a) $p(x) * q(x) = \int_0^1 |p(x)q(x)| \, dx$
   (b) $p(x) * q(x) = p(0)q(0) + p(1)q(1) + p(2)q(2)$

Problems 5 and 6 refer to the vector space of all 2×2 matrices. Decide which of the suggested operations $A * B$ on

$$A = \begin{bmatrix} a_1 & a_2 \\ a_3 & a_4 \end{bmatrix}, \qquad B = \begin{bmatrix} b_1 & b_2 \\ b_3 & b_4 \end{bmatrix}$$

define an inner product on the space.

5. (a) $A * B = a_1 b_1 + a_2 b_2 + a_3 b_3 + a_4 b_4$
   (b) $A * B = a_1 b_1 + a_4 b_4$

6. (a) $A * B = \det(AB)$
   (b) $A * B = \det(A^t B)$

Problems 7 and 8 refer to the vector space of all complex numbers. Decide which of the suggested operations $\mathbf{x} * \mathbf{y}$ on $\mathbf{x} = x_1 + ix_2$ and $\mathbf{y} = y_1 + iy_2$ define an inner product on the space.

7. (a) $\mathbf{x} * \mathbf{y} = x_1 y_1 + x_2 y_2$
   (b) $\mathbf{x} * \mathbf{y} = |x||y| = \sqrt{x_1^2 + x_2^2} \sqrt{y_1^2 + y_2^2}$

8. (a) $\mathbf{x} * \mathbf{y} = |xy|$
   (b) $\mathbf{x} * \mathbf{y} = (x_1 + ix_2)(y_1 - iy_2)$

C 9. In P_2, the space of polynomials of degree 2 or less, an inner product is defined by

$$\langle p(x), q(x) \rangle = \int_{-1}^1 p(x)q(x) \, dx$$

(a) Show that $p_1(x) = x$ and $p_2(x) = x^2 - \frac{1}{3}$ are orthogonal.
(b) Show that $p_2(x) = x^2 - \frac{1}{3}$ and $p_3(x) = x^3 - \frac{3}{5}x$ are orthogonal.

C 10. In C, the space of continuous functions, an inner product is defined by

$$\langle f(x), g(x) \rangle = \int_{-\pi}^{\pi} f(x)g(x) \, dx$$

(a) Show that $s_1(x) = \sin x$ and $s_2(x) = \sin 2x$ are orthogonal.
(b) Show that $s_2(x) = \sin 2x$ and $c_2(x) = \cos 2x$ are orthogonal.

C 11. Using the inner product defined in Problem 9, find the "distance" between $p(x) = x^4 - 2x^2 + 1$ and $q(x) = 1 - x^2$. Draw a graph.

C 12. Using the inner product defined in Problem 10, find the "distance" between $f(x) = \sin x$ and $g(x) = x^3 - x$. Draw a graph.

Problems 13–16 state properties that are true for any inner product space. Prove them, using Properties 1 through 4 of an inner product space and the definition of $\|\mathbf{u}\|$.

13. $\langle \mathbf{0}, \mathbf{v} \rangle = 0$ for any vector $\mathbf{v}$. *Hint:* Use Property 2 on $\langle \mathbf{0} + \mathbf{0}, \mathbf{v} \rangle$.

14. $\langle \mathbf{u} + \mathbf{v}, \mathbf{u} + \mathbf{v} \rangle = \|\mathbf{u}\|^2 + 2\langle \mathbf{u}, \mathbf{v} \rangle + \|\mathbf{v}\|^2$

15. $\langle \mathbf{u}, \mathbf{v} \rangle = \frac{1}{4}\|\mathbf{u} + \mathbf{v}\|^2 - \frac{1}{4}\|\mathbf{u} - \mathbf{v}\|^2$

16. $\|\mathbf{u} + \mathbf{v}\|^2 + \|\mathbf{u} - \mathbf{v}\|^2 = 2\|\mathbf{u}\|^2 + 2\|\mathbf{v}\|^2$ (This is called the **parallelogram law**. Draw a diagram that explains the name.)

17. Let $\mathbf{u}$ and $\mathbf{v}$ be any two vectors in an inner product space. Prove the CBS inequality as follows:
    (a) Using Properties 1 through 4 for an inner product, show that for any two scalars r and s,

    $$0 \le \langle r\mathbf{u} \pm s\mathbf{v}, r\mathbf{u} \pm s\mathbf{v} \rangle = r^2\langle \mathbf{u}, \mathbf{u} \rangle + s^2\langle \mathbf{v}, \mathbf{v} \rangle \pm 2rs\langle \mathbf{u}, \mathbf{v} \rangle$$

    (b) Show that the particular choice of $r = \|\mathbf{u}\|$ and $s = \|\mathbf{v}\|$ in part (a) leads to the conclusion that

    $$2\|\mathbf{u}\|\|\mathbf{v}\|\langle \mathbf{u}, \mathbf{v} \rangle \le 2\|\mathbf{u}\|^2\|\mathbf{v}\|^2$$

    (c) If $\mathbf{u} \ne \mathbf{0}$ and $\mathbf{v} \ne \mathbf{0}$, the last inequality establishes the CBS inequality. What about the case where either $\mathbf{u} = \mathbf{0}$ or $\mathbf{v} = \mathbf{0}$?

18. Refer to Problem 9. Verify the CBS inequality for the functions in part (a). C

19. Refer to Problem 9. Verify the CBS inequality for the functions in part (b). C

20. Prove that equality holds in the CBS inequality if and only if $\mathbf{u}$ and $\mathbf{v}$ are linearly dependent.

ORTHONORMAL SETS   4-7

Two nonzero vectors $\mathbf{u}$ and $\mathbf{v}$ in an inner product space V have been called orthogonal if $\langle \mathbf{u}, \mathbf{v} \rangle = 0$. A set $B = \{\mathbf{v}_1, \ldots, \mathbf{v}_k\}$ of nonzero vectors in V is said to be an **orthogonal set** if $\langle \mathbf{v}_i, \mathbf{v}_j \rangle = 0$ whenever $i \ne j$. In R^2 or R^3, an orthogonal set is a set of mutually perpendicular vectors. The vectors in an orthogonal set are always linearly independent; for if

$$r_1\mathbf{v}_1 + \cdots + r_j\mathbf{v}_j + \cdots + r_k\mathbf{v}_k = \mathbf{0}$$

then for any j,

$$r_1\langle \mathbf{v}_1, \mathbf{v}_j \rangle + \cdots + r_j\langle \mathbf{v}_j, \mathbf{v}_j \rangle + \cdots + r_k\langle \mathbf{v}_k, \mathbf{v}_j \rangle = \langle \mathbf{0}, \mathbf{v}_j \rangle$$

Since $\langle \mathbf{v}_i, \mathbf{v}_j \rangle = 0$ when $i \ne j$ and $\langle \mathbf{v}_j, \mathbf{v}_j \rangle = \|\mathbf{v}_j\|^2$, this equation reduces to $r_j\|\mathbf{v}_j\|^2 = 0$. Having required that vectors $\mathbf{v}_j$ in B be nonzero, it follows that $r_j = 0$. Since j was arbitrary, we conclude that $r_j = 0$ for all j.

Let W be the subspace of V spanned by the vectors of an orthogonal set B. Since the vectors of B are linearly independent, it is clear that B is a basis for W. Any orthogonal set in V is, therefore, a basis of a subspace W of V. It is called an **orthogonal basis** of W.

DEFINITION

A basis $B = \{v_1, \ldots, v_k\}$ of a subspace W of V is said to be **orthogonal** if $\langle v_i, v_j \rangle = 0$ whenever $i \neq j$. If, in addition, each vector has length 1; that is, $\langle v_i, v_i \rangle = 1$ for each i, then the basis is said to be **orthonormal**.

If a vector w is in the space for which $B = \{v_1, \ldots, v_k\}$ is a basis, then w can be uniquely represented in the form

(1)
$$w = r_1 v_1 + r_2 v_2 + \cdots + r_k v_k$$

That's easy to see; it follows from the definition of a basis. Generally speaking, it is not so easy to determine the actual values of the r_i. That requires solving a system of k equations having k variables. But now, *when our basis is orthonormal, it is also easy to determine the values of the r_i.* Consider the inner product of w as given in (1) with v_1. We get

(2)
$$\langle w, v_1 \rangle = r_1 \langle v_1, v_1 \rangle + r_2 \langle v_2, v_1 \rangle + \cdots + r_k \langle v_k, v_1 \rangle$$

From the definition of an orthonormal set, $\langle v_1, v_1 \rangle = 1$ and $\langle v_j, v_1 \rangle = 0$ for each $j \neq 1$. We are left with $\langle w, v_1 \rangle = r_1$. The same trick can be used to determine all the scalars, leading to the conclusion that in (1),

$$r_1 = \langle w, v_1 \rangle, \qquad r_2 = \langle w, v_2 \rangle, \ldots, r_k = \langle w, v_k \rangle$$

Example A

In our discussion of rotation of axes in Section 4-5, we noted that a basis for R^2 is given by

$$v_1 = (\cos \theta, \sin \theta)$$
$$v_2 = (-\sin \theta, \cos \theta)$$

This basis is orthonormal for any choice of θ. Set $\theta = \dfrac{\pi}{6}$ and express $w = (3, 4)$ in the form $w = r_1 v_1 + r_2 v_2$.

It is geometrically evident from the way they were obtained (rotating $(1, 0)$ and $(0, 1)$ each through the angle θ) that $\{v_1, v_2\}$ would be orthonormal. It is also evident directly from the definition since

$$\langle v_1, v_2 \rangle = -\sin \theta \cos \theta + \sin \theta \cos \theta = 0$$
$$\langle v_1, v_1 \rangle = \langle v_2, v_2 \rangle = \sin^2 \theta + \cos^2 \theta = 1$$

When $\theta = \dfrac{\pi}{6}$ and $w = (3, 4)$,

$$r_1 = \langle w, v_1 \rangle = (3, 4) \cdot \left(\frac{\sqrt{3}}{2}, \frac{1}{2} \right) = \frac{3\sqrt{3} + 4}{2} \approx 4.60$$

$$r_2 = \langle w, v_2 \rangle = (3, 4) \cdot \left(-\frac{1}{2}, \frac{\sqrt{3}}{2} \right) = \frac{-3 + 4\sqrt{3}}{2} \approx 1.96$$

Vectors v_1, v_2, and w are shown in Figure 4-14. □

⊂ A STEP TOWARD FOURIER SERIES

It is not difficult to show that in C, the space of continuous functions, the set

$$B = \left\{ f_0(x) = \frac{1}{\sqrt{2\pi}}, f_1(x) = \frac{\sin x}{\sqrt{\pi}}, f_2(x) = \frac{\cos x}{\sqrt{\pi}}, \right.$$
$$\left. f_3(x) = \frac{\sin 2x}{\sqrt{\pi}}, f_4(x) = \frac{\cos 2x}{\sqrt{\pi}} \right\}$$

is an orthonormal set when the inner product is defined by

$$\langle f(x), g(x) \rangle = \int_{-\pi}^{\pi} f(x)g(x)\, dx$$

Suppose we look for the vector in the subspace spanned by B that is closest to the function $g(x) = |x|$. Stripped of the language of vector spaces, this same question asks how to choose the coefficients $r_0, \ldots, r_4$ in

$$F(x) = r_0 \frac{1}{\sqrt{2\pi}} + r_1 \frac{\sin x}{\sqrt{\pi}} + r_2 \frac{\cos x}{\sqrt{\pi}} + r_3 \frac{\sin 2x}{\sqrt{\pi}} + r_4 \frac{\cos 2x}{\sqrt{\pi}}$$

so as to minimize $\left[\int_{-\pi}^{\pi} [g(x) - F(x)]^2 \right]^{1/2}$

This is a problem in approximation. It is much the same idea as one has when trying to find the Taylor polynomial that approximates a given function. The approximation of $g(x)$ with a Taylor polynomial depends, you may recall, on the function g being differentiable at the point x_0 around which the expansion is developed. The method outlined above does not require differentiability; indeed, the function $g(x) = |x|$ is not differentiable at $x = 0$. The above method can also be used to get approximations to *sawtooth* functions (Figure 4-H) and even to *discontinuous* functions (Figure 4-I).

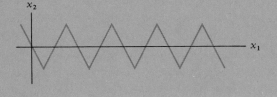

x_2

x_1

Figure 4-H

Figure 4-I

Coming back to the example at hand and the theory developed in this section, we can find the coefficients from $r_i = \langle g(x), f_i(x) \rangle$. For instance,

$$r_2 = \langle g(x), f_2(x) \rangle = \frac{1}{\sqrt{\pi}} \int_{-\pi}^{\pi} |x| \cos x \, dx = \frac{2}{\sqrt{\pi}} \int_0^{\pi} x \cos x \, dx$$

$$= \frac{2}{\sqrt{\pi}} (\cos x + x \sin x) \Big|_0^{\pi} = -\frac{4}{\sqrt{\pi}}$$

Similar computations give $r_0 = \dfrac{\pi^2}{\sqrt{2\pi}}$ and $r_1 = r_3 = r_5 = 0$.

The best approximation is given by

$$F(x) = \frac{\pi^2}{\sqrt{2\pi}} \frac{1}{\sqrt{2\pi}} - \frac{4}{\sqrt{\pi}} \frac{\cos x}{\sqrt{\pi}} = \frac{\pi}{2} - \frac{4}{\pi} \cos x$$

The graphs of $g(x) = |x|$ and $F(x)$ are drawn on the same axes in Figure 4-J.

This same process, carried out for an infinite orthonormal basis of which B is a finite subset, leads to the Fourier series corresponding to $g(x) = |x|$.

Figure 4-J

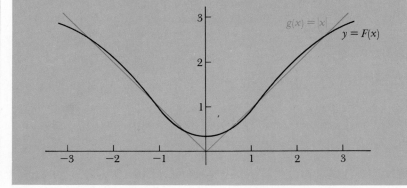

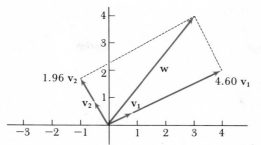

Figure 4-14

There is another advantage to having an orthonormal basis for a subspace W of an inner product space V. In order to have a picture in mind, let us consider a 2-dimensional subspace (a plane through the origin) in R^3. Suppose that $B = \{v_1, v_2\}$ is an orthonormal basis of W, that u, drawn with its initial point at the origin, is a vector of R^3 which is not in W, and that we wish to know the shortest distance from the terminal point of u to W (Figure 4-15).

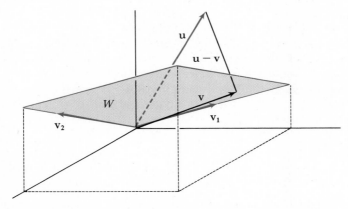

Figure 4-15

Numerous methods are available to us so long as we work in R^3, but we wish to solve the problem in a way that can be used for any vector space V and any proper subspace W. We therefore rephrase our problem slightly. We seek that vector

$$v = r_1 v_1 + r_2 v_2$$

for which $\|u - v\|$ is a minimum, or, what amounts to the same thing but relieves us of working with square roots, for which $\|u - v\|^2$ is a minimum. Now the properties of the inner product can be used together with the facts that $\langle v_1, v_1 \rangle = \langle v_2, v_2 \rangle = 1$ and $\langle v_1, v_2 \rangle = 0$ to show that

$$\|u - v\|^2 = \langle u - r_1 v_1 - r_2 v_2, u - r_1 v_1 - r_2 v_2 \rangle$$
$$= \langle u, u \rangle + r_1^2 + r_2^2 - 2r_1 \langle u, v_1 \rangle - 2r_2 \langle u, v_2 \rangle$$

If we complete the square by adding and subtracting the terms $\langle u, v_1 \rangle^2$ and $\langle u, v_2 \rangle^2$ to this expression,

$$\|u - v\|^2 = \|u\|^2 + [r_1 - \langle u, v_1 \rangle]^2 + [r_2 - \langle u, v_2 \rangle]^2$$
$$- \langle u, v_1 \rangle^2 - \langle u, v_2 \rangle^2$$

Our goal, remember, is to choose $v = r_1 v_1 + r_2 v_2$ so as to minimize $\|u - v\|^2$. Since $\|u - v\|^2 \geq 0$, this expression will clearly be minimized by choosing $r_1 = \langle u, v_1 \rangle$ and $r_2 = \langle u, v_2 \rangle$. That is, choose

(3)
$$v = \langle u, v_1 \rangle v_1 + \langle u, v_2 \rangle v_2$$

In R^3 where we have a picture (Figure 4-15) to guide us, it is geometrically evident that we should choose v so that $u - v$ is perpendicular to the plane that represents W. A quick check shows that

$$\langle u - v, v_1 \rangle = \langle u, v_1 \rangle - \langle v, v_1 \rangle$$
$$= \langle u, v_1 \rangle - r_1 \langle v_1, v_1 \rangle - r_2 \langle v_1, v_2 \rangle$$
$$= \langle u, v_1 \rangle - r_1(1) - r_2(0) = 0$$

because $r_1 = \langle u, v_1 \rangle$. Similarly, $\langle u - v, v_2 \rangle = 0$, confirming that $u - v$ is orthogonal to the plane.

It is natural enough in R^3 to call the vector v defined in (3) the projection of u onto W, and as is our established custom, we carry the terminology over to a general inner product space V. If $B = \{v_1, \ldots, v_k\}$ is an orthonormal set in V that spans the subset W, and if u is a vector in V, then the **projection of u onto W**, written $\text{proj}_W u$, is defined to be

(4)
$$\text{proj}_W u = \langle u, v_1 \rangle v_1 + \cdots + \langle u, v_k \rangle v_k$$

The vector $u - \text{proj}_W u$ is, as in the 2-dimensional case just illustrated, orthogonal to every vector in W, or more succinctly, $u - \text{proj}_W u$ is orthogonal to W.

Our definition of the projection of u does not require that u not be in W, but if u is in W, then (4) reduces to (2) and we see that $u = \text{proj}_W u$ (as geometric intuition tells us that it should).

The results we have proved can be summarized as follows:

THEOREM A

Let $B = \{v_1, \ldots, v_k\}$ be an orthonormal set in the inner product space V. Then

(a) The set B is linearly independent and thus forms a basis for the subspace W which it spans.

(b) The projection of vector u onto W is defined to be

$$\text{proj}_W u = \langle u, v_1 \rangle v_1 + \cdots + \langle u, v_k \rangle v_k$$

(c) If u is in W, $w = \text{proj}_W u$ and the expression in (b) is simply the unique representation of u as a linear combination of the basis vectors.

(d) If $\mathbf{u}$ is not in W, $\mathbf{u} - \mathrm{proj}_W\mathbf{u}$ is orthogonal to W and $\|\mathbf{u} - \mathrm{proj}_W\mathbf{u}\|$ is the distance from $\mathbf{u}$ to W.

The set

Example B

$$B = \left\{ \mathbf{v}_1 = \left(\frac{1}{3\sqrt{2}}, -\frac{1}{3\sqrt{2}}, \frac{4}{3\sqrt{2}}\right), \mathbf{v}_2 = \left(\frac{1}{\sqrt{2}}, \frac{1}{\sqrt{2}}, 0\right) \right\}$$

is an orthonormal set in R^3. Find the distance from the point $P(1, 4, 3)$ to the plane W spanned by B.

We begin by finding the projection of $\overrightarrow{OP} = (1, 4, 3)$ onto W. Since $\langle \overrightarrow{OP}, \mathbf{v}_1 \rangle = 9/3\sqrt{2}$ and $\langle \overrightarrow{OP}, \mathbf{v}_2 \rangle = 5/\sqrt{2}$,

$$\mathrm{proj}_W \overrightarrow{OP} = \frac{3}{\sqrt{2}}\left(\frac{1}{3\sqrt{2}}, -\frac{1}{3\sqrt{2}}, \frac{4}{3\sqrt{2}}\right) + \frac{5}{\sqrt{2}}\left(\frac{1}{\sqrt{2}}, \frac{1}{\sqrt{2}}, 0\right) = (3, 2, 2)$$

$$\overrightarrow{OP} - \mathrm{proj}_W \overrightarrow{OP} = (1, 4, 3) - (3, 2, 2) = (-2, 2, 1)$$

The required distance is, therefore $\|\overrightarrow{OP} - \mathrm{proj}_W \overrightarrow{OP}\| = \sqrt{4 + 4 + 1} = 3$. For a check on our work, we can verify that

$$\langle \overrightarrow{OP} - \mathrm{proj}_W \overrightarrow{OP}, \mathbf{v}_1 \rangle = \langle \overrightarrow{OP} - \mathrm{proj}_W \overrightarrow{OP}, \mathbf{v}_2 \rangle = 0 \quad \square$$

THE GRAM-SCHMIDT PROCESS

It is clear that there are advantages to having an orthonormal basis to a subspace. This raises an obvious question. Suppose that $B = \{\mathbf{u}_1, \ldots, \mathbf{u}_k\}$ is an arbitrary basis for a subspace W of an inner product space V. Can we use the basis vectors of B to get a second basis of W that is orthonormal? The answer is yes. The method used to find the second basis is called the **Gram-Schmidt orthogonalization process,** and it is to a description of this process that we now turn.

If $k = 1$, the problem is easy since all we need to do is **normalize** the single vector $\mathbf{u}_1$; that is, we need to find a unit vector that spans the same subspace as does $\mathbf{u}_1$. Quite obviously

$$\mathbf{v}_1 = \frac{1}{\|\mathbf{u}_1\|}\mathbf{u}_1$$

will do the job.

When $k > 1$, we get to make use of what we know about projections. Let W_1 be the subspace spanned by $\mathbf{v}_1$. Then form $\mathbf{w}_2 = \mathbf{u}_2 - \mathrm{proj}_{W_1}\mathbf{u}_2$, a vector that we know will be orthogonal to every vector in W_1 (which certainly includes $\mathbf{v}_1$). Since we want an orthonormal set, let $\mathbf{v}_2 = \frac{\mathbf{w}_2}{\|\mathbf{w}_2\|}$.

Thus,

$$\mathbf{v}_2 = \frac{\mathbf{w}_2}{\|\mathbf{w}_2\|} \quad \text{where} \quad \mathbf{w}_2 = \mathbf{u}_2 - \langle \mathbf{u}_2, \mathbf{v}_1 \rangle \mathbf{v}_1$$

This process can now be repeated endlessly. We will run through it once more. Let W_2 be the subspace spanned by $\mathbf{v}_1$ and $\mathbf{v}_2$. Then form $\mathbf{w}_3 = \mathbf{u}_3 - \text{proj}_{W_2}\mathbf{u}_3$, a vector that will be orthogonal to W_2, so certainly to $\mathbf{v}_1$ and $\mathbf{v}_2$. Finally, set $\mathbf{v}_3 = \mathbf{w}_3/\|\mathbf{w}_3\|$. Thus,

$$\mathbf{v}_3 = \frac{\mathbf{w}_3}{\|\mathbf{w}_3\|} \quad \text{where} \quad \mathbf{w}_3 = \mathbf{u}_3 - \langle \mathbf{u}_3, \mathbf{v}_1 \rangle \mathbf{v}_1 - \langle \mathbf{u}_3, \mathbf{v}_2 \rangle \mathbf{v}_2$$

Example C   The subspace W of R^5 is spanned by

$$\mathbf{u}_1 = (1, 0, -1, 1, 1)$$
$$\mathbf{u}_2 = (0, 1, -1, 1, 2)$$
$$\mathbf{u}_3 = (0, -1, 2, 0, 4)$$

Find an orthonormal basis for W.

Following the Gram-Schmidt orthogonalization process, we begin by setting

$$\mathbf{v}_1 = \frac{\mathbf{u}_1}{\|\mathbf{u}_1\|} = (\tfrac{1}{2}, 0, -\tfrac{1}{2}, \tfrac{1}{2}, \tfrac{1}{2})$$

Then

$$\mathbf{w}_2 = \mathbf{u}_2 - \langle \mathbf{u}_2, \mathbf{v}_1 \rangle \mathbf{v}_1$$
$$\mathbf{w}_2 = (0, 1, -1, 1, 2) - 2(\tfrac{1}{2}, 0, -\tfrac{1}{2}, \tfrac{1}{2}, \tfrac{1}{2}) = (-1, 1, 0, 0, 1)$$
$$\mathbf{v}_2 = \frac{\mathbf{w}_2}{\|\mathbf{w}_2\|} = \left(-\frac{1}{\sqrt{3}}, \frac{1}{\sqrt{3}}, 0, 0, \frac{1}{\sqrt{3}} \right)$$

Finally,

$$\mathbf{w}_3 = \mathbf{u}_3 - \langle \mathbf{u}_3, \mathbf{v}_1 \rangle \mathbf{v}_1 - \langle \mathbf{u}_3, \mathbf{v}_2 \rangle \mathbf{v}_2$$
$$= (0, -1, 2, 0, 4) - 1\left(\frac{1}{2}, 0, -\frac{1}{2}, \frac{1}{2}, \frac{1}{2} \right) - \sqrt{3}\left(-\frac{1}{\sqrt{3}}, \frac{1}{\sqrt{3}}, 0, 0, \frac{1}{\sqrt{3}} \right)$$
$$= (\tfrac{1}{2}, -2, \tfrac{5}{2}, -\tfrac{1}{2}, \tfrac{5}{2})$$
$$\mathbf{v}_3 = \frac{\mathbf{w}_3}{\|\mathbf{w}_3\|} = \frac{1}{2\sqrt{17}}(1, -4, 5, -1, 5)$$

You should satisfy yourself that $\{\mathbf{v}_1, \mathbf{v}_2, \mathbf{v}_3\}$ really is an orthogonal set by verifying that $\langle \mathbf{v}_1, \mathbf{v}_2 \rangle = \langle \mathbf{v}_1, \mathbf{v}_3 \rangle = \langle \mathbf{v}_2, \mathbf{v}_3 \rangle = 0.$ $\square$

PROBLEM
SET 4-7   *Two orthonormal bases of R^2 are*

$$B = \left\{ \mathbf{u}_1 = \left(\frac{1}{\sqrt{5}}, \frac{2}{\sqrt{5}} \right), \mathbf{u}_2 = \left(-\frac{2}{\sqrt{5}}, \frac{1}{\sqrt{5}} \right) \right\}$$
$$C = \left\{ \mathbf{v}_1 = \left(-\frac{3}{5}, \frac{4}{5} \right), \mathbf{v}_2 = \left(-\frac{4}{5}, -\frac{3}{5} \right) \right\}$$

In Problems 1–4, you are given a vector **w**. Find the coordinates of **w** with respect to each basis and draw a diagram showing the two bases and **w**. Use Theorem A of this section.

1. $\mathbf{w} = (5, -3)$ $\qquad\qquad\qquad$ 2. $\mathbf{w} = (-6, 5)$

3. $\mathbf{w} = (-7, 4)$ $\qquad\qquad\qquad$ 4. $\mathbf{w} = (3, -5)$

In Problems 5–8, find the coordinates of the given vector with respect to the basis of R^3 given by

$$\mathbf{u}_1 = \left(\frac{2}{3}, \frac{2}{3}, -\frac{1}{3}\right), \qquad \mathbf{u}_2 = \left(\frac{1}{\sqrt{2}}, -\frac{1}{\sqrt{2}}, 0\right), \qquad \mathbf{u}_3 = \left(\frac{1}{3\sqrt{2}}, \frac{1}{3\sqrt{2}}, \frac{4}{3\sqrt{2}}\right)$$

5. $\mathbf{w} = (-1, 2, 5)$ $\qquad\qquad\qquad$ 6. $\mathbf{w} = (4, 2, -3)$

7. $\mathbf{w} = (5, -2, 3)$ $\qquad\qquad\qquad$ 8. $\mathbf{w} = (-1, 2, -3)$

In Problems 9–14 you are given a set B of vectors which span a subspace W of one of the inner product spaces we have studied, together with a vector **u** in the same inner product space. In each problem,

(a) Show that the set B is orthonormal.
(b) Find $\text{proj}_W \mathbf{u}$.
(c) Verify that $\mathbf{u} - \text{proj}_W \mathbf{u}$ is orthogonal to each vector in B.

9. Let V be the inner product space of 2×2 matrices in which the inner product of

$$\mathbf{x} = \begin{bmatrix} x_1 & x_2 \\ x_3 & x_4 \end{bmatrix}, \qquad \mathbf{y} = \begin{bmatrix} y_1 & y_2 \\ y_3 & y_4 \end{bmatrix}$$

is

$$\langle \mathbf{x}, \mathbf{y} \rangle = x_1 y_1 + x_2 y_2 + x_3 y_3 + x_4 y_4$$

$$B = \left\{ \mathbf{u}_1 = \begin{bmatrix} \frac{1}{\sqrt{2}} & 0 \\ 0 & -\frac{1}{\sqrt{2}} \end{bmatrix}, \mathbf{u}_2 = \begin{bmatrix} \frac{1}{\sqrt{6}} & 0 \\ \frac{2}{\sqrt{6}} & \frac{1}{\sqrt{6}} \end{bmatrix}, \mathbf{u}_3 = \begin{bmatrix} \frac{1}{\sqrt{3}} & 0 \\ -\frac{1}{\sqrt{3}} & \frac{1}{\sqrt{3}} \end{bmatrix} \right\}$$

$$\mathbf{u} = \begin{bmatrix} 2 & 3 \\ 0 & -1 \end{bmatrix}$$

10. Use the inner product space of Problem 9.

$$B = \left\{ \mathbf{u}_1 = \begin{bmatrix} \frac{2}{\sqrt{15}} & \frac{1}{\sqrt{15}} \\ \frac{3}{\sqrt{15}} & -\frac{1}{\sqrt{15}} \end{bmatrix}, \mathbf{u}_2 = \begin{bmatrix} -\frac{2}{\sqrt{10}} & -\frac{1}{\sqrt{10}} \\ \frac{2}{\sqrt{10}} & \frac{1}{\sqrt{10}} \end{bmatrix}, \mathbf{u}_3 = \begin{bmatrix} 0 & \frac{1}{\sqrt{2}} \\ 0 & \frac{1}{\sqrt{2}} \end{bmatrix} \right\}$$

$$\mathbf{u} = \begin{bmatrix} -1 & 2 \\ 1 & 0 \end{bmatrix}$$

[C] 11. Let P_1 be the vector space of all polynomials in which the inner product of $p(x)$ and $q(x)$ is

$$\langle p(x), q(x)\rangle = \int_{-1}^{1} p(x)\, q(x)\, dx$$

$$B = \left\{ u_0(x) = \frac{1}{\sqrt{2}}, u_1(x) = \frac{\sqrt{3}}{\sqrt{2}} x, u_2(x) = \frac{3\sqrt{5}}{2\sqrt{2}}\left(x^2 - \frac{1}{3}\right) \right\}$$

$$u(x) = x^3 + x$$

The polynomials in set B are called Legendre polynomials.

[C] 12. Let P_2 be the vector space of all polynomials in which the inner product of $p(x)$ and $q(x)$ is

$$\langle p(x), q(x)\rangle = \int_{0}^{1} p(x)q(x)\, dx$$

$$B = \{u_0(x) = 1,\ u_1(x) = \sqrt{3}(2x - 1),\ u_2(x) = \sqrt{5}(6x^2 - 6x + 1)\}$$

$$u(x) = x^3 + x$$

[C] 13. In the space C of all continuous functions, define the inner product by

$$\langle f(x), g(x)\rangle = \int_{0}^{\pi} f(x)g(x)\, dx$$

$$B = \left\{ u_0(x) = \frac{1}{\sqrt{\pi}}, u_1(x) = \sqrt{\frac{2}{\pi}} \cos x, u_2(x) = \sqrt{\frac{2}{\pi}} \cos 2x \right\}$$

$$u(x) = \tfrac{1}{2}$$

[C] 14. Use the inner product space of Problem 13.

$$B = \left\{ u_0(x) = \sqrt{\frac{2}{\pi}} \sin x, u_1(x) = \sqrt{\frac{2}{\pi}} \sin 2x, u_2(x) = \sqrt{\frac{2}{\pi}} \sin 3x \right\}$$

$$u(x) = \tfrac{1}{2}$$

In 15–18, find an orthonormal basis for the subspaces of R^4 having the given basis.

15. $\mathbf{u}_1 = (1, 2, 0, -2)$
    $\mathbf{u}_2 = (2, 0, -1, 2)$

16. $\mathbf{u}_1 = (2, 1, 0, -1)$
    $\mathbf{u}_2 = (1, 0, -2, 1)$

17. $\mathbf{u}_1 = (0, 3, 0, 3)$
    $\mathbf{u}_2 = (1, 1, 0, 0)$
    $\mathbf{u}_3 = (3, 1, 2, 2)$

18. $\mathbf{u}_1 = (1, -1, 0, 1)$
    $\mathbf{u}_2 = (-1, 0, 1, -1)$
    $\mathbf{u}_3 = (1, 3, -1, 1)$

19. Find an orthogonal basis for R^4 that includes the vectors

$$\mathbf{v}_1 = (1, 1, -1, 1), \quad \mathbf{v}_2 = (1, 0, 2, 1), \quad \mathbf{v}_3 = (-1, 0, 0, 1)$$

20. Find an orthogonal basis for R^4 that includes the vectors

$$\mathbf{v}_1 = (2, 0, 1, 1), \quad \mathbf{v}_2 = (1, 3, -1, -1), \quad \mathbf{v}_3 = (1, -1, 0, -2)$$

21. Sometimes calculations in the Gram-Schmidt process can be relieved by first finding an orthogonal basis for the given subspace, and then normaliz-

ing all vectors as the final step. Prove that given a basis $\{\mathbf{u}_1, \mathbf{u}_2, \mathbf{u}_3\}$, we can find an orthogonal basis $\{\mathbf{v}_1, \mathbf{v}_2, \mathbf{v}_3\}$ from the formulas

$$\mathbf{v}_1 = \mathbf{u}_1$$

$$\mathbf{v}_2 = \mathbf{u}_2 - \frac{\langle \mathbf{v}_1, \mathbf{u}_2 \rangle}{\langle \mathbf{v}_1, \mathbf{v}_1 \rangle} \mathbf{v}_1$$

$$\mathbf{v}_3 = \mathbf{u}_3 - \frac{\langle \mathbf{v}_1, \mathbf{u}_3 \rangle}{\langle \mathbf{v}_1, \mathbf{v}_1 \rangle} \mathbf{v}_1 - \frac{\langle \mathbf{v}_2, \mathbf{u}_3 \rangle}{\langle \mathbf{v}_2, \mathbf{v}_2 \rangle} \mathbf{v}_2$$

22. Use the formulas given in Problem 21 to find an orthogonal set, then an orthonormal set, in R^4 spanning the subspace spanned by

$$\mathbf{u}_1 = (2, 1, 0, 0), \qquad \mathbf{u}_2 = (3, 1, 3, -1), \qquad \mathbf{u}_3 = (1, 3, -3, 1)$$

23. Suppose that B is an orthonormal basis of a proper subspace W in an inner product space V. Show that every vector $\mathbf{v}$ in V can be uniquely expressed in the form $\mathbf{v} = \mathbf{w} + \mathbf{n}$ where $\mathbf{w}$ is in W and $\mathbf{n}$ is orthogonal to W. *Hint:* Begin by setting $\mathbf{n} = \mathbf{v} - \text{proj}_W \mathbf{v}$.

1. Explain why the following sets do not form vector spaces:
   (a) The set of all 2×2 nonsingular matrices, addition defined as usual for matrices.
   (b) The set of infinite sequences of non-negative entries, addition defined by

   $$(a_1, a_2, \ldots) + (b_1, b_2, \ldots) = (a_1 + b_1, a_2 + b_2, \ldots)$$

   (c) The set of all second degree polynomials.
   (d) The ordered triples (x_1, x_2, x_3) for which $x_1 + x_2 + x_3 = 1$.

2. Prove that in the vector space C' of functions that are everywhere differentiable, $f_1(x) = e^x$, $f_2(x) = e^{2x}$, and $f_3(x) = e^{3x}$ are linearly independent.

3. Give your reason for marking the following statements true or false.
   (a) A subset of a linearly independent set must be linearly independent.
   (b) A subset of a linearly dependent set must be linearly dependent.
   (c) The span of a set of k elements must be a subspace of dimension k or more.
   (d) The projection of a vector $\mathbf{w}$ in an inner product space V onto a subspace S spanned by the orthonormal set $S = \{\mathbf{v}_1, \ldots, \mathbf{v}_k\}$ will be orthogonal to each $\mathbf{v}_i$ in S.

4. In the vector space C of functions everywhere continuous, find the dimension of the space spanned by

$$s_1(x) = \sin x \qquad s_2(x) = \cos x \qquad s_3(x) = \sin^2 x$$
$$s_4(x) = \cos^2 x \qquad s_5(x) = \sin 2x \qquad s_6(x) = \cos 2x$$

5. Show that a vector space $\mathcal{P}$ can be formed by 2×2 matrices in which entries in the matrix are polynomials of degree 2 or less, and the addition and scalar multiplication of matrices by real numbers proceeds as usual.

6. In the space $\mathcal{P}$ defined in Problem 5, show that

$$\mathbf{v}_1 = \begin{bmatrix} 1 & 1 \\ 1 & 1 \end{bmatrix}, \qquad \mathbf{v}_2 = \begin{bmatrix} x & x \\ x & x \end{bmatrix}, \qquad \mathbf{v}_3 = \begin{bmatrix} x^2 & x^2 \\ x^2 & x^2 \end{bmatrix}$$

are linearly independent. Do they form a basis for $\mathcal{P}$? What is the dimension of the vector space $\mathcal{P}$?

7. Consider only those matrices in $\mathcal{P}$ for which all four polynomials have a root at $x = 0$. Do these matrices form a subspace? If so, what is the dimension of the subspace?

8. In the space $\mathcal{P}$, decide whether the following four matrices are linearly independent.

$$\mathbf{u}_1 = \begin{bmatrix} x^2 & x^2 + 1 \\ 0 & -x \end{bmatrix}, \qquad\qquad \mathbf{u}_2 = \begin{bmatrix} x^2 - 1 & 3 \\ 3x & 0 \end{bmatrix},$$

$$\mathbf{u}_3 = \begin{bmatrix} 4x^2 + 2 & 6x^2 + x \\ -x^2 - 6x & -1 \end{bmatrix}, \qquad \mathbf{u}_4 = \begin{bmatrix} 0 & -x \\ x^2 & -6x + 1 \end{bmatrix}$$

9. Suppose that $\mathbf{v}_1$ and $\mathbf{v}_2$ are linearly independent vectors in a vector space V that has a basis $\{\mathbf{u}_1, \mathbf{u}_2, \mathbf{u}_3\}$. Prove that you can choose some i so that $\{\mathbf{v}_1, \mathbf{v}_2, \mathbf{u}_i\}$ forms a basis.

10. The set $B_1 = \{\mathbf{u}_1 = (1, 2), \mathbf{u}_2 = (-1, 1)\}$ forms a basis for R^2, as does the set $B_2 = \{\mathbf{v}_1, \mathbf{v}_2\}$ where

$$\mathbf{v}_1 = \mathbf{u}_1 + \mathbf{u}_2 = (0, 3), \qquad \mathbf{v}_2 = \mathbf{u}_1 - \mathbf{u}_2 = (2, 1)$$

Write down (no calculations are necessary) the transition matrix P that relates the coordinates $(x_1, x_2)_{B_1}$ and $(x'_1, x'_2)_{B_2}$ of an arbitrary point by

$$\begin{bmatrix} x_1 \\ x_2 \end{bmatrix} = P \begin{bmatrix} x'_1 \\ x'_2 \end{bmatrix}$$

Given that $\mathbf{w} = (3, -4)_{B_2}$, find $\mathbf{w}_{B_1}$. How can you check your answer (besides looking in the back of the book)?

LINEAR TRANSFORMATIONS 5

A function *f* may be likened to a cannon on a target range, except that instead of shooting shells, it propels points or vectors from one set *D* into another set, say *E* (Figure 5-1). The set from which the ammunition can be drawn is called the **domain** of *f*. If **x** is a member of the domain, it is always "fired" into a unique point $f(\mathbf{x})$ that is called the **image** of **x** under *f*. Since it is possible for different members of the domain to be fired onto the same target point, the **preimage** of **y** in *E*, defined to be the set of all points **x** such that $f(\mathbf{x}) = \mathbf{y}$, may be empty, contain just one point, or contain many points. Those members of *E* that have a nonempty preimage form the **range** of *f*. Thus, **y** is in the range of *f* if there exists at least one **x** in *D* such that $f(\mathbf{x}) = \mathbf{y}$.

The functions most familiar from elementary mathematics have domains and ranges that are sets of real numbers, and the destination of a number *x* in the domain is usually described by a formula.

$$R(x) = \sqrt{1 - x} \quad \text{has domain } (-\infty, 1], \text{ range } [0, \infty) \tag{1}$$
$$S(x) = \sin x \quad \text{has domain } (-\infty, \infty), \text{ range } [-1, 1]$$
$$T(x) = 2x \quad \text{has domain } (-\infty, \infty), \text{ range } (-\infty, \infty)$$

Figure 5-1

x •

f

• $f(x)$

D

E

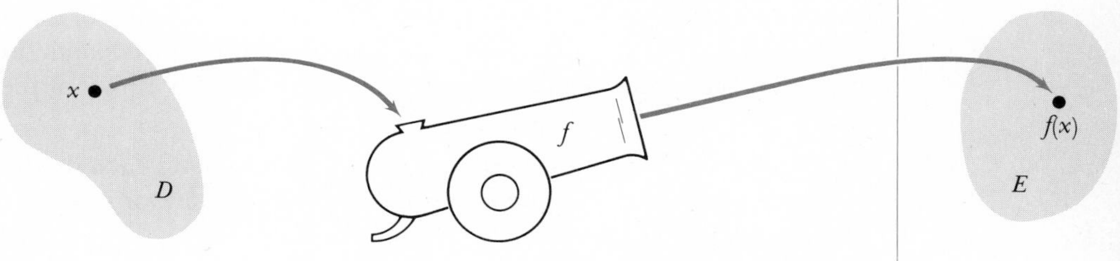

It is also likely that a student who has studied calculus will have at least briefly considered real-valued functions of several variables. We may use the vocabulary of Chapter 4 in describing the domain of such functions. The domain of $G(x_1, x_2) = \sqrt{4 - (x_1^2 + x_2^2)}$ is a subset of the vector space R^2, and the domain of

(2)
$$L(x_1, x_2) = -x_1 + 3x_2$$

is the entire space R^2. Similarly, the ranges of G and L are respectively a subset of R^1 and the entire 1-dimensional vector space R^1, often designated by either R or $(-\infty, \infty)$. See Figure 5-2.

Figure 5-2

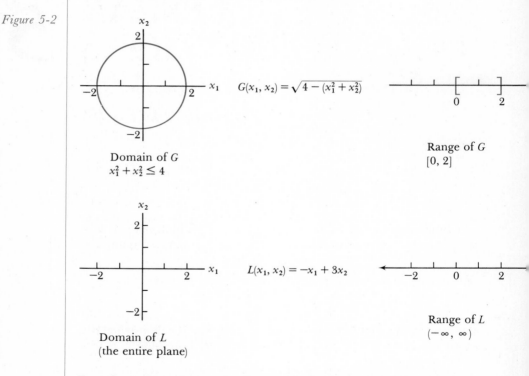

Domain of G
$x_1^2 + x_2^2 \le 4$

$G(x_1, x_2) = \sqrt{4 - (x_1^2 + x_2^2)}$

Range of G
[0, 2]

Domain of L
(the entire plane)

$L(x_1, x_2) = -x_1 + 3x_2$

Range of L
$(-\infty, \infty)$

Functions whose domain and range are both vector spaces are sometimes given special names, such as **operator, mapping,** or **transformation.** We shall generally prefer the last one, although we will also use the common expression of saying (when $y = T(x)$) that T "maps" or "carries" x into y. Thus, a transformation is a function having its domain in one vector space and its range in another (which may be a copy of the first).

You are familiar with other examples of transformations, though you have probably not called them by that name. Consider, for example, the operation of differentiation which may be indicated by the so-called

differential operator D. Thus,

$$D(x^3 - 5x + 1) = 3x^2 - 5$$
$$D(\sin x + e^{-2x}) = \cos x - 2e^{-2x}$$

and, more generally,

$$Df = f'$$

is defined on the vector space $C'(a, b)$ of all functions continuously differentiable on (a, b). From the definition of $C'(a, b)$, it follows that the derivatives f' are continuous on (a, b); that is, they are members of $C(a, b)$. We may think of D then as an operator that fires vectors from the vector space $C'(a, b)$ into the vector space $C(a, b)$ (Figure 5-3).

$C'[0, 1]$

$f(x) = x^3 - 5x + 1$

D

$C[0, 1]$

$Df(x) = 3x^2 - 5$

Figure 5-3

Vector space of functions
having a continuous
derivative on [0, 1].

Vector space of functions
continuous on [0, 1].

The functions T and L described in (1) and (2) are both referred to in the elementary calculus as linear functions. If we write them in the form

$$y_1 = [2][x_1]$$

$$y_1 = \begin{bmatrix} -1 & 3 \end{bmatrix} \begin{bmatrix} x_1 \\ x_2 \end{bmatrix}$$

they are seen to be simple examples of transformations from the vector space R^n to the vector space R^m that can be described by a matrix equation of the form

$$\begin{bmatrix} y_1 \\ \vdots \\ y_m \end{bmatrix} = \begin{bmatrix} a_{11} & \cdots & a_{1n} \\ \vdots & & \vdots \\ a_{m1} & \cdots & a_{mn} \end{bmatrix} \begin{bmatrix} x_1 \\ \vdots \\ x_n \end{bmatrix} \tag{3}$$

In this chapter we will extend the definition of a transformation so that any transformation that can be described in the form of the matrix equation (3) will be called *linear*. We will in fact extend the definition of a linear transformation so that a surprising number of mathematical operations (like differentiation) can be included in a study of linear transformations. But we will also attach particular importance to those that can be written in the form of (3).

5-2 | DEFINITIONS

There are just two operations that can be performed on a vector **u** in an arbitrary vector space V. It can be multiplied by a scalar r, giving $r\mathbf{u}$, or it may be added to a vector **v**, giving $\mathbf{u} + \mathbf{v}$. Consider now a transformation T defined on V, having its range in a second vector space W, a situation commonly indicated by writing

$$T: V \longrightarrow W$$

This transformation, being defined on all of V, is defined for the vectors mentioned above; $T(\mathbf{u})$, $T(\mathbf{v})$, $T(r\mathbf{u})$, and $T(\mathbf{u} + \mathbf{v})$ are all in W (Figure 5-4).

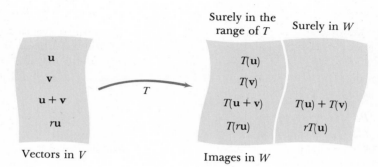

Figure 5-4

Since W is a vector space, the scalar product $rT(\mathbf{u})$ and the sum $T(\mathbf{u}) + T(\mathbf{v})$ must also be in W. There is no guarantee, of course, that any of these will be in the range of T. If T is nicely behaved, however, then we could hope not only that $rT(\mathbf{u})$ and $T(\mathbf{u}) + T(\mathbf{v})$ are in the range of T, but that $T(r\mathbf{u}) = rT(\mathbf{u})$ and $T(\mathbf{u} + \mathbf{v}) = T(\mathbf{u}) + T(\mathbf{v})$.

These are exactly the properties that distinguish the transformation now to be studied.

DEFINITION $T: V \to W$ is called a **linear transformation** if for every **u** and **v** in V and for every scalar r,

(4)
$$T(r\mathbf{u}) = rT(\mathbf{u})$$
$$T(\mathbf{u} + \mathbf{v}) = T(\mathbf{u}) + T(\mathbf{v})$$

We have already said that the real-valued function of a real variable defined by $T(x) = 2x$ is linear, and that is easy to verify from the definition. For any two members x_1 and x_2 of R,

$$T(x_1 + x_2) = 2(x_1 + x_2) = 2x_1 + 2x_2 = T(x_1) + T(x_2)$$

Since the scalar r as well as the "vector" x comes from R in this simple case, it is trivial almost to the point of confusion to note that

$$T(rx) = 2(rx) = r(2x) = rT(x)$$

There is a potential for worse confusion. You have been taught from your youth that $f(x) = mx + b$ is, for any fixed m and b, to be called linear. Yet, in this case

$$f(rx) = mrx + b \neq r(mx + b) = rf(x).$$

Therefore, according to our definition, this function is not linear! It is unfortunate to have to ask students to abandon a familiar meaning for a word and adopt a new one, but no amount of fulminating here will change common usage.

Since being familiar can sometimes be the enemy of being correct, it may be helpful to see some examples of linear transformations in settings where we have no previous associations with the word *linear*. We will look, therefore, at examples of linear transformations defined on each of the four vector spaces introduced as examples in Section 4-2: the spaces

P—the set of all polynomials
M_{mn}—the set of all $m \times n$ matrices
$C'(a, b)$—the set of all continuously differentiable functions
S—the set of all infinite sequences

The use of these names for the indicated vector spaces, together with the use of R for the space R^1 of real numbers, allows us to concisely specify the domain and range for each of our examples.

A linear transformation $I: P \to R$ is defined by

Example A

$$I(\mathbf{p}) = \int_0^1 p(x)\, dx$$

The linearity of I follows from familiar properties of the integral, namely

$$I(\mathbf{p} + \mathbf{q}) = \int_0^1 [p(x) + q(x)]\, dx = \int_0^1 p(x)\, dx + \int_0^1 q(x)\, dx$$
$$= I(\mathbf{p}) + I(\mathbf{q})$$
$$I(r\mathbf{p}) = \int_0^1 rp(x)\, dx = r\int_0^1 p(x)\, dx = rI(\mathbf{p}) \quad \square$$

A linear transformation $T: M_{mn} \to M_{nm}$ is defined by letting $T(A) = A^t$, the transpose of the matrix A.

Example B

We have only to note that

$$T(A + B) = (A + B)^t = A^t + B^t = T(A) + T(B)$$

and

$$T(rA) = (rA)^t = rA^t = rT(A) \quad \square$$

Example C

A linear transformation $D: C'(a, b) \to C(a, b)$ is defined by the differential operator D.

The linearity of D is again a consequence of what we know about calculus.

$$D(\mathbf{f} + \mathbf{g}) = D\mathbf{f} + D\mathbf{g}$$
$$D(r\mathbf{f}) = rD\mathbf{f} \quad \square$$

Example D

A linear transformation $L: S \to S$ is defined by the left-shift operator which has the effect of moving every entry of a given sequence one position to the left (and dropping the first entry altogether).

It may be well to begin with an illustration of this unfamiliar transformation.

$$L(1, \tfrac{1}{2}, \tfrac{1}{3}, \tfrac{1}{4}, \ldots) = (\tfrac{1}{2}, \tfrac{1}{3}, \tfrac{1}{4}, \ldots)$$

In general,

$$L(x_1, x_2, x_3, x_4, \ldots) = (x_2, x_3, x_4, \ldots)$$
$$L(y_1, y_2, y_3, y_4, \ldots) = (y_2, y_3, y_4, \ldots)$$

Then

$$L(\mathbf{x} + \mathbf{y}) = L(x_1 + y_1, x_2 + y_2, x_3 + y_3, \ldots)$$
$$= (x_2 + y_2, x_3 + y_3, \ldots)$$
$$= (x_2, x_3, \ldots) + (y_2, y_3, \ldots)$$
$$= L(\mathbf{x}) + L(\mathbf{y})$$

We also need to check

$$L(r\mathbf{x}) = L(rx_1, rx_2, rx_3, \ldots)$$
$$= (rx_2, rx_3, \ldots)$$
$$= r(x_2, x_3, \ldots)$$
$$= rL(\mathbf{x}) \quad \square$$

Perhaps we are now ready to see that our definition does indeed imply linearity in a more familiar setting, the setting most important to our purposes in this text.

Let A be an $m \times n$ matrix. Then $T: R^n \rightarrow R^m$ defined by $T(\mathbf{x}) = A\mathbf{x}$ is   *Example E*
a linear transformation.

First let us note that we depend here (as we shall depend in the future)
on the context to make it clear that $\mathbf{x}$ in R^n is being written as a column
matrix. Thus, with $\mathbf{u}$ and $\mathbf{v}$ represented by column matrices, the fact
that matrix multiplication is distributive enables us to write

$$T(\mathbf{u} + \mathbf{v}) = A(\mathbf{u} + \mathbf{v}) = A\mathbf{u} + A\mathbf{v} = T(\mathbf{u}) + T(\mathbf{v})$$

And if r is a scalar,

$$T(r\mathbf{u}) = \begin{bmatrix} a_{11} & \cdots & a_{1n} \\ \vdots & & \vdots \\ a_{m1} & \cdots & a_{mn} \end{bmatrix} \begin{bmatrix} ru_1 \\ \vdots \\ ru_n \end{bmatrix} = \begin{bmatrix} a_{11}ru_1 + \cdots + a_{1n}ru_n \\ \vdots & & \vdots \\ a_{m1}ru_1 + \cdots + a_{mn}ru_n \end{bmatrix}$$

and so

$$T(r\mathbf{u}) = r \begin{bmatrix} a_{11}u_1 + \cdots + a_{1n}u_n \\ \vdots & & \vdots \\ a_{m1}u_1 + \cdots + a_{mn}u_n \end{bmatrix} = rA\mathbf{u} = rT(\mathbf{u}) \quad \square$$

Two other examples of linear transformations must be mentioned. For
any vector space V, the transformation $T: V \rightarrow V$ is called the **identity**
transformation if $T(\mathbf{v}) = \mathbf{v}$ for every $\mathbf{v}$, and it is called the **zero** transformation if $T(\mathbf{v}) = \mathbf{0}$ for every $\mathbf{v}$. Both of these transformations are easily
seen to be linear.

There is one property shared by all linear transformations, no matter
what their domain or range, that is of central importance. Let $\mathbf{0}_V$ be the
zero vector in the domain V of a linear transformation T, and let $\mathbf{0}_W$ be
the zero vector in the range W. It follows from the linearity of T that

$$T(\mathbf{0}_V) = T(\mathbf{0}_V + \mathbf{0}_V) = T(\mathbf{0}_V) + T(\mathbf{0}_V)$$

Subtraction of $T(\mathbf{0}_V)$ from each side gives $\mathbf{0}_W = T(\mathbf{0}_V)$. We state this as
our first theorem, using the notation just defined.

Let $T:V \rightarrow W$ be a linear transformation; then $T(\mathbf{0}_V) = \mathbf{0}_W$.   *THEOREM A*

If we had not noticed it before, it would have been obvious from this
theorem that functions of the form $f(x) = mx + b$ are linear if and only
if $b = 0$, since $f(0) = b$. If a real-valued function of one or two variables
is linear, its graph (a line or a plane) must pass through the origin.

We are always helped in our understanding of a mathematical concept
if we can associate it with a diagram. For this reason, it is helpful to ask
what effect a linear transformation $T: R^2 \rightarrow R^2$ has on certain geometric figures in the plane. Numerous questions of this type are asked in the
following problem set, and we conclude this section with one example
of this type.

Example F

Every nonzero linear transformation $T: R^2 \to R^2$ takes parallel line segments into parallel line segments.

You may wish to look at a specific example (Figure 5-5a). Note that the segment joining $A(-3, 1)$ to $B(1, 3)$ and the segment joining

Figure 5-5

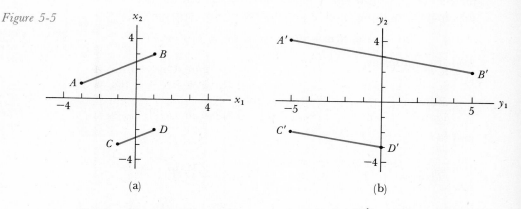

(a)   (b)

$C(-1, -3)$ to $D(1, -2)$ are parallel; in fact $\overrightarrow{AB} = (4, 2) = 2(2, 1) = 2\overrightarrow{CD}$. Now under the linear transformation described by

$$\begin{bmatrix} y_1 \\ y_2 \end{bmatrix} = \begin{bmatrix} 2 & 1 \\ -1 & 1 \end{bmatrix} \begin{bmatrix} x_1 \\ y_1 \end{bmatrix}$$

we find the images $T(\overrightarrow{AB}) = \overrightarrow{A'B'}$ and $T(\overrightarrow{CD}) = \overrightarrow{C'D'}$ from

$$\begin{bmatrix} 2 & 1 \\ -1 & 1 \end{bmatrix} \begin{bmatrix} 4 \\ 2 \end{bmatrix} = \begin{bmatrix} 10 \\ -2 \end{bmatrix} \quad \text{and} \quad \begin{bmatrix} 2 & 1 \\ -1 & 1 \end{bmatrix} \begin{bmatrix} 2 \\ 1 \end{bmatrix} = \begin{bmatrix} 5 \\ -1 \end{bmatrix}$$

We see that $\overrightarrow{A'B'} = (10, -2) = 2(5, -1) = 2\overrightarrow{C'D'}$. These vectors can be drawn (Figure 5-5b) between the image points $A'(-5, 4)$, $B'(5, 2)$, $C'(-5, -2)$, and $D'(0, -3)$.

A general proof proceeds as follows: If the segments joining A to B and C to D are parallel, then $\overrightarrow{AB} = r\overrightarrow{CD}$ for some scalar r. If we let $\overrightarrow{A'B'} = T(\overrightarrow{AB})$ and $\overrightarrow{C'D'} = T(\overrightarrow{CD})$, then it follows from the linearity of T that

$$\overrightarrow{A'B'} = T(\overrightarrow{AB}) = T(r\overrightarrow{CD}) = rT(\overrightarrow{CD}) = r\overrightarrow{C'D'}$$

This says that the segments joining A' to B' and C' to D' are also parallel. □

*PROBLEM
SET 5-2*

Problems 1–8 test your understanding of the definition of a linear transformation. Each problem describes two vector spaces V and W and a transformation $T: V \to W$ that maps vectors from V into W. You are to decide whether T is linear.

1. V is the vector space of all 2×2 matrices; $W = R^1$, the space of all real numbers.

$$T\left(\begin{bmatrix} a & b \\ c & d \end{bmatrix}\right) = \det \begin{bmatrix} a & b \\ c & d \end{bmatrix}$$

(*Problem set continues on p. 204*)

TRANSFORMED AREAS

Let $T: R^2 \to R^2$ be a linear transformation defined by $\mathbf{y} = A\mathbf{x}$, A being a 2×2 matrix. It is easy to show (Problem 19) that a parallelogram P will be transformed by T into a parallelogram P' (Figure 5-A). We wish to find the relationship between the areas of the two parallelograms.

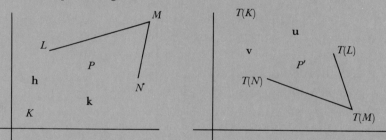

Figure 5-A

Use the vertices $KLMN$ of parallelogram P to form the column vectors $\mathbf{h} = \overrightarrow{KL}$ and $\mathbf{k} = \overrightarrow{KN}$. Then $\mathbf{u} = A\mathbf{h}$ and $\mathbf{v} = A\mathbf{k}$ are nonparallel edges of P'. Since the determinant of a matrix equals the determinant of its transpose, it follows (see Section 3-1) that

$$\text{area } P = \pm \det \begin{bmatrix} h_1 & k_1 \\ h_2 & k_2 \end{bmatrix} = \pm \det [\mathbf{h} \quad \mathbf{k}]$$

$$\text{area } P' = \pm \det \begin{bmatrix} u_1 & v_1 \\ u_2 & v_2 \end{bmatrix} = \pm \det [\mathbf{u} \quad \mathbf{v}]$$

Note that $[\mathbf{u} \quad \mathbf{v}] = [A\mathbf{h} \quad A\mathbf{k}] = A[\mathbf{h} \quad \mathbf{k}]$. Since the determinant of a product is the product of the determinants, $\det [\mathbf{u} \quad \mathbf{v}] = (\det A)(\det [\mathbf{h} \quad \mathbf{k}])$. Thus,

$$\text{area } P' = \pm (\det A)(\text{area } P) \qquad (i)$$

The potential value of this result is illustrated by the very simple linear transformation defined by

$$\begin{bmatrix} y_1 \\ y_2 \end{bmatrix} = \begin{bmatrix} a & 0 \\ 0 & b \end{bmatrix} \begin{bmatrix} x_1 \\ x_2 \end{bmatrix}$$

This transforms the circle $x_1^2 + x_2^2 = 1$ into the ellipse $y_1^2/a^2 + y_2^2/b^2 = 1$. The area enclosed by the circle is defined by a limiting process involving rectangles that cover the circle, and these rectangles are transformed into rectangles that cover the ellipse. You are circling in on the truth if you guess that in analogy with (i),

$$\text{area of the ellipse} = \det \begin{bmatrix} a & 0 \\ 0 & b \end{bmatrix} (\text{area of the circle}) = ab\pi$$

2. $V = R^2$, $W = R^1$, and $T(x_1, x_2) = \det \begin{bmatrix} 3 & x_1 \\ -2 & x_2 \end{bmatrix}$

3. $V = R^3$, $W = R^1$, and $T(x_1, x_2, x_3) = \det \begin{bmatrix} 1 & -1 & 2 \\ 0 & 1 & -1 \\ x_1 & x_2 & x_3 \end{bmatrix}$.

4. $V = R^4$, $W = R^1$, and $T(\mathbf{x}) = \mathbf{x} \cdot \mathbf{x}$.

5. $V = R^2$, $W = R^2$, and $T(x_1, x_2) = (|x_1|, |x_2|)$

6. $V = W = P_2$, the space of all polynomials of degree two or less.

$$T(a_2 x^2 + a_1 x + a_0) = (a_0 + a_2)x^2 + (a_1 + a_2)x + a_1 + a_0$$

7. $V = W = R^2$; $T(x_1, x_2) = (\sin(x_1 + x_2), \cos(x_1 - x_2))$.

8. $V = W = R^2$; $T(x_1, x_2) = (\sin x_1 + \sin x_2, \cos x_1 + \cos x_2)$.

Points $J(2, 8)$, $K(4, 4)$, and $L(10, 7)$ are vertices of a right triangle, and points K and L, together with $M(8, -1)$ and $N(2, -4)$, are vertices of a parallelogram. In Problems 9–16, you are given a matrix A which can be used to define a linear transformation $T: R^2 \rightarrow R^2$ in the usual way, $T(\mathbf{x}) = A\mathbf{x}$. For each problem, answer the following questions:

(a) Is the image of $\triangle JKL$ again a right triangle?

(b) Is the image of $\square KLMN$ again a parallelogram?

(c) Do the images of the segments from K to M and L to N again bisect each other?

9. $\begin{bmatrix} 0 & 1 \\ -1 & 0 \end{bmatrix}$

10. $\begin{bmatrix} \frac{3}{5} & -\frac{4}{5} \\ \frac{4}{5} & \frac{3}{5} \end{bmatrix}$

11. $\begin{bmatrix} \frac{5}{13} & -\frac{12}{13} \\ \frac{12}{13} & \frac{5}{13} \end{bmatrix}$

12. $\begin{bmatrix} -1 & 0 \\ 0 & 1 \end{bmatrix}$

13. $\begin{bmatrix} 1 & 2 \\ -1 & 1 \end{bmatrix}$

14. $\begin{bmatrix} -1 & 1 \\ 2 & 1 \end{bmatrix}$

15. $\begin{bmatrix} -1 & 2 \\ 2 & -4 \end{bmatrix}$

16. $\begin{bmatrix} 1 & -2 \\ -\frac{1}{2} & 1 \end{bmatrix}$

Problems 17–24 all refer to a linear transformation $T: R^3 \to R^3$.

17. Suppose $\mathbf{u}$ and $\mathbf{v}$ are parallel, meaning of course that $\mathbf{u} = r\mathbf{v}$ for some scalar r. Must $T(\mathbf{u})$ and $T(\mathbf{v})$ be parallel?

18. Suppose $\mathbf{u}$ and $\mathbf{v}$ are perpendicular, meaning of course that $\mathbf{u} \cdot \mathbf{v} = 0$. Must $T(\mathbf{u})$ and $T(\mathbf{v})$ be perpendicular?

19. Suppose points $\mathbf{x}_1$, $\mathbf{x}_2$, $\mathbf{x}_3$, and $\mathbf{x}_4$ are vertices of a parallelogram. Must the points $\mathbf{y}_i = T(\mathbf{x}_i)$ be vertices of a parallelogram?

20. Suppose points $\mathbf{x}_1$, $\mathbf{x}_2$, and $\mathbf{x}_3$ are vertices of a right triangle in R^3. Must the points $\mathbf{y}_i = T(\mathbf{x}_i)$ be vertices of a right triangle?

21. Suppose points $\mathbf{x}_1$, $\mathbf{x}_2$, and $\mathbf{x}_3$ are collinear. Must $\mathbf{y}_i = T(\mathbf{x}_i)$ be collinear?

22. Let points $\mathbf{x}_1$, $\mathbf{x}_2$, and $\mathbf{x}_3$ be vertices of a triangle having medians that intersect at $\mathbf{x}_4$. If $\mathbf{y}_i = T(\mathbf{x}_i)$, do the medians of the triangle with vertices at $\mathbf{y}_1$, $\mathbf{y}_2$, and $\mathbf{y}_3$ intersect at $\mathbf{y}_4$?

23. Suppose $\mathbf{x}_3$ is a point that divides the line segment joining points $\mathbf{x}_1$ and $\mathbf{x}_2$ into the ratio of 1 to 3. Does $\mathbf{y}_3 = T(\mathbf{x}_3)$ divide the line segment joining points $\mathbf{y}_1 = T(\mathbf{x}_1)$ and $\mathbf{y}_2 = T(\mathbf{x}_2)$ in the same ratio?

24. If points $\mathbf{x}_1$, $\mathbf{x}_2$, $\mathbf{x}_3$, and $\mathbf{x}_4$ are vertices of a trapezoid, must the points $\mathbf{y}_i = T(\mathbf{x}_i)$ also be vertices of a trapezoid?

25. Suppose that $T: U \to V$ is linear, and that $S: V \to W$ is also linear. Show that $S \circ T: U \to W$, defined by $S \circ T(x) = S(T(x))$, is linear.

26. The projection of a vector $\mathbf{u}$ onto $\mathbf{v}$ was defined in Section 2-3 by

$$\operatorname{proj}_{\mathbf{v}} \mathbf{u} = \frac{\mathbf{u} \cdot \mathbf{v}}{\mathbf{v} \cdot \mathbf{v}} \mathbf{v}$$

Let $\mathbf{v}$ be a fixed vector of R^3 and let K be the subspace spanned by $\mathbf{v}$. Define $P: R^3 \to K$ by $P(\mathbf{x}) = \operatorname{proj}_{\mathbf{v}} \mathbf{x}$. Prove that P is linear.

*27. The linear transformation P described in Problem 26 is sometimes called an *orthogonal* projection to distinguish it from the following generalization of the concept. Suppose that K and L are proper subspaces of a vector space V, that $\mathbf{0}$ is the only vector common to K and L, and that any vector $\mathbf{v}$ in V can be written in the form $\mathbf{v} = \mathbf{k} + \mathbf{l}$ where $\mathbf{k}$ is in K, $\mathbf{l}$ is in L. Then $P: V \to K$, defined by $P(\mathbf{v}) = \mathbf{k}$, is called a **projection** of V onto K. Prove that a projection is a linear transformation.

*28. Let $\{\mathbf{u}_1, \ldots, \mathbf{u}_n\}$ be an orthonormal basis for R^n, and suppose $T: R^n \to R^n$ is a linear transformation. Show that if

$$M = \max\{|T(\mathbf{u}_1)|, \ldots, |T(\mathbf{u}_n)|\}$$

then for any $\mathbf{x}$ in R^n, $|T(\mathbf{x})| < nM|\mathbf{x}|$.

29. The discussion of transformed areas can be illustrated with the parallelogram used in Problems 9–16. In this case,

$$\mathbf{h} = \overrightarrow{KL} = \begin{bmatrix} 6 \\ 3 \end{bmatrix}, \qquad \mathbf{k} = \overrightarrow{KN} = \begin{bmatrix} -2 \\ -8 \end{bmatrix}$$

and the area of $P = \Box KLMN$ is

$$\text{area } P = \pm \det \begin{bmatrix} 6 & -2 \\ 3 & -8 \end{bmatrix} = 42$$

Use the matrix A of Problem 13 to define a linear transformation $T: R^2 \rightarrow R^2$, find $\mathbf{u} = A\mathbf{h}$, $\mathbf{v} = A\mathbf{k}$, and verify that

$$\text{area } P' = \pm (\det A)(\text{area } P)$$

5-3 | *MATRIX REPRESENTATIONS OF A LINEAR TRANSFORMATION*

Many transformations of R^2 into itself which can be described geometrically turn out to be linear transformations. A number of them have names and are described in the following examples. In each case, you should verify that the transformation T is linear; that is, verify that

$$T(\mathbf{u} + \mathbf{v}) = T(\mathbf{u}) + T(\mathbf{v})$$
$$T(r\mathbf{u}) = rT(\mathbf{u})$$

Example A   Define $T: R^2 \rightarrow R^2$ to be the transformation that reflects every point in the line $x_2 = -x_1$ (Figure 5-6). T is called a **reflection.**   □

Figure 5-6

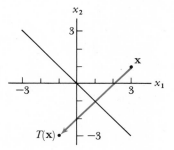

Example B   A transformation $T: R^2 \rightarrow R^2$ defined by $T(\mathbf{x}) = r\mathbf{x}$ is called a **dilation** if $r > 1$, and a **contraction** if $0 < r < 1$ (Figure 5-7).   □

Figure 5-7

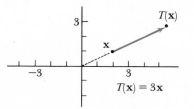

Example C

A **rotation** is a transformation $T: R^2 \to R^2$ that moves each point $\mathbf{x}$ around the circle of radius $|\mathbf{x}|$ through some fixed angle θ (Figure 5-8). ☐

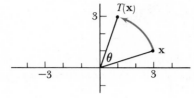

Figure 5-8

The transformation that takes (x_1, x_2) into $(x_1 + 2x_2, x_2)$ is called a **shearing.** Each point moves along a horizontal line, the direction of motion determined by whether the horizontal line is above or below the origin, and the distance moved is proportional to the distance of the horizontal line from the origin (Figure 5-9). ☐

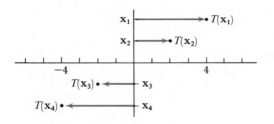

Figure 5-9

The transformation $T: R^2 \to R^2$ that projects every point onto the line $x_2 = 2x_1$ is called a projection (Figure 5-10). ☐

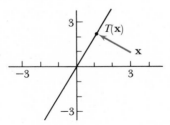

Figure 5-10

We know that if A is an $m \times n$ matrix, then $T(\mathbf{x}) = A\mathbf{x}$ defines a linear transformation from R^n to R^m. The remarkable fact is that no matter how a linear transformation from R^n to R^m is described, it too can be described by a matrix multiplication. Thus, each of the five examples given above can (and will, in due time) be associated with a 2×2 matrix.

We can do even better. Let U and V be any two vector spaces of dimensions n and m, respectively. Any basis of U will have n vectors, of course, and any basis of V will have m vectors. We are not particular about how the bases are chosen, but for our purposes it is very important that we choose bases for each space, and then stick with them throughout the discussion. Suppose, therefore, that

$$U \text{ has basis } B_n = \{\mathbf{u}_1, \ldots, \mathbf{u}_n\}$$
$$V \text{ has basis } B_m = \{\mathbf{v}_1, \ldots, \mathbf{v}_m\}$$

Then any linear transformation $T: U \to V$ can be uniquely represented by an $m \times n$ matrix A; that is, we can find a matrix $\mathbf{x}$ such that $T(\mathbf{x}) = A\mathbf{x}$.

To see how this is done while at the same time holding notational problems down, let us consider the case where $n = 2$ and $m = 3$. Then $T(\mathbf{u}_1)$ and $T(\mathbf{u}_2)$, being in V, can be written in the form

$$T(\mathbf{u}_1) = a_{11}\mathbf{v}_1 + a_{21}\mathbf{v}_2 + a_{31}\mathbf{v}_3$$
$$T(\mathbf{u}_2) = a_{12}\mathbf{v}_1 + a_{22}\mathbf{v}_2 + a_{32}\mathbf{v}_3$$

Consider now any $\mathbf{w} = x_1\mathbf{u}_1 + x_2\mathbf{u}_2$ in U. Then

$$T(\mathbf{w}) = T(x_1\mathbf{u}_1 + x_2\mathbf{u}_2)$$

and, because T is linear,

$$T(\mathbf{w}) = x_1 T(\mathbf{u}_1) + x_2 T(\mathbf{u}_2)$$
$$= x_1[a_{11}\mathbf{v}_1 + a_{21}\mathbf{v}_2 + a_{31}\mathbf{v}_3]$$
$$+ x_2[a_{12}\mathbf{v}_1 + a_{22}\mathbf{v}_2 + a_{32}\mathbf{v}_3]$$

After collecting coefficients of the $\mathbf{v}_i$, we get

$$T(\mathbf{w}) = [a_{11}x_1 + a_{12}x_2]\mathbf{v}_1$$
$$+ [a_{21}x_1 + a_{22}x_2]\mathbf{v}_2$$
$$+ [a_{31}x_1 + a_{32}x_2]\mathbf{v}_3$$

If we let y_i be the coefficient of $\mathbf{v}_i$ in $T(\mathbf{w})$, then

$$\begin{bmatrix} y_1 \\ y_2 \\ y_3 \end{bmatrix} = \begin{bmatrix} a_{11} & a_{12} \\ a_{21} & a_{22} \\ a_{31} & a_{32} \end{bmatrix} \begin{bmatrix} x_1 \\ x_2 \end{bmatrix}$$

This basic result, extended to any $m, n > 0$, may be stated this way:

THEOREM A

Let U and V be n-dimensional and m-dimensional vector spaces with bases

$$B_n = \{\mathbf{u}_1, \ldots, \mathbf{u}_n\} \quad \text{and} \quad B_m = \{\mathbf{v}_1, \ldots, \mathbf{v}_m\}$$

and let $T: U \to V$ be a linear transformation for which

$$T(\mathbf{u}_i) = a_{1i}\mathbf{v}_1 + a_{2i}\mathbf{v}_2 + \cdots + a_{mi}\mathbf{v}_m$$

Then T may be represented by the matrix product $\mathbf{y} = A\mathbf{x}$ *where*

$$A = \begin{bmatrix} a_{11} & a_{12} & \cdots & a_{1n} \\ a_{21} & a_{22} & \cdots & a_{2n} \\ \vdots & \vdots & & \vdots \\ a_{m1} & a_{m2} & \cdots & a_{mn} \end{bmatrix}$$

It cannot be overemphasized that the matrix A depends upon the bases chosen for U and V. Change one or both of the bases, and the matrix A will change.

The theorem can be remembered this way: The coordinates of the image of the ith basis vector in the domain become the entries in the ith column of the matrix.

Evidently we can find the matrices of the transformations of Examples A through E by finding the images of the basis vectors. We will illustrate the procedure in each case, using the standard basis $\mathbf{e}_1 = (1, 0)$ and $\mathbf{e}_2 = (0, 1)$.

It is geometrically evident (Figure 5-11) that under reflection in the line $x_2 = -x_1$,

$$T(\mathbf{e}_1) = 0\mathbf{e}_1 - 1\mathbf{e}_2$$
$$T(\mathbf{e}_2) = -1\mathbf{e}_1 + 0\mathbf{e}_2$$

Example A'

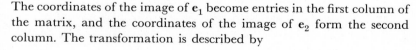

Figure 5-11

The coordinates of the image of $\mathbf{e}_1$ become entries in the first column of the matrix, and the coordinates of the image of $\mathbf{e}_2$ form the second column. The transformation is described by

$$\begin{bmatrix} y_1 \\ y_2 \end{bmatrix} = \begin{bmatrix} 0 & -1 \\ -1 & 0 \end{bmatrix} \begin{bmatrix} x_1 \\ x_2 \end{bmatrix}$$

Try it on $\mathbf{x} = (3, 1)$. □

Since $T(r\mathbf{x}) = r\mathbf{x}$,

$$T(\mathbf{e}_1) = r\mathbf{e}_1$$
$$T(\mathbf{e}_2) = r\mathbf{e}_2$$

The transformation is described by

$$\begin{bmatrix} y_1 \\ y_2 \end{bmatrix} = \begin{bmatrix} r & 0 \\ 0 & r \end{bmatrix}\begin{bmatrix} x_1 \\ x_2 \end{bmatrix} \quad \square$$

Example C'

It is evident from Figure 5-12 that in a rotation through θ,

$$T(1, 0) = (\cos \theta, \sin \theta)$$
$$T(0, 1) = (-\sin \theta, \cos \theta)$$

Figure 5-12

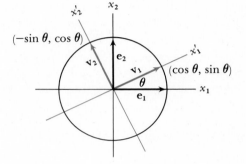

In agreement with equations (3) of Section 4-5, obtained when this same problem was viewed as a change of basis, we obtain

$$\begin{bmatrix} y_1 \\ y_2 \end{bmatrix} = \begin{bmatrix} \cos \theta & -\sin \theta \\ \sin \theta & \cos \theta \end{bmatrix}\begin{bmatrix} x_1 \\ x_2 \end{bmatrix} \quad \square$$

Example D'

Since (x_1, x_2) goes into $(x_1 + 2x_2, x_2)$, $(1, 0)$ goes into $(1, 0)$ and $(0, 1)$ goes into $(2, 1)$.

The transformation is therefore described by

$$\begin{bmatrix} y_1 \\ y_2 \end{bmatrix} = \begin{bmatrix} 1 & 2 \\ 0 & 1 \end{bmatrix}\begin{bmatrix} x_1 \\ x_2 \end{bmatrix}$$

In this case, the multiplication which shows $y_1 = x_1 + 2x_2$ and $y_2 = x_2$ is seen to be exactly what was to be expected from the definition. $\square$

Example E'

The projections of $\mathbf{e}_1$ and $\mathbf{e}_2$ onto the line $x_2 = 2x_1$ can be found geometrically (Figure 5-13), but it is easier to find the projections onto $\mathbf{v} = (1, 2)$ using the formula

$$\text{proj}_{\mathbf{v}}\mathbf{e}_i = \frac{\mathbf{e}_i \cdot \mathbf{v}}{|\mathbf{v}|^2}\,\mathbf{v}$$

$$T(\mathbf{e}_1) = \text{proj}_{\mathbf{v}}\mathbf{e}_1 = \tfrac{1}{5}(1, 2) = (\tfrac{1}{5}, \tfrac{2}{5})$$
$$T(\mathbf{e}_2) = \text{proj}_{\mathbf{v}}\mathbf{e}_2 = \tfrac{2}{5}(1, 2) = (\tfrac{2}{5}, \tfrac{4}{5})$$

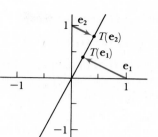

Figure 5-13

The transformation is described by

$$\begin{bmatrix} y_1 \\ y_2 \end{bmatrix} = \begin{bmatrix} \frac{1}{5} & \frac{2}{5} \\ \frac{2}{5} & \frac{4}{5} \end{bmatrix} \begin{bmatrix} x_1 \\ x_2 \end{bmatrix}$$

Try it on $\mathbf{x} = (3, 1)$. □

If $T: R^n \to R^m$ is a linear transformation, Theorem A can be used to determine the matrix representation of T if we know the images of any n linearly independent vectors in R^n.

Suppose that $T: R^3 \to R^4$ maps

Example F'

$$\mathbf{u}_1 = (1, 3, -2), \quad \mathbf{u}_2 = (-2, -5, 4) \quad \text{and} \quad \mathbf{u}_3 = (3, 7, -5)$$

into the indicated vectors in R^4.

$$T(\mathbf{u}_1) = (-1, 2, 1, -1) = \mathbf{v}_1$$
$$T(\mathbf{u}_2) = (5, -3, -1, 4) = \mathbf{v}_2$$
$$T(\mathbf{u}_3) = (1, 4, 0, 2) = \mathbf{v}_3$$

Find the matrix representation of T.

Begin by writing down a matrix having $\mathbf{u}_1$, $\mathbf{u}_2$, and $\mathbf{u}_3$ as row vectors. Adjacent to this matrix, separated for convenience by a vertical dotted line, list the coordinates of the image vectors.

$$\begin{bmatrix} 1 & 3 & -2 & -1 & 2 & 1 & -1 \\ -2 & -5 & 4 & 5 & -3 & -1 & 4 \\ 3 & 7 & -5 & 1 & 4 & 0 & 2 \end{bmatrix}$$

Let us take the first step in reducing the matrix on the left to row echelon form.

$$\begin{bmatrix} 1 & 3 & -2 & -1 & 2 & 1 & -1 \\ 0 & 1 & 0 & 3 & 1 & 1 & 2 \\ 0 & -2 & 1 & 4 & -2 & -3 & 5 \end{bmatrix}$$

first row copied
2(first row) + (second row)
−3(first row) + (third row)

In the second row, we have $2\mathbf{u}_1 + \mathbf{u}_2$ in the left-hand matrix, $2\mathbf{v}_1 + \mathbf{v}_2$

in the right-hand matrix. Moreover, since

$$T(2\mathbf{u}_1 + \mathbf{u}_2) = 2T(\mathbf{u}_1) + T(\mathbf{u}_2) = 2\mathbf{v}_1 + \mathbf{v}_2$$

the row vector on the right is still the image of the one on the left under T. And since the one on the left happens to be the basis vector $\mathbf{e}_2$ of R^3, we now know the coordinates of the image of $\mathbf{e}_2$.

If the given vectors $\mathbf{u}_1$, $\mathbf{u}_2$, and $\mathbf{u}_3$ are linearly independent, the row echelon form of the matrix on the left will be I (meaning the row vectors are the standard basis vectors for R^3). The reasoning above shows us that if the operations used in bringing the left matrix to row echelon form I are simultaneously performed on the right matrix, the image of each basis vector will be known. We continue, therefore, with this process.

$$\begin{bmatrix} 1 & 0 & -2 & | & -10 & -1 & -2 & -7 \\ 0 & 1 & 0 & | & 3 & 1 & 1 & 2 \\ 0 & 0 & 1 & | & 10 & 0 & -1 & 9 \end{bmatrix} \begin{matrix} -3(\text{second row}) + (\text{first row}) \\ \text{second row copied} \\ 2(\text{second row}) + (\text{third row}) \end{matrix}$$

$$\begin{bmatrix} 1 & 0 & 0 & | & 10 & -1 & -4 & 11 \\ 0 & 1 & 0 & | & 3 & 1 & 1 & 2 \\ 0 & 0 & 1 & | & 10 & 0 & -1 & 9 \end{bmatrix} \begin{matrix} 2(\text{third row}) + (\text{first row}) \\ \text{second row copied} \\ \text{third row copied} \end{matrix}$$

The images of the basis vectors $\mathbf{e}_1$, $\mathbf{e}_2$, and $\mathbf{e}_3$ now appear in the respective rows of the right-hand matrix. Using them as column vectors, write down the matrix representation of T with respect to the standard bases in R^3 and R^4.

$$A = \begin{bmatrix} 10 & 3 & 10 \\ -1 & 1 & 0 \\ -4 & 1 & -1 \\ 11 & 2 & 9 \end{bmatrix}$$

The matrix can be checked, verifying that

$$A\mathbf{u}_1 = \begin{bmatrix} -1 \\ 2 \\ 1 \\ -1 \end{bmatrix}, \qquad A\mathbf{u}_2 = \begin{bmatrix} 5 \\ -3 \\ -1 \\ 4 \end{bmatrix}, \qquad A\mathbf{u}_3 = \begin{bmatrix} 1 \\ 4 \\ 0 \\ 2 \end{bmatrix} \quad \Box$$

COMPOSITION OF FUNCTIONS

The composition of two linear transformations $T: R^n \to R^k$ and $S: R^k \to R^m$ is sometimes used to motivate the definition for matrix multiplication. For us, it certainly confirms that we have a useful definition. The general procedure is illustrated here by finding the result of following $T: R^3 \to R^2$ by $S: R^2 \to R^4$. Let T be described by

$$T: \begin{aligned} y_1 &= 2x_1 - 3x_2 + x_3 \\ y_2 &= x_1 + x_2 - x_3 \end{aligned}$$

and let S be described by

$$S: \begin{aligned} z_1 &= y_1 - y_2 \\ z_2 &= 3y_1 + y_2 \\ z_3 &= -y_1 + 2y_2 \\ z_4 &= y_1 + 3y_2 \end{aligned}$$

Then direct substitution gives, for $S \circ T$, the equations

$$\begin{aligned} z_1 &= (2x_1 - 3x_2 + x_3) - (x_1 + x_2 - x_3) = x_1 - 4x_2 + 2x_3 \\ z_2 &= 3(2x_1 - 3x_2 + x_3) + (x_1 + x_2 - x_3) = 7x_1 - 8x_2 + 2x_3 \\ z_3 &= -(2x_1 - 3x_2 + x_3) + 2(x_1 + x_2 - x_3) = 5x_2 - 3x_3 \\ z_4 &= (2x_1 - 3x_2 + x_3) + 3(x_1 + x_2 - x_3) = 5x_1 - 2x_3 \end{aligned}$$

The same result is obtained by multiplying the appropriate matrices:

$$\begin{bmatrix} z_1 \\ z_2 \\ z_3 \\ z_4 \end{bmatrix} = \begin{bmatrix} 1 & -1 \\ 3 & 1 \\ -1 & 2 \\ 1 & 3 \end{bmatrix} \begin{bmatrix} 2 & -3 & 1 \\ 1 & 1 & -1 \end{bmatrix} \begin{bmatrix} x_1 \\ x_2 \\ x_3 \end{bmatrix} = \begin{bmatrix} 1 & -4 & 2 \\ 7 & -8 & 2 \\ 0 & 5 & -3 \\ 5 & 0 & -2 \end{bmatrix} \begin{bmatrix} x_1 \\ x_2 \\ x_3 \end{bmatrix}$$

*PROBLEM
SET 5-3*

1. $T: R^3 \to R^2$ is a linear transformation that maps

$$\mathbf{e}_1 \longrightarrow \begin{bmatrix} 5 \\ 3 \end{bmatrix}, \quad \mathbf{e}_2 \longrightarrow \begin{bmatrix} -1 \\ 2 \end{bmatrix}, \quad \mathbf{e}_3 \longrightarrow \begin{bmatrix} -3 \\ -2 \end{bmatrix}$$

Find the matrix representation of T, and find $T(\mathbf{v})$ where $\mathbf{v} = (-1, 3, -2)$.

2. $T: R^4 \to R^2$ is a linear transformation that maps

$$\mathbf{e}_1 \longrightarrow \begin{bmatrix} 1 \\ 3 \end{bmatrix}, \quad \mathbf{e}_2 \longrightarrow \begin{bmatrix} -1 \\ 1 \end{bmatrix}, \quad \mathbf{e}_3 \longrightarrow \begin{bmatrix} -2 \\ -3 \end{bmatrix}, \quad \mathbf{e}_4 \longrightarrow \begin{bmatrix} 3 \\ -1 \end{bmatrix}$$

Find the matrix representation of T, and find $T(\mathbf{v})$ where $\mathbf{v} = (5, -1, 0, 3)$.

3. $S: R^2 \to R^4$ is a linear transformation that maps

$$\mathbf{e}_1 \longrightarrow \begin{bmatrix} 1 \\ 0 \\ -1 \\ 2 \end{bmatrix}, \quad \mathbf{e}_2 \longrightarrow \begin{bmatrix} -1 \\ 4 \\ 3 \\ 0 \end{bmatrix}$$

Find the matrix representation of S, and find $S(\mathbf{w})$ where $\mathbf{w} = (-2, 7)$.

4. $S: R^2 \to R^3$ is a linear transformation that maps

$$\mathbf{e}_1 \longrightarrow \begin{bmatrix} 1 \\ -1 \\ 0 \end{bmatrix}, \qquad \mathbf{e}_2 \longrightarrow \begin{bmatrix} 0 \\ 4 \\ -3 \end{bmatrix}$$

Find the matrix representation of S, and find $S(\mathbf{w})$ where $\mathbf{w} = (15, 11)$.

Problems 5 and 6 refer to the basis of R^2 given by

$$D = \{\mathbf{u}_1 = (3, 2), \mathbf{u}_2 = (-1, 2)\}$$

5. A linear transformation $T: R^2 \to R^2$ maps

$$\mathbf{u}_1 \longrightarrow 3\mathbf{u}_1 - 4\mathbf{u}_2, \qquad \mathbf{u}_2 \longrightarrow -\mathbf{u}_1 + 2\mathbf{u}_2$$

Find the matrix of T with respect to the basis D. If $\mathbf{w} = (2, 1)_D$, what is $T(\mathbf{w})_D$?

6. The linear transformation $S: R^2 \to R^2$ maps

$$\mathbf{u}_1 \longrightarrow -3\mathbf{u}_1 + 2\mathbf{u}_2, \qquad \mathbf{u}_2 \longrightarrow 4\mathbf{u}_1 - \mathbf{u}_2$$

Find the matrix of S with respect to the basis D. If $\mathbf{w} = (2, 1)_D$, find $S(\mathbf{w})_D$.

7. Referring to Problem 5, find the matrix of T with respect to the standard basis. Find the coordinates of $\mathbf{w}$ and $T(\mathbf{w})$ with respect to the standard basis. Can you think of a way to check your answers—other than looking in the back of the book?

8. Referring to Problem 6, find the matrix of S with respect to the standard basis. Find the coordinates of $\mathbf{w}$ and $T(\mathbf{w})$ with respect to the standard basis. Can you devise a check?

9. The linear transformation $T: R^3 \to R^4$ maps

$$\begin{bmatrix} -1 \\ 2 \\ 1 \end{bmatrix} \longrightarrow \begin{bmatrix} -4 \\ -3 \\ 1 \\ 3 \end{bmatrix}, \qquad \begin{bmatrix} -2 \\ 0 \\ 3 \end{bmatrix} \longrightarrow \begin{bmatrix} 1 \\ -8 \\ -5 \\ 3 \end{bmatrix}, \qquad \begin{bmatrix} 4 \\ 1 \\ -2 \end{bmatrix} \longrightarrow \begin{bmatrix} 0 \\ 8 \\ -1 \\ -1 \end{bmatrix}$$

Find $T(\mathbf{v})$ where $\mathbf{v} = (4, 1, -3)$.

10. The linear transformation $S: R^4 \to R^3$ maps

$$\begin{bmatrix} 1 \\ 0 \\ 1 \\ 2 \end{bmatrix} \longrightarrow \begin{bmatrix} 4 \\ 7 \\ 1 \end{bmatrix}, \qquad \begin{bmatrix} -2 \\ 1 \\ 0 \\ 3 \end{bmatrix} \longrightarrow \begin{bmatrix} 4 \\ 10 \\ 7 \end{bmatrix},$$

$$\begin{bmatrix} 4 \\ 1 \\ -2 \\ 1 \end{bmatrix} \longrightarrow \begin{bmatrix} 8 \\ 2 \\ -1 \end{bmatrix}, \qquad \begin{bmatrix} 1 \\ -2 \\ 1 \\ 0 \end{bmatrix} \longrightarrow \begin{bmatrix} 0 \\ -1 \\ -5 \end{bmatrix}$$

Find $S(\mathbf{w})$ where $\mathbf{w} = (-1, 2, 1, 3)$.

11. Referring to Problems 9 and 10, find the matrix representation of $S \circ T: R^3 \to R^3$. Then find $S \circ T(\mathbf{v})$.

(Problem set continues on p. 216)

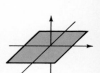

THE DERIVATIVE AS A LINEAR TRANSFORMATION

In elementary calculus, one learns that the derivative may be used to give an approximation for $f(x_0 + h) - f(x_0)$ when $f(x_0)$ is known and h is small. The idea is to use (Figure 5-B)

$$f(x_0 + h) - f(x_0) \approx [f'(x_0)]h \qquad (i)$$

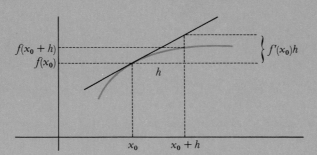

Figure 5-B

Since x_0 is fixed, the so-called **differential** $[f'(x_0)]h$ is linear in h. A key observation for our purposes is to note that the derivative $[f'(x_0)]$ may be viewed as a 1×1 matrix.

How shall we define the derivative of a function of two variables? In analogy with (i), we would like it to be a linear transformation that gives us an approximation for $f(\mathbf{x}_0 + \mathbf{h}) - f(\mathbf{x}_0)$ when $f(\mathbf{x}_0)$ is known and $\mathbf{h} = (h_1, h_2)$ is small. Now a real-valued linear transformation on R^2 is represented by a 1×2 matrix. Again one learns in calculus the correct matrix to use, the so-called **gradient,** obtained by evaluating at $\mathbf{x}_0$ the partial derivatives f_1 and f_2. This is the matrix that we will call the derivative

$$f'(\mathbf{x}_0) = [f_1(\mathbf{x}_0) \quad f_2(\mathbf{x}_0)] \qquad (ii)$$

With this definition of $[f'(\mathbf{x}_0)]$, we have in analog with (i),

$$f(\mathbf{x}_0 + \mathbf{h}) - f(\mathbf{x}_0) \approx [f'(\mathbf{x}_0)] \begin{bmatrix} h_1 \\ h_2 \end{bmatrix}$$

One more example points the way to the general rule. Suppose $T: R^2 \to R^2$ is a transformation represented by

$$T: \quad \begin{aligned} y_1 &= f(\mathbf{x}) = f(x_1, x_2) \\ y_2 &= g(\mathbf{x}) = g(x_1, x_2) \end{aligned}$$

Again we want to define $[T'(\mathbf{x}_0)]$ to be a linear transformation affording an approximation to $T(\mathbf{x}_0 + \mathbf{h}) - T(\mathbf{x}_0)$ when $T(\mathbf{x}_0)$ and $\mathbf{h} = (h_1, h_2)$ are known. Now a linear transformation from R^2 to R^2 is represented by a 2×2 matrix. Which one? It is the so-called **Jacobian matrix**

$$[T'(\mathbf{x}_0)] = \begin{bmatrix} f_1(\mathbf{x}_0) & f_2(\mathbf{x}_0) \\ g_1(\mathbf{x}_0) & g_2(\mathbf{x}_0) \end{bmatrix} \qquad (iii)$$

With this definition of $[T'(\mathbf{x}_0)]$, we have in analogy with (i),

$$T(\mathbf{x}_0 + \mathbf{h}) - T(\mathbf{x}_0) \approx [T'(\mathbf{x}_0)] \begin{bmatrix} h_1 \\ h_2 \end{bmatrix}$$

12. Referring to Problems 9 and 10, find the matrix representation of $T \circ S \colon R^4 \to R^4$. Then find $T \circ S(\mathbf{w})$.

13. Using T and S as defined in Problems 1 and 3, find the matrix representation of $S \circ T \colon R^3 \to R^4$. Then find $S \circ T(\mathbf{v})$ where $\mathbf{v}$ is the vector defined in Problem 1.

14. Using T and S as defined in Problems 2 and 4, find the matrix representation of $S \circ T \colon R^4 \to R^3$. Then find $S \circ T(\mathbf{v})$ where $\mathbf{v}$ is the vector defined in Problem 2.

15. $T \colon R^2 \to R^2$ reflects every vector in the line $x_1 = x_2$. Show that T is linear and find the matrix representation.

16. $T \colon R^2 \to R^2$ reflects every vector in the line $x_2 = 0$. Show that T is linear and find the matrix representation.

17. $P \colon R^3 \to R^2$ projects every vector (x_1, x_2, x_3) into (x_1, x_2) in the $x_1 x_2$ plane. Find the matrix representation of P.

18. $S \colon R^3 \to R^2$ by multiplying by 2 the projection of every vector (x_1, x_2, x_3) into the $x_1 x_3$ plane. Find the matrix representation for S.

*19. $T \colon R^3 \to R^3$ reflects every vector in the plane $x_1 + x_2 - 2x_3 = 0$. Show that T is linear and find the matrix representation.

*20. $S \colon R^3 \to R^3$ reflects every vector in the plane $x_1 - x_2 + x_3 = 0$. Show that S is linear and find the matrix representation.

*21. Let $f(x_1, x_2) = \sqrt{2x_1 + x_2}$. Using equation (ii) from the preceding vignette, approximate $f(2.2, 5.3)$ with $\mathbf{x}_0^t = [2 \quad 5]$, $\mathbf{h}^t = [0.2 \quad 0.3]$.

*22. Using the function of Problem 21, approximate $f(1.7, 5.2)$.

*23. Let $T \colon R^2 \to R^2$ be described by

$$y_1 = x_1 + \sqrt{x_1 + 2x_2}$$
$$y_2 = x_2 - \sqrt{x_1 - 2x_2}$$

Using equation (iii) from the preceding vignette, approximate $T(5.3, 1.8)$ with $\mathbf{x}_0^t = [5 \quad 2]$, $\mathbf{h}^t = [0.3 \quad -0.2]$.

*24. Using the transformation of Problem 23, approximate $T(5.2, 2.3)$.

*25. Matrices can be used to represent a linear transformation between any two finite-dimensional vector spaces once the bases are determined. Consider for example the set $\{x^2 + x, x^2 - x, x^3 + 1, x^3 - 1\}$ which forms a basis for the linear space P_3 of all polynomials of degree three or less. What 4×4 matrix could be used to represent the differential operator $D: P_3 \to P_3$ that is defined by $Dp(x) = p'(x)$?

CHANGE OF BASIS | 5-4

The central fact of the previous section is easily stated: if for an n-dimensional vector space U and an m-dimensional vector space V, bases B_n and B_m are specified, then a linear transformation $T: U \to V$ is uniquely represented by an $m \times n$ matrix A. It was emphasized that the matrix depends on the choice of the bases B_n and B_m. Our purpose in this section is to see how a change in the bases changes the matrix A.

If $m = n$, then the same basis B can be used in both the domain and the range of T. We can simplify our work and at the same time deal with the case of greatest importance by beginning with this special case.

A word is in order about a possible source of confusion in the way we think and talk (the preferred order, by the way) about linear transformations of the form $T: R^n \to R^n$. On the one hand, we think of equations such as

$$y_1 = \tfrac{3}{5}x_1 - \tfrac{4}{5}x_2$$
$$y_2 = \tfrac{4}{5}x_1 + \tfrac{3}{5}x_2$$

(1)

as mappings that take a point from the domain and carry it into a different point in the range (Figure 5-14). On the other hand, the equa-

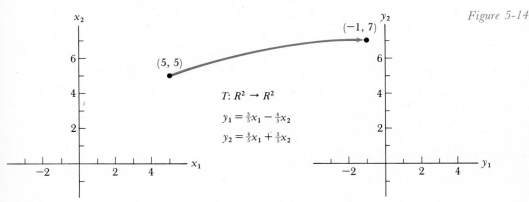

Figure 5-14

tions (1) are of the form

$$(2) \qquad \begin{aligned} y_1 &= (\cos\theta)x_1 - (\sin\theta)x_2 \\ y_2 &= (\sin\theta)x_1 + (\cos\theta)x_2 \end{aligned}$$

where $\cos\theta = \frac{3}{5}$, and $\sin\theta = \frac{4}{5}$. These are the equations that are used to relate the coordinates (y_1, y_2) of a point with respect to a given set of axes and the coordinates (x_1, x_2) of the same point with respect to rotated axes (Figure 5-15). We think of the equations (2) as a means of giving different names to the same point; the point is the same but is named with respect to two different frames of reference.

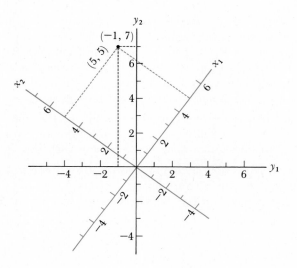

Figure 5-15

In our previous discussion of rotation of axes (Section 4-5), the equations (2) were written in the form

$$\begin{aligned} x_1 &= (\cos\theta)x_1' - (\sin\theta)x_2' \\ x_2 &= (\sin\theta)x_1' + (\cos\theta)x_2' \end{aligned}$$

The notation was chosen to remind us that (x_1, x_2) and (x_1', x_2') give the "address" of the same point, still in the "**x** plane," with respect to two different bases. We will continue with this notational reminder:

$\mathbf{y} = A\mathbf{x}$ indicates that we are thinking of a transformation that carries a point from the "**x** space" to the "**y** space."

$\mathbf{x} = P\mathbf{x}'$ indicates that we are thinking of a change of variables that gives the relationship between the coordinates of a point with respect to two different bases, say B and B', in the "**x** space."

THE SPECIAL CASE

Figure 5-16, with $V = R^2$, nicely pictures the problem that we have set for ourselves.

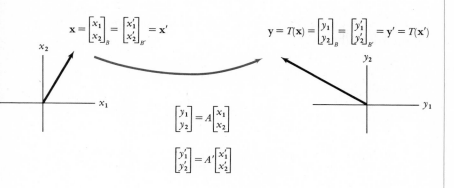

Figure 5-16

The n-dimensional vector space V has two bases $B = \{\mathbf{u}_1, \ldots, \mathbf{u}_n\}$ and $B' = \{\mathbf{v}_1, \ldots, \mathbf{v}_n\}$. $T: V \to V$ is a linear transformation. A point $\mathbf{x}$ in the domain of T has coordinates $(x_1, \ldots, x_n)$ and $(x'_1, \ldots, x'_n)$ with respect to the two bases, and the image of $\mathbf{x}$ in the range of T is $\mathbf{y}$, having coordinates $(y_1, \ldots, y_n)$ and $(y'_1, \ldots, y'_n)$ with respect to the same two bases. The linear transformation T is represented with respect to the basis B by the matrix A such that $\mathbf{y} = A\mathbf{x}$. It is represented with respect to basis B' by the matrix A' such that $\mathbf{y}' = A'\mathbf{x}'$.

Our problem is to determine how matrices A and A' are related.

Now is the time to review Theorem A of Section 4-5 which was tailored to our present situation. We begin by using the basis vectors of B and B' as columns to form two $n \times n$ matrices

$$U = [\mathbf{u}_1 \quad \cdots \quad \mathbf{u}_n] \quad \text{and} \quad V = [\mathbf{v}_1 \quad \cdots \quad \mathbf{v}_n]$$

Then this theorem tells us that there is an $n \times n$ matrix P such that $V = UP$, and the coordinates $(x_1, \ldots, x_n)_B$ and $(x'_1, \ldots, x'_n)_{B'}$ of an arbitrary point with respect to the two bases are related by

$$\begin{bmatrix} x_1 \\ \vdots \\ x_n \end{bmatrix} = P \begin{bmatrix} x'_1 \\ \vdots \\ x'_n \end{bmatrix} \tag{3}$$

Since the same bases are being used in the range of T, we similarly have

$$\begin{bmatrix} y_1 \\ \vdots \\ y_n \end{bmatrix} = P \begin{bmatrix} y'_1 \\ \vdots \\ y'_n \end{bmatrix} \tag{4}$$

Substitution of (3) and (4) in $\mathbf{y} = A\mathbf{x}$ gives us

$$P\begin{bmatrix} y'_1 \\ \vdots \\ y'_n \end{bmatrix} = AP\begin{bmatrix} x'_1 \\ \vdots \\ x'_n \end{bmatrix}$$

Since P is nonsingular (Problem 14, Problem Set 4-5), both sides can be multiplied on the left by P^{-1}.

$$\begin{bmatrix} y'_1 \\ \vdots \\ y'_n \end{bmatrix} = P^{-1}AP\begin{bmatrix} x'_1 \\ \vdots \\ x'_n \end{bmatrix}$$

T is represented by the matrix $P^{-1}AP$, and $A' = P^{-1}AP$.

THEOREM A

If under the conditions of Hypothesis A, $T: V \to V$ is represented by the two matrices A and A', then there exists a nonsingular matrix P such that $[\mathbf{v}_1 \quad \cdots \quad \mathbf{v}_n] = [\mathbf{u}_1 \quad \cdots \quad \mathbf{u}_n]P$, $\mathbf{x} = P\mathbf{x}'$, and $A' = P^{-1}AP$.

Example A

Two bases for R^2 are

$$B = \left\{ \mathbf{u}_1 = \begin{bmatrix} -3 \\ -2 \end{bmatrix}, \mathbf{u}_2 = \begin{bmatrix} -1 \\ 2 \end{bmatrix} \right\},$$

$$B' = \left\{ \mathbf{v}_1 = \begin{bmatrix} -1 \\ -6 \end{bmatrix}, \mathbf{v}_2 = \begin{bmatrix} 2 \\ 4 \end{bmatrix} \right\}$$

The transformation $T: R^2 \to R^2$ is represented with respect to the basis B by the matrix

$$A = \begin{bmatrix} 5 & 2 \\ 7 & 3 \end{bmatrix}$$

and the coordinates of $\mathbf{w}$ with respect to basis B are $(1, -4)_B$. Find the matrix representing T with respect to basis B', and the coordinates of $T(\mathbf{w})$ with respect to both bases.

After forming the matrices $U = [\mathbf{u}_1 \quad \mathbf{u}_2]$ and $V = [\mathbf{v}_1 \quad \mathbf{v}_2]$, our next task is to determine P such that $V = UP$. A little work can be saved by noting that P was found for this particular problem in Example A of Section 4-5, but it's not much more work to find P from the fact that $P = U^{-1}V$. Either way we find P, and then go on to find P^{-1}.

$$P = \begin{bmatrix} 1 & -1 \\ -2 & 1 \end{bmatrix}, \quad P^{-1} = \begin{bmatrix} -1 & -1 \\ -2 & -1 \end{bmatrix}$$

According to Theorem A, T is represented with respect to basis B' by

$$P^{-1}AP = \begin{bmatrix} -1 & -1 \\ -2 & -1 \end{bmatrix}\begin{bmatrix} 5 & 2 \\ 7 & 3 \end{bmatrix}\begin{bmatrix} 1 & -1 \\ -2 & 1 \end{bmatrix} = \begin{bmatrix} -2 & 7 \\ -3 & 10 \end{bmatrix}$$

The coordinates (w_1, w_2) and (w'_1, w'_2) are related by $\mathbf{w} = P\mathbf{w}'$. We were given $\mathbf{w} = (1, -4)_B$, so $\mathbf{w}'$ is found from $\mathbf{w}' = P^{-1}\mathbf{w}$.

$$\mathbf{w}' = \begin{bmatrix} -1 & -1 \\ -2 & -1 \end{bmatrix} \begin{bmatrix} 1 \\ -4 \end{bmatrix} = \begin{bmatrix} 3 \\ 2 \end{bmatrix}_{B'}$$

Now $T(\mathbf{w})$ can be found with respect to either coordinate system by using the appropriate coordinates for $\mathbf{w}$ and the appropriate matrix for T.

$$T(\mathbf{w})_B = \begin{bmatrix} 5 & 2 \\ 7 & 3 \end{bmatrix} \begin{bmatrix} 1 \\ -4 \end{bmatrix} = \begin{bmatrix} -3 \\ -5 \end{bmatrix}_B$$

$$T(\mathbf{w})_{B'} = \begin{bmatrix} -2 & 7 \\ -3 & 10 \end{bmatrix} \begin{bmatrix} 3 \\ 2 \end{bmatrix} = \begin{bmatrix} 8 \\ 11 \end{bmatrix}_{B'}$$

Several checks on our work are easy to make. For example, we can find the coordinates of $T(\mathbf{w})$ with respect to the standard basis from either $T(\mathbf{w})_B$ or $T(\mathbf{w})_{B'}$. We thus verify that

$$-3\begin{bmatrix} -3 \\ -2 \end{bmatrix} - 5\begin{bmatrix} -1 \\ 2 \end{bmatrix} = 8\begin{bmatrix} -1 \\ -6 \end{bmatrix} + 11\begin{bmatrix} 2 \\ 4 \end{bmatrix}$$

You should also verify that $T(\mathbf{w})_B = PT(\mathbf{w})_{B'}$. $\square$

THE GENERAL CASE

The way is now paved for a solution to the more general problem posed at the beginning of this section, pictured in Figure 5-17.

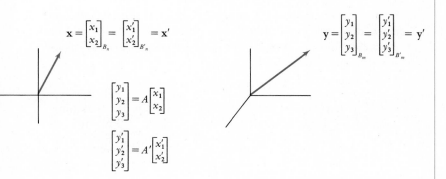

$$\mathbf{x} = \begin{bmatrix} x_1 \\ x_2 \end{bmatrix}_{B_n} = \begin{bmatrix} x'_1 \\ x'_2 \end{bmatrix}_{B'_n} = \mathbf{x}'$$

$$\mathbf{y} = \begin{bmatrix} y_1 \\ y_2 \\ y_3 \end{bmatrix}_{B_m} = \begin{bmatrix} y'_1 \\ y'_2 \\ y'_3 \end{bmatrix}_{B'_m} = \mathbf{y}'$$

$$\begin{bmatrix} y_1 \\ y_2 \\ y_3 \end{bmatrix} = A\begin{bmatrix} x_1 \\ x_2 \end{bmatrix}$$

$$\begin{bmatrix} y'_1 \\ y'_2 \\ y'_3 \end{bmatrix} = A'\begin{bmatrix} x'_1 \\ x'_2 \end{bmatrix}$$

Figure 5-17

A point in the domain U of a linear transformation $T: U \to V$ has coordinates $(x_1, \ldots, x_n)$ and $(x'_1, \ldots, x'_n)$ with respect to the two bases $B_n = \{\mathbf{u}_1, \ldots, \mathbf{u}_n\}$ and $B'_n = \{\mathbf{v}_1, \ldots, \mathbf{v}_n\}$. A point in the range of T has coordinates $(y_1, \ldots, y_m)$ and $(y'_1, \ldots, y'_m)$ with respect to the two bases $B_m = \{\mathbf{r}_1, \ldots, \mathbf{r}_m\}$ and $B'_m = \{\mathbf{s}_1, \ldots, \mathbf{s}_m\}$. The linear

HYPOTHESIS B

transformation T is represented with respect to the bases B_n and B_m by the matrix A where $\mathbf{y} = A\mathbf{x}$. It is represented with respect to the bases B'_n and B'_m by the matrix A' where $\mathbf{y}' = A'\mathbf{x}'$.

Our problem is again to determine how matrices A and A' are related.

This time we use Theorem A of Section 4-5 twice. It tells us that there exists an $n \times n$ nonsingular matrix P and an $m \times m$ nonsingular matrix Q such that

(5)
$$[\mathbf{v}_1 \quad \cdots \quad \mathbf{v}_n] = [\mathbf{u}_1 \quad \cdots \quad \mathbf{u}_n]P, \qquad \mathbf{x} = P\mathbf{x}'$$

and

(6)
$$[\mathbf{s}_1 \quad \cdots \quad \mathbf{s}_m] = [\mathbf{r}_1 \quad \cdots \quad \mathbf{r}_m]Q, \qquad \mathbf{y} = Q\mathbf{y}'$$

Substitution of (5) and (6) in $\mathbf{y} = A\mathbf{x}$ gives us $Q\mathbf{y}' = AP\mathbf{x}'$. The same argument used in the special case for matrix P can be used to show that Q has an inverse, giving us $\mathbf{y}' = Q^{-1}AP\mathbf{x}'$ and our second theorem.

THEOREM B

If under the conditions of Hypothesis B, $T: U \to V$ is represented by the two matrices A and A', then there exist nonsingular matrices P and Q such that (5) and (6) both hold and $A' = Q^{-1}AP$.

We conclude with a theorem that will be useful in the next section.

THEOREM C

If a transformation $T: U \to V$ has, with respect to different bases in either the domain or range or both, the matrix representations A and A', then the rank of A equals the rank of A'.

Proof: We know from Theorem B that there are nonsingular matrices P and Q such that $A' = Q^{-1}AP$. Set $K = Q^{-1}A$ so that $A' = KP$. Since the rank of a product never exceeds the rank of either factor (Problem 16, Problem Set 2-8),

$$\text{rank } A' \leq \text{rank } K \leq \text{rank } A$$

But if $A' = Q^{-1}AP$, then $A = QA'P^{-1}$ and similar reasoning leads to the conclusion that rank $A \leq$ rank A'. The two inequalities give us rank $A =$ rank A'. $\quad \square$

SIMILAR MATRICES

The matrix A is said to be **similar** to matrix B if there exists a nonsingular matrix P such that $A = P^{-1}BP$. Note that if A is similar to B, there is a nonsingular matrix $Q = P^{-1}$ such that $B = Q^{-1}AQ$ so that B is also similar to A. We may therefore simply refer to A and B as similar matrices, a relationship we express by writing $A \sim B$.

STANDARD BUT NOT BEST

Let B and B' be two bases of R^2 described by

$$B = \left\{ \mathbf{u}_1 = \begin{bmatrix} 1 \\ 0 \end{bmatrix}, \ \mathbf{u}_2 = \begin{bmatrix} 0 \\ 1 \end{bmatrix} \right\},$$

$$B' = \left\{ \mathbf{v}_1 = \begin{bmatrix} 1 \\ -1 \end{bmatrix}, \ \mathbf{v}_2 = \begin{bmatrix} 2 \\ 1 \end{bmatrix} \right\}$$

and let $T: R^2 \to R^2$ be described with respect to B by

$$\begin{bmatrix} y_1 \\ y_2 \end{bmatrix} = \begin{bmatrix} 3 & 2 \\ 1 & 2 \end{bmatrix} \begin{bmatrix} x_1 \\ x_2 \end{bmatrix}$$

To find the matrix representation of T with respect to B', we first seek P such that $[\mathbf{v}_1 \ \ \mathbf{v}_2] = [\mathbf{u}_1 \ \ \mathbf{u}_2]P$. This is no challenge, however, since $[\mathbf{u}_1 \ \ \mathbf{u}_2] = I$. Therefore,

$$P = \begin{bmatrix} 1 & 2 \\ -1 & 1 \end{bmatrix} \quad \text{and} \quad P^{-1} = \begin{bmatrix} \frac{1}{3} & -\frac{2}{3} \\ \frac{1}{3} & \frac{1}{3} \end{bmatrix}$$

The matrix representation for T with respect to B' is, therefore,

$$\begin{bmatrix} \frac{1}{3} & -\frac{2}{3} \\ \frac{1}{3} & \frac{1}{3} \end{bmatrix} \begin{bmatrix} 3 & 2 \\ 1 & 2 \end{bmatrix} \begin{bmatrix} 1 & 2 \\ -1 & 1 \end{bmatrix} = \begin{bmatrix} 1 & 0 \\ 0 & 4 \end{bmatrix}$$

It is easy now to describe the effect that T has on vectors if we use the basis B'. The vector $(x_1', x_2')_{B'}$ is mapped into $(x_1', 4x_2')_{B'}$. For instance,

$$\text{if} \quad \mathbf{w} = \begin{bmatrix} 2 \\ 1 \end{bmatrix}_{B'}, \quad \text{then} \quad T(\mathbf{w})_{B'} = \begin{bmatrix} 1 & 0 \\ 0 & 4 \end{bmatrix} \begin{bmatrix} 2 \\ 1 \end{bmatrix}_{B'} = \begin{bmatrix} 2 \\ 4 \end{bmatrix}_{B'}$$

To express the same vector $\mathbf{w}$ and its image $T(\mathbf{w})$ with respect to the standard basis B, we first note that

$$\mathbf{w} = 2\mathbf{v}_1 + \mathbf{v}_2 = 2(1, -1) + (2, 1) = (4, -1) = 4\mathbf{u}_1 - \mathbf{u}_2$$

Then

$$\mathbf{w} = \begin{bmatrix} 4 \\ -1 \end{bmatrix}_B \quad \text{and} \quad T(\mathbf{w})_B = \begin{bmatrix} 3 & 2 \\ 1 & 2 \end{bmatrix} \begin{bmatrix} 4 \\ -1 \end{bmatrix}_B = \begin{bmatrix} 10 \\ 2 \end{bmatrix}_B$$

See Figure 5-C. Now B is, of course, the standard basis of R^2, but B' is a better basis to use if we wish to study T.

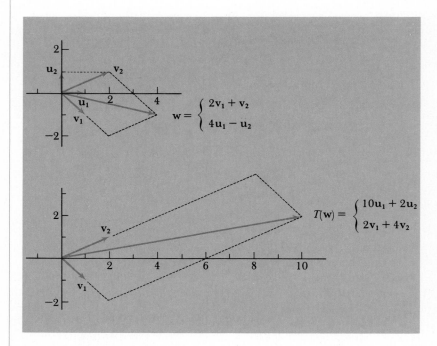

Figure 5-C

Attention is naturally drawn to similar matrices by Theorem A which says that if $T: R^n \to R^n$ is represented by two matrices A and A', then $A \sim A'$. The converse is also true. If $A \sim A'$, then the equations $\mathbf{y} = A\mathbf{x}$ and $\mathbf{y}' = A'\mathbf{x}$ represent the same transformation with respect to different bases (Problem 21, Problem Set 5-4).

The proof of Theorem C shows that if $A \sim B$, then rank A equals rank B. Also, if $A \sim B$, then det A equals det B. To see this, appeal twice to the rule for the determinant of a product of matrices, $\det(MN) = (\det M)(\det N)$. Then if $A \sim B$ so that $A = P^{-1}BP$,

$$\det A = (\det P^{-1})(\det B)(\det P)$$

Now, while the product of the matrices P^{-1} and B may not commute, the product of the numbers det P^{-1} and det B surely commutes. That, together with another appeal to the rule for the determinant of a product, gives the result

$$\det A = (\det B)(\det P^{-1})(\det P)$$
$$= (\det B)(\det P^{-1}P)$$
$$= (\det B)(\det I) = \det B$$

PROBLEM
SET 5-4

In Problems 1–4 you are given two bases B and B' of R^2; the matrix A representing with respect to basis B a transformation $T: R^2 \to R^2$; and the coordinates with respect to basis B' of a vector $\mathbf{w}$.

(a) *Illustrate equation (3) of this section by finding P such that* $[\mathbf{v}_1 \quad \mathbf{v}_2] = [\mathbf{u}_1 \quad \mathbf{u}_2]P.$
(b) *Use* $\mathbf{w} = P\mathbf{w}'$ *to find coordinates with respect to basis B of the vector* $\mathbf{w}.$
(c) *Use* $\mathbf{y} = A\mathbf{x}$ *to find the coordinates with respect to basis B of* $T(\mathbf{w}).$
(d) *Find* $P^{-1}AP$, *the matrix representation with respect to the basis B' of the transformation T.*
(e) *Use* $\mathbf{y}' = P^{-1}AP\mathbf{x}'$ *to find the coordinates with respect to basis B' of* $T(\mathbf{w}).$
(f) *Check your answers to (c) and (e) by verifying that your answers satisfy* $T(\mathbf{w})_B = PT(\mathbf{w})_{B'}.$

1. $B = \left\{ \mathbf{u}_1 = \begin{bmatrix} -1 \\ 2 \end{bmatrix}, \mathbf{u}_2 = \begin{bmatrix} 1 \\ 1 \end{bmatrix} \right\}, \quad B' = \left\{ \mathbf{v}_1 = \begin{bmatrix} -5 \\ 4 \end{bmatrix}, \mathbf{v}_2 = \begin{bmatrix} -1 \\ 5 \end{bmatrix} \right\}$

$$A = \begin{bmatrix} 2 & -1 \\ 3 & 5 \end{bmatrix}, \quad \mathbf{w} = \begin{bmatrix} 4 \\ 3 \end{bmatrix}_{B'}$$

2. $B = \left\{ \mathbf{u}_1 = \begin{bmatrix} -1 \\ 1 \end{bmatrix}, \mathbf{u}_2 = \begin{bmatrix} 2 \\ -1 \end{bmatrix} \right\}, \quad B' = \left\{ \mathbf{v}_1 = \begin{bmatrix} 1 \\ 1 \end{bmatrix}, \mathbf{v}_2 = \begin{bmatrix} -3 \\ 2 \end{bmatrix} \right\}$

$$A = \begin{bmatrix} 2 & 1 \\ -3 & -2 \end{bmatrix}, \quad \mathbf{w} = \begin{bmatrix} -7 \\ 8 \end{bmatrix}_{B'}$$

3. $B = \left\{ \mathbf{u}_1 = \begin{bmatrix} 1 \\ -2 \end{bmatrix}, \mathbf{u}_2 = \begin{bmatrix} -1 \\ 1 \end{bmatrix} \right\}, \quad B' = \left\{ \mathbf{v}_1 = \begin{bmatrix} 0 \\ -1 \end{bmatrix}, \mathbf{v}_2 = \begin{bmatrix} 2 \\ -5 \end{bmatrix} \right\}$

$$A = \begin{bmatrix} 5 & 2 \\ 3 & 1 \end{bmatrix}, \quad \mathbf{w} = \begin{bmatrix} 4 \\ -1 \end{bmatrix}_{B'}$$

4. $B = \left\{ \mathbf{u}_1 = \begin{bmatrix} -2 \\ 3 \end{bmatrix}, \mathbf{u}_2 = \begin{bmatrix} 5 \\ 4 \end{bmatrix} \right\}, \quad B' = \left\{ \mathbf{v}_1 = \begin{bmatrix} 1 \\ 10 \end{bmatrix}, \mathbf{v}_2 = \begin{bmatrix} 7 \\ 1 \end{bmatrix} \right\}$

$$A = \begin{bmatrix} -1 & 2 \\ 3 & 1 \end{bmatrix}, \quad \mathbf{w} = \begin{bmatrix} -1 \\ 1 \end{bmatrix}_{B'}$$

5–8. *Using the data of Problems 1–4, answer the following questions:*
   (a) What are the coordinates with respect to the standard basis of $\mathbf{w}$ and $T(\mathbf{w})$? That is, find $\mathbf{w}_S$ and $T(\mathbf{w})_S$.
   (b) Find the matrix representation M with respect to the standard basis of the transformation T.
   (c) Verify that $T(\mathbf{w})_S = M\mathbf{w}_S$.

In Problems 9–12 you are given two bases B_n *and* B_n' *of* R^n; *two bases* B_m *and* B_m' *of* R^m; *the matrix A representing with respect to the bases* B_n *and* B_m *a transformation* $T: R^n \to R^m$; *and the coordinates with respect to basis* B_n' *of a vector* $\mathbf{w}.$
(a) *Find a matrix P to illustrate equation (5).*
(b) *Find a matrix Q to illustrate equation (6).*
(c) *Use equation (5) to find the coordinates with respect to basis* B_n *of the vector* $\mathbf{w}.$
(d) *Use* $\mathbf{y} = A\mathbf{x}$ *to find the coordinates with respect to basis* B_m *of the vector* $T(\mathbf{w}).$
(e) *Find* $Q^{-1}AP$, *the matrix representation with respect to the bases* B_n' *and* B_m' *of the transformation T.*

(f) *Use* $\mathbf{y}' = Q^{-1}AP\mathbf{x}'$ *to find the coordinates with respect to basis* B'_m *of the vector* $T(\mathbf{w})$.

(g) *Check your answers to* (d) *and* (f) *by verifying that your answers satisfy* $T(\mathbf{w})_{B_m} = QT(\mathbf{w})_{B'_m}$.

9. $B_2 = \left\{ \mathbf{u}_1 = \begin{bmatrix} 1 \\ -1 \end{bmatrix}, \mathbf{u}_2 = \begin{bmatrix} -1 \\ 2 \end{bmatrix} \right\}, \quad B'_2 = \left\{ \mathbf{v}_1 = \begin{bmatrix} -2 \\ 5 \end{bmatrix}, \mathbf{v}_2 = \begin{bmatrix} 1 \\ -3 \end{bmatrix} \right\}$

$B_3 = \left\{ \mathbf{r}_1 = \begin{bmatrix} 1 \\ 0 \\ 1 \end{bmatrix}, \mathbf{r}_2 = \begin{bmatrix} -1 \\ 1 \\ 0 \end{bmatrix}, \mathbf{r}_3 = \begin{bmatrix} 0 \\ -1 \\ 1 \end{bmatrix} \right\},$

$B'_3 = \left\{ \mathbf{s}_1 = \begin{bmatrix} 0 \\ 0 \\ 2 \end{bmatrix}, \mathbf{s}_2 = \begin{bmatrix} 2 \\ 1 \\ 1 \end{bmatrix}, \mathbf{s}_3 = \begin{bmatrix} -1 \\ 2 \\ -1 \end{bmatrix} \right\}$

$T: R^2 \to R^3$ is represented by $A = \begin{bmatrix} -1 & 2 \\ 0 & 1 \\ 3 & -1 \end{bmatrix}$ and $\mathbf{w}_{B_2} = \begin{bmatrix} 2 \\ 1 \end{bmatrix}$.

10. $B_2 = \left\{ \mathbf{u}_1 = \begin{bmatrix} 2 \\ 3 \end{bmatrix}, \mathbf{u}_2 = \begin{bmatrix} -1 \\ -2 \end{bmatrix} \right\}, \quad B'_2 = \left\{ \mathbf{v}_1 = \begin{bmatrix} -1 \\ -3 \end{bmatrix}, \mathbf{v}_2 = \begin{bmatrix} 0 \\ 1 \end{bmatrix} \right\}$

$B_3 = \left\{ \mathbf{r}_1 = \begin{bmatrix} 2 \\ -1 \\ 0 \end{bmatrix}, \mathbf{r}_2 = \begin{bmatrix} 1 \\ 0 \\ -2 \end{bmatrix}, \mathbf{r}_3 = \begin{bmatrix} 0 \\ 1 \\ 3 \end{bmatrix} \right\},$

$B'_3 = \left\{ \mathbf{s}_1 = \begin{bmatrix} 3 \\ 0 \\ 1 \end{bmatrix}, \mathbf{s}_2 = \begin{bmatrix} 4 \\ -3 \\ -3 \end{bmatrix}, \mathbf{s}_3 = \begin{bmatrix} 1 \\ -1 \\ -5 \end{bmatrix} \right\}$

$T: R^2 \to R^3$ is represented by $A = \begin{bmatrix} 3 & -2 \\ 1 & -3 \\ -2 & 1 \end{bmatrix}$ and $\mathbf{w}_{B_2} = \begin{bmatrix} 3 \\ 4 \end{bmatrix}$.

11. $B_3 = \left\{ \mathbf{u}_1 = \begin{bmatrix} -1 \\ 0 \\ 3 \end{bmatrix}, \mathbf{u}_2 = \begin{bmatrix} 0 \\ -1 \\ 1 \end{bmatrix}, \mathbf{u}_3 = \begin{bmatrix} 2 \\ 0 \\ -1 \end{bmatrix} \right\},$

$B'_3 = \left\{ \mathbf{v}_1 = \begin{bmatrix} -3 \\ -1 \\ 5 \end{bmatrix}, \mathbf{v}_2 = \begin{bmatrix} 3 \\ 0 \\ -4 \end{bmatrix}, \mathbf{v}_3 = \begin{bmatrix} 2 \\ -1 \\ 0 \end{bmatrix} \right\}$

$B_2 = \left\{ \mathbf{r}_1 = \begin{bmatrix} 5 \\ 3 \end{bmatrix}, \mathbf{r}_2 = \begin{bmatrix} -3 \\ -1 \end{bmatrix} \right\}, \quad B'_2 = \left\{ \mathbf{s}_1 = \begin{bmatrix} 9 \\ 7 \end{bmatrix}, \mathbf{s}_2 = \begin{bmatrix} 11 \\ 9 \end{bmatrix} \right\}$

$T: R^3 \to R^2$ is represented by $A = \begin{bmatrix} 1 & -1 & 2 \\ 3 & 2 & -4 \end{bmatrix}$ and $\mathbf{w}_{B_3} = \begin{bmatrix} -1 \\ 1 \\ 2 \end{bmatrix}$.

12. $B_3 = \left\{ \mathbf{u}_1 = \begin{bmatrix} 1 \\ 2 \\ 1 \end{bmatrix}, \mathbf{u}_2 = \begin{bmatrix} -2 \\ 0 \\ -1 \end{bmatrix}, \mathbf{u}_3 = \begin{bmatrix} -1 \\ 2 \\ 0 \end{bmatrix} \right\},$

$B_3' = \mathbf{v}_1 = \begin{bmatrix} 2 \\ 0 \\ 0 \end{bmatrix}, \mathbf{v}_2 = \begin{bmatrix} -2 \\ 0 \\ -1 \end{bmatrix}, \mathbf{v}_3 = \begin{bmatrix} -3 \\ 2 \\ -1 \end{bmatrix}$

$B_2 = \left\{ \mathbf{r}_1 = \begin{bmatrix} 2 \\ -3 \end{bmatrix}, \mathbf{r}_2 = \begin{bmatrix} -5 \\ 4 \end{bmatrix} \right\}, \qquad B_2' = \left\{ \mathbf{s}_1 = \begin{bmatrix} -4 \\ -1 \end{bmatrix}, \mathbf{s}_2 = \begin{bmatrix} -7 \\ 0 \end{bmatrix} \right\}$

$T: R^3 \to R^2$ is represented by $A = \begin{bmatrix} 2 & 0 & 3 \\ 1 & -1 & 0 \end{bmatrix}$ and $\mathbf{w}_{B_3'} = \begin{bmatrix} 4 \\ 3 \\ -2 \end{bmatrix}.$

13–16. *Using the data of Problems 9–12, answer the following questions:*
  (a) What are the coordinates with respect to the standard bases of $\mathbf{w}$ and $T(\mathbf{w})$? That is, find $\mathbf{w}_S$ and $T(\mathbf{w})_S$.
  (b) Find the matrix representation M with respect to the standard bases of $T: R^n \to R^m$.
  (c) Verify that $T(\mathbf{w})_S = M\mathbf{w}_S$.

In Problems 17–20, use the given matrix A to form the similar matrix $B = P^{-1}AP$ where

$$P = \begin{bmatrix} 1 & -1 & 0 & 1 \\ -4 & 1 & 1 & -1 \\ -2 & 0 & 1 & 0 \\ 1 & 1 & 0 & -2 \end{bmatrix}, \qquad P^{-1} = \begin{bmatrix} -1 & -1 & 1 & 0 \\ -5 & -3 & 3 & -1 \\ -2 & -2 & 3 & 0 \\ -3 & -2 & 2 & -1 \end{bmatrix}$$

Then verify (a) rank A = rank B, *and* (b) det A = det B.

17. $A = \begin{bmatrix} 1 & 0 & 0 & -1 \\ 0 & 2 & 1 & 0 \\ 0 & 0 & -1 & 0 \\ 0 & 0 & 0 & 3 \end{bmatrix}$ 18. $A = \begin{bmatrix} -1 & 0 & 0 & 0 \\ 0 & 3 & 1 & 0 \\ 0 & 0 & -2 & 4 \\ 0 & 0 & 0 & 1 \end{bmatrix}$

19. $A = \begin{bmatrix} 3 & -2 & 1 & 5 \\ -2 & 4 & 2 & -3 \\ 0 & 8 & 8 & 1 \\ 1 & 2 & 3 & 2 \end{bmatrix}$ 20. $A = \begin{bmatrix} 4 & 5 & 7 & -3 \\ -1 & -2 & -2 & 5 \\ 5 & 4 & 8 & 9 \\ 11 & 13 & 19 & -4 \end{bmatrix}$

21. Show that if $A \sim A'$, then $\mathbf{y} = A\mathbf{x}$ and $\mathbf{y}' = A'\mathbf{x}'$ represent the same linear transformation with respect to different bases. *Hint:* Choose a basis $\{\mathbf{u}_1, \ldots, \mathbf{u}_n\}$ and define $T: R^n \to R^n$ with respect to this basis by $\mathbf{y} = A\mathbf{x}$.

22. Show that if $A \sim B$, then $A^2 \sim B^2$.

23. Show that if A is an $n \times n$ nonsingular matrix, then for any other $n \times n$ matrix B, $AB \sim BA$.

24. If $A \sim B$ and $B \sim C$, is $A \sim C$?

25. If $A \sim B$, is $A^t \sim B^t$?

V 26. Let $T: R^2 \rightarrow R^2$ be described with respect to the standard basis by the matrix equation $\mathbf{y} = A\mathbf{x}$, matrix A being

$$A = \begin{bmatrix} 7 & 3 \\ -9 & -5 \end{bmatrix}$$

Show, in the sense of *Standard but Not Best*, that this transformation is better described with respect to the basis

$$B' = \left\{ \mathbf{v}_1 = \begin{bmatrix} 1 \\ -1 \end{bmatrix}, \ \mathbf{v}_2 = \begin{bmatrix} -2 \\ 1 \end{bmatrix} \right\}$$

5-5 THE NULL SPACE AND THE RANGE

Let $T: V \rightarrow W$ be a linear transformation of the vector space V into the vector space W. Let the zero vector in the two spaces be, respectively, $\mathbf{0}_V$ and $\mathbf{0}_W$. We know that $T(\mathbf{0}_V) = \mathbf{0}_W$, so it is clear that the preimage of $\mathbf{0}_W$ is never empty. The preimage of the zero vector $\mathbf{0}_W$ is called the **kernel** of the transformation.

THEOREM A

The kernel of a linear transformation $T: V \rightarrow W$ is a subspace of V.

Proof: Suppose that $\mathbf{u}$ and $\mathbf{v}$ are in the kernel. Then for any two scalars r and s,

$$\begin{aligned} T(r\mathbf{u} + s\mathbf{v}) &= T(r\mathbf{u}) + T(s\mathbf{v}) \\ &= rT(\mathbf{u}) + sT(\mathbf{v}) \\ &= r\mathbf{0}_W + s\mathbf{0}_W = \mathbf{0}_W \end{aligned}$$

Thus, $r\mathbf{u} + s\mathbf{v}$ is also in the kernel, and this is enough (Theorem A of Section 4-2) to show that the kernel is a subspace. $\square$

Because the kernel turns out to be a vector space, a vector space that maps into $\mathbf{0}$, it is sometimes referred to as the **null space** of T. We use the terms kernel and null space interchangeably.

It is frequently useful to describe the null space of a linear transformation, but no general rules can be given. Each situation provides a new problem in which one must ask how to describe the subspace of vectors that a particular transformation will carry into the zero vector. By way of illustration, we describe the null space of each of the transformations first introduced in Section 5-2. Recall that these examples referred to

P—the set of all polynomials
M_{mn}—the set of all $m \times n$ matrices
$C'(a, b)$—the set of continuously differentiable functions
S—the set of all infinite sequences
R—the space R^1 of real numbers

The null space of $I: P \rightarrow R$ defined by

Example A
C

$$I(\mathbf{p}) = \int_0^1 p(x)\, dx$$

is the set of all polynomials which, like $p(x) = 2x - 1$, bound equal areas above and below the x axis between 0 and 1 (Figure 5-18). □

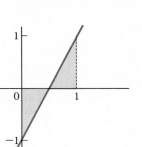

Figure 5-18

The null space of $T: M_{mn} \rightarrow M_{nm}$ defined by $T(A) = A^t$ is the $m \times n$ Θ matrix, the matrix having all zero entries. The null space of T is the trivial subspace of M_{mn} consisting of just the element Θ. □

Example B

The differential operator $D: C'(a, b) \rightarrow C(a, b)$ has as its null space the set of all functions that are constant on (a, b). □

Example C
C

The left-shift operator $L: S \rightarrow S$ sends into the zero vector any sequence having all zero entries except possibly in the first position, since

Example D

$$L(x_1, 0, 0, 0, \ldots) = (0, 0, 0, \ldots) \quad \square$$

The null space of the transformation $T: R^n \rightarrow R^m$, defined by the matrix equation $\mathbf{y} = A\mathbf{x}$ is, of course, the set of all vectors $\mathbf{x}$ for which $A\mathbf{x} = \mathbf{0}$. This is the same set in R^n that we have previously described as the solution space for the homogeneous system $A\mathbf{x} = \mathbf{0}$. □

Example E

It is reasonable to wonder why we should be so interested in the kernel of a transformation. To answer this question, let us for a few moments personify the points in the domain V and in the space W that contains the range of the linear transformation $T: V \rightarrow W$. Each point $\mathbf{x}$ in V is going to be taken somewhere by T. There is not much to ask, except,

"Where will I be taken?" A point $\mathbf{y}$ in W, however, may have several questions: "Will anyone come to my location? Will I receive just a single visitor, or will there be many?"

It turns out that the answers to $\mathbf{y}$'s questions are all related to the dimension of the null space of the transformation T. It also turns out, since every linear transformation from an n-dimensional space V to an m-dimensional space W can be represented by an $m \times n$ matrix, that we will be greatly helped by our knowledge of matrices.

Let us begin with the observation that all the $\mathbf{y}$'s that are to be visited constitute a subspace of T.

THEOREM B

Let $T:V \rightarrow W$ be a linear transformation, and suppose K is a subspace of V. Then K is mapped into a subspace of W. In particular, the range of T is a subspace of W.

Proof: T maps K into some subset L contained in W. If $\mathbf{y}_1$ and $\mathbf{y}_2$ are arbitrary members of L, then there must be two vectors $\mathbf{x}_1$ and $\mathbf{x}_2$ in K such that $T(\mathbf{x}_1) = \mathbf{y}_1$, $T(\mathbf{x}_2) = \mathbf{y}_2$. Now let r and s be arbitrary scalars. Since K is a vector space, $r\mathbf{x}_1 + s\mathbf{x}_2$ is in K. But then $T(r\mathbf{x}_1 + s\mathbf{x}_2)$ must be in L. That is,

$$T(r\mathbf{x}_1 + s\mathbf{x}_2) = rT(\mathbf{x}_1) + sT(\mathbf{x}_2) = r\mathbf{y}_1 + s\mathbf{y}_2$$

is in L, which is enough to show (by Theorem A of Section 4-2) that L is a subspace of W. If $K = V$, this says that the range of T is a subspace of W. $\square$

Suppose that bases for an n-dimensional vector space V and an m-dimensional vector space W have been chosen, allowing us to identify $T: V \rightarrow W$ with an $m \times n$ matrix A, and to describe T by the matrix equation $\mathbf{y} = A\mathbf{x}$. We can then describe both the null space and the range of T in terms of the matrix A. An example will remind you of how this is done.

Example F

Describe the null space and the range of a transformation represented with respect to some fixed bases by the matrix

$$A = \begin{bmatrix} -1 & 2 & 0 & -1 & 3 \\ 3 & 8 & -4 & -5 & 13 \\ 3 & 1 & -2 & -1 & 2 \\ -9 & 4 & 4 & -1 & 5 \end{bmatrix}$$

Proceeding in the usual way, we find the row echelon form of matrix A:

$$\begin{bmatrix} 1 & 0 & -\frac{4}{7} & -\frac{1}{7} & \frac{1}{7} \\ 0 & 1 & -\frac{2}{7} & -\frac{4}{7} & \frac{11}{7} \\ 0 & 0 & 0 & 0 & 0 \\ 0 & 0 & 0 & 0 & 0 \end{bmatrix}$$

It is clear that to satisfy $A\mathbf{x} = \mathbf{0}$, we must have

$$x_1 = \tfrac{4}{7}x_3 + \tfrac{1}{7}x_4 - \tfrac{1}{7}x_5$$
$$x_2 = \tfrac{2}{7}x_3 + \tfrac{4}{7}x_4 - \tfrac{11}{7}x_5$$

Vectors in the solution space are

$$(\tfrac{4}{7}x_3 + \tfrac{1}{7}x_4 - \tfrac{1}{7}x_5, \tfrac{2}{7}x_3 + \tfrac{4}{7}x_4 - \tfrac{11}{7}x_5, x_3, x_4, x_5)$$
$$= x_3(\tfrac{4}{7}, \tfrac{2}{7}, 1, 0, 0) + x_4(\tfrac{1}{7}, \tfrac{4}{7}, 0, 1, 0) + x_5(-\tfrac{1}{7}, -\tfrac{11}{7}, 0, 0, 1)$$

The null space of T has a basis of

$$\mathbf{v}_1 = (\tfrac{4}{7}, \tfrac{2}{7}, 1, 0, 0), \qquad \mathbf{v}_2 = (\tfrac{1}{7}, \tfrac{4}{7}, 0, 1, 0), \qquad \mathbf{v}_3 = (-\tfrac{1}{7}, -\tfrac{11}{7}, 0, 0, 1)$$

Since $A\mathbf{x}$ is, for any $\mathbf{x}$, a linear combination of the column vectors of A, the range of T is simply the space spanned by the column vectors of A. This is the same space spanned by the row vectors of A^t, so we can find a basis for the range by finding the row echelon form of A^t. It turns out to be

$$\begin{bmatrix} 1 & 0 & -\tfrac{3}{2} & 6 \\ 0 & 1 & \tfrac{1}{2} & -1 \\ 0 & 0 & 0 & 0 \\ 0 & 0 & 0 & 0 \\ 0 & 0 & 0 & 0 \end{bmatrix}$$

Basis vectors for the range of T are

$$\mathbf{u}_1 = (1, 0, -\tfrac{3}{2}, 6), \qquad \mathbf{u}_2 = (0, 1, \tfrac{1}{2}, -1)$$

The dimension of the range of the linear transformation represented by A is 2, the (column) rank of A. The dimension of the null space is 3, consistent with Theorem B of Section 2-8 which asserts that a homogeneous system $A\mathbf{x} = \mathbf{0}$ of m equations in n variables has a solution space of dimension $n - r$ where r is the rank of A. In this example, $n - r = 5 - 2 = 3$. $\square$

You, as a careful reader, will worry for a moment about the fact that had different bases for V or W been used, we would have had a different matrix A' representing T. But, being a careful reader, you will quickly see the usefulness of Theorem C of Section 5-4. The rank of A' must be the same as the rank of A.

The same methods enable us to analyze any linear transformation having both domain and range in finite-dimensional vector spaces.

> *Let $T: V \to W$ be a linear transformation from the n-dimensional vector space V into the m-dimensional vector space W, and let A be any matrix representation of T.*
> (a) *The dimension of the range of T is r, the rank of A.*

THEOREM C

(b) *The dimension of the null space of T is $n - r$.*
Obviously, the dimension of the null space plus the dimension of the range is n.

Consider now a particular $\mathbf{y} = \mathbf{b}$ in W. The question of how many, if any, $\mathbf{x}$ in V will be sent to $\mathbf{b}$ comes down to this: Let bases for V and W be chosen, express $\mathbf{b}$ in terms of the basis of W and, for the corresponding matrix A, ask if there are solutions to $A\mathbf{x} = \mathbf{b}$.

We happen to know all about the solution set $\mathcal{S}$ to such a system; review the summary at the conclusion of Secton 2-8. $\mathcal{S}$ may be empty, meaning that $\mathbf{b}$ is not in the range of T; there will be no visitors to $\mathbf{b}$ coming from V. If $\mathcal{S}$ contains at least one member $\mathbf{p}$, then the entire set can be described in the form

$$\mathcal{S} = \{\mathbf{p} + \mathbf{s} : \mathbf{s} \text{ is a solution to } A\mathbf{x} = \mathbf{0}\}$$

If $A\mathbf{x} = \mathbf{0}$ has just one solution (the dimension of the null space is 0), then just the one visitor $\mathbf{p}$ will be sent to $\mathbf{b}$. If $A\mathbf{x} = \mathbf{0}$ has more than one solution, it will have infinitely many, but they will all come from a subspace of dimension $n - r$. This gives us some measure, then, of how many points get sent to $\mathbf{b}$.

*PROBLEM
SET 5-5*

In Problems 1–6, describe the null space of the given linear transformation.

1. T takes the space of 2×2 matrices into itself, T defined by

$$T\left(\begin{bmatrix} a_{11} & a_{12} \\ a_{21} & a_{22} \end{bmatrix}\right) = \begin{bmatrix} a_{11} & a_{12} \\ 0 & a_{22} \end{bmatrix}$$

2. T takes the space of 2×2 matrices into R^2, T defined by

$$T(A) = A\begin{bmatrix} 1 \\ 1 \end{bmatrix}$$

C

3. T is defined on the space C'' of twice-differentiable functions by $T(f) = f''$.

4. T is defined on the space C of continuous functions by $T(f)(x) = xf(x)$; for example, if $f(x) = \sin x$, $T(f)(x) = x \sin x$.

5. T takes the space P_2 of second degree polynomials into itself, and is defined on the Lagrange polynomials $q_1(x) = \frac{1}{2}x(1 - x)$, $q_2(x) = -x^2 + 1$, $q_3(x) = \frac{1}{2}x(1 + x)$ by

$$T(q_1(x)) = x^2 + x, \qquad T(q_2(x)) = x^2 - 1, \qquad T(q_3(x)) = x + 1$$

6. T takes the space P_2 into itself, and is defined on the basis $\{p_2(x) = x^2,$ $p_1(x) = x,\ p_0(x) = 1\}$ by

$$T(p_2(x)) = x^2 + 1, \qquad T(p_1(x)) = x^2 - 1, \qquad T(p_0(x)) = x$$

In Problems 7–10, you are given a matrix which defines a linear transformation $T: R^2 \rightarrow R^2$. In each case, answer the following questions:

(a) What is the range of T?

(b) What is the kernel of T?

(c) Let K be the subspace of all vectors $\mathbf{x} = (x_1, x_2)$ in which $x_1 = x_2$. According to Theorem B, T maps K into a subspace of R^2. Describe the subspace.

7. $\begin{bmatrix} 1 & 2 \\ -1 & 1 \end{bmatrix}$

8. $\begin{bmatrix} -1 & 1 \\ 2 & 1 \end{bmatrix}$

9. $\begin{bmatrix} -1 & 2 \\ 2 & -4 \end{bmatrix}$

10. $\begin{bmatrix} 1 & -2 \\ -\frac{1}{2} & 1 \end{bmatrix}$

Problems 11 and 12 refer to a linear transformation $T: V \rightarrow W$ that takes the 3-dimensional vector space V into the 5-dimensional vector space W. In each case, from the information given, find the dimension of the range of T.

11. (a) If $\mathbf{x}_1 \neq \mathbf{x}_2$, then $T(\mathbf{x}_1) \neq T(\mathbf{x}_2)$.

(b) A basis $\{\mathbf{v}_1, \mathbf{v}_2, \mathbf{v}_3\}$ can be chosen for V so that

$$T(\mathbf{v}_1) = 2T(\mathbf{v}_2) = 3T(\mathbf{v}_3) \neq \mathbf{0}$$

(c) The dimension of the null space of T is 1.

12. (a) If $\mathbf{x}_1$ and $\mathbf{x}_2$ are linearly independent, then so are $T(\mathbf{x}_1)$ and $T(\mathbf{x}_2)$.

(b) Bases $\{\mathbf{v}_1, \mathbf{v}_2, \mathbf{v}_3\}$ and $\{\mathbf{w}_1, \mathbf{w}_2, \mathbf{w}_3, \mathbf{w}_4, \mathbf{w}_5\}$ can be chosen so that $T(\mathbf{v}_i) = \mathbf{w}_i$ for $i = 1, 2, 3$.

(c) T can be represented by a matrix of rank 2.

13. A linear transformation $T: V \rightarrow W$ is represented by a 6×9 matrix A. Give the dimension of the range of T in each of the following cases:

(a) The rank of A is 5.

(b) $A\mathbf{x} = \mathbf{0}$ has a 2-dimensional solution space.

14. A linear transformation $T: V \rightarrow W$ is represented by a 5×7 matrix A. Give the dimension of the range of T in each of the following cases:

(a) A has the maximal possible rank.

(b) The row vectors span a subspace of dimension 3.

15. A linear transformation $T: V \rightarrow W$ is said to be **1–1 (one-to-one)** if, whenever $\mathbf{x}_1 \neq \mathbf{x}_2$, $T(\mathbf{x}_1) \neq T(\mathbf{x}_2)$. Prove the following:

(a) If T is 1–1, the null space of T consists of $\mathbf{0}$ alone.

(b) If the null space of T consists of $\mathbf{0}$ alone, then T is 1–1.

(c) If T is 1–1, then the dimension of $W \geq$ the dimension of V.

16. A linear transformation T is represented by an $n \times n$ matrix A. Prove the following:

(a) The null space of T is $\{\mathbf{0}\}$ if $\det A \neq 0$.

(b) If $\det A \neq 0$, then the null space of T is $\{\mathbf{0}\}$.

(c) If the rank of A is less than n, then there is a nonzero vector $\mathbf{x}$ such that $A\mathbf{x} = \mathbf{0}$.

17. Let $T: V \to W$ be linear, and suppose that $K = \{x_1, \ldots, x_k\}$ is a set of vectors in V. If the set $L = \{T(x_1), \ldots, T(x_k)\}$ is linearly independent, must K be linearly independent?

18. What about the converse of Problem 17: If K is linearly independent, must L be linearly independent?

In Problems 19–22, you are given an $m \times n$ matrix which is to be used to define a linear transformation $T: R^n \to R^m$. For each transformation,
(a) *Find a basis for the null space.*
(b) *Find a basis for the range.*
(c) *Verify Theorem C of this section.*

19. $\begin{bmatrix} 2 & -1 & 3 & 4 \\ 8 & -7 & 7 & 6 \\ -1 & 2 & 1 & 3 \end{bmatrix}$

20. $\begin{bmatrix} 1 & 3 & -1 \\ 2 & -1 & 4 \end{bmatrix}$

21. $\begin{bmatrix} -1 & 5 & 2 & -3 & 1 \\ 0 & 1 & -1 & 1 & 0 \\ 1 & 6 & -2 & 2 & -2 \\ 2 & -2 & -1 & 2 & -3 \end{bmatrix}$

22. $\begin{bmatrix} 1 & 0 & 0 & -1 & 2 \\ 1 & 2 & 5 & 4 & 4 \\ 2 & 0 & 1 & -1 & 0 \\ -1 & 1 & 2 & 3 & 1 \end{bmatrix}$

CHAPTER 5
SELF-TEST

1. Let K be the set of functions in 3 variables having continuous partial derivatives $f_1, f_2,$ and f_3. Define $T: K \to R^3$ by $T(f) = (f_1(0), f_2(0), f_3(0))$. Is T linear?

2. Define $T: R^2 \to R$ by

$$T(x_1, x_2) = \det \begin{bmatrix} x_1 & 1 \\ x_2 & 1 \end{bmatrix}$$

Is T linear? If not, why not? If so, what is its kernel?

3. Define $T: C'' \to C$ by $T(f) = f'' - f$. Show that T is linear, and decide which of the following are in its kernel: $\cos x, \sin x, e^x, e^{-x}, \cosh x, \sinh x$.

4. With P_2 being the vector space of polynomials of degree 2 or less, let $T: P_2 \to P_2$ be a linear transformation defined on the basis $B = \{1, x, x^2\}$ by

$$T(1) = x^2 + 1, \qquad T(x) = x + x^2, \qquad T(x^2) = 1 + x$$

Find the matrix of T with respect to the basis B.

5. Let T be a linear transformation taking the space of 2×2 matrices into itself, defined on the basis

$$S = \left\{ A = \begin{bmatrix} 1 & 0 \\ 0 & 0 \end{bmatrix}, B = \begin{bmatrix} 0 & 1 \\ 0 & 0 \end{bmatrix}, C = \begin{bmatrix} 0 & 0 \\ 1 & 0 \end{bmatrix}, D = \begin{bmatrix} 0 & 0 \\ 0 & 1 \end{bmatrix} \right\}$$

by

$$T(A) = 2A + B, \qquad T(B) = A - C + D,$$
$$T(C) = B + 2D, \qquad T(D) = B - C$$

Find the matrix of T with respect to the basis S. Careful! The desired matrix is 4×4.

6. Give a geometric description of the possible null spaces of a linear transformation $T: R^3 \to R^2$.

7. A linear transformation $T: R^3 \to R^3$ is defined with respect to the standard basis by $\mathbf{y} = A\mathbf{x}$ where

$$A = \begin{bmatrix} 1 & -1 & 2 \\ 0 & -1 & 1 \\ -1 & 0 & 1 \end{bmatrix}$$

Find bases for the kernel and for the range. Do your answers "add up?"

8. Two bases for R^2 are

$$B_1 = \{\mathbf{u}_1 = (-1, 1), \mathbf{u}_2 = (0, -1)\}, \quad B_2 = \{\mathbf{v}_1 = (1, -1), \mathbf{v}_2 = (1, 1)\}$$

The matrix representation of $T: R^2 \to R^2$ with respect to B_1 is

$$\begin{bmatrix} 1 & 2 \\ -1 & 1 \end{bmatrix}$$

Find the matrix representation of T with respect to B_2.

9. Prove, without reference to matrices, that if V is an n-dimensional vector space, and $T: V \to W$ is linear, then the dimension of the range of T is at most n.

10. Prove that the only $n \times n$ matrix A similar to the $n \times n$ matrix I is I itself.

6 SQUARE MATRICES

6-1 ## INTRODUCTION

Our study of linear transformations has been oriented to transformations that can be represented by $m \times n$ matrices. If we further restrict ourselves to the case where $m = n$, then these matrix representations are square, and this is the context in which we shall begin our discussion in Section 6-2.

Square matrices also arise naturally when we study the relationships between two bases of R^n. If $\{\mathbf{u}_1, \ldots, \mathbf{u}_n\}$ is one such basis, the matrix $U = [\mathbf{u}_1 \quad \cdots \quad \mathbf{u}_n]$ is a square matrix. A second basis for R^n will give rise to a second square matrix $V = [\mathbf{v}_1 \quad \cdots \quad \mathbf{v}_n]$, in which case there is yet a third important square matrix P that relates the bases by the matrix equation $V = UP$. This last matrix P also relates the coordinates $(x_1, \ldots, x_n)$ of a point with respect to the column vectors of U and the coordinates $(x'_1, \ldots, x'_n)$ of a point with respect to the column vectors of V by a matrix equation $\mathbf{x} = P\mathbf{x}'$ that is sometimes described as a change of variables.

It is to be stressed, however, that square matrices also arise in numerous places not yet mentioned. They are important, for example, in the study of quadratic forms. A **quadratic form** in n variables consists of the second degree terms of a second degree polynomial in n variables. Straightforward matrix multiplication shows that the general quadratic forms in 2 and in 3 variables can be written

(1)

$$\begin{bmatrix} x_1 & x_2 \end{bmatrix} \begin{bmatrix} a & \dfrac{b}{2} \\ \dfrac{b}{2} & c \end{bmatrix} \begin{bmatrix} x_1 \\ x_2 \end{bmatrix} = ax_1^2 + bx_1x_2 + cx_2^2$$

$$[x_1 \quad x_2 \quad x_3] \begin{bmatrix} a & \dfrac{d}{2} & \dfrac{f}{2} \\ \dfrac{d}{2} & b & \dfrac{e}{2} \\ \dfrac{f}{2} & \dfrac{e}{2} & c \end{bmatrix} \begin{bmatrix} x_1 \\ x_2 \\ x_3 \end{bmatrix}$$

$$= ax_1^2 + bx_2^2 + cx_3^2 + dx_1x_2 + ex_2x_3 + fx_1x_3$$

The general quadratic form in n variables can similarly be written using an $n \times n$ matrix.

One of our notes on applications shows that $n \times n$ matrices are used in the study of graphs with n nodes, and we shall shortly see in another such note that they occur in writing the second derivative of a real-valued function in n variables. In Chapter 8 they are central to the study of Markov chains; in Chapter 9 we encounter them in the study of nth-order differential equations. In short, square matrices occur in many applications.

Square matrices also play a special part in matrix theory. You are already familiar with numerous matrix-related concepts which make sense only if the matrices under consideration are square. If A is a square matrix, then det A is defined, A^{-1} may exist, and there are matrices similar to A. It is our purpose in this chapter to study still other concepts that can be defined when attention is confined to square matrices.

EIGENVALUES AND EIGENVECTORS 6-2

In discussing a transformation $T: R^n \to R^n$, we could use one basis in the domain and a different basis in the range. That's what we could do, but with a high regard for making things simple for ourselves, there is no question about what we should do: we should use the same basis in both the domain and the range.

This same passion for making things simple does raise a question, however, when a particular transformation $T: R^n \to R^n$ is given. The matrix representation of T depends on the choice of a basis for R^n. How, we ask, can we choose that basis which will result in the most simple matrix representation of T?

Since the concept of a simple matrix is not defined, the question we have asked is (deliberately) vague. There would probably be rather broad agreement, however, that if T could be represented by a *diagonal matrix* (a matrix in which all entries not on the main diagonal are zero), that would be a simple representation. You might, in this regard, turn back to Section 5-4 and read the vignette *Standard but Not Best*.

Suppose, to begin with, that T is represented by the matrix A with respect to the standard basis (or any convenient basis). If T could also be represented by a diagonal matrix D then, according to Theorem A of Section 5-4, matrices A and D would be similar. That is, we would be able to find a nonsingular matrix P such that $P^{-1}AP = D$. It would follow that $AP = PD$, an equation which in the case of $n = 2$ takes the form

$$AP = PD = \begin{bmatrix} p_{11} & p_{12} \\ p_{21} & p_{22} \end{bmatrix} \begin{bmatrix} r_1 & 0 \\ 0 & r_2 \end{bmatrix} = \begin{bmatrix} r_1 p_{11} & r_2 p_{12} \\ r_1 p_{21} & r_2 p_{22} \end{bmatrix}$$

Let us rewrite this in a way that focuses attention on the column vectors of matrix P,

(1)
$$A[\mathbf{P}_{\cdot 1} \quad \mathbf{P}_{\cdot 2}] = [r_1 \mathbf{P}_{\cdot 1} \quad r_2 \mathbf{P}_{\cdot 2}]$$

Looking at the columns individually, we have

$$A\mathbf{P}_{\cdot 1} = r_1 \mathbf{P}_{\cdot 1}, \qquad A\mathbf{P}_{\cdot 2} = r_2 \mathbf{P}_{\cdot 2}$$

Consider the first of these matrix equations. It may be written in the form

$$(A - r_1 I)\mathbf{P}_{\cdot 1} = \mathbf{0}$$

Since matrix P is to be nonsingular, column vector $\mathbf{P}_{\cdot 1}$ cannot be the $\mathbf{0}$ vector. Thus, the homogeneous system $(A - r_1 I)\mathbf{x} = \mathbf{0}$ would have a nontrivial solution. This can only happen, according to the 8th listed property of determinants (Section 3-3), if $\det (A - r_1 I) = 0$.

The 2-dimensional case was considered only for convenience of notation. The same reasoning, applied to a transformation $T: R^n \to R^n$ represented by a matrix A, would lead us to begin our search for a diagonal representation D by looking for real numbers r that satisfy $\det (A - rI) = 0$.

Example A

Suppose $T: R^3 \to R^3$ is represented with respect to the standard basis by

$$A = \begin{bmatrix} 9 & -7 & 7 \\ 3 & -1 & 3 \\ -5 & 5 & -3 \end{bmatrix}$$

Can we find a nonsingular matrix P such that $P^{-1}AP$ is a diagonal matrix?

Begin by looking for choices of r for which

(2)
$$\det (A - rI) = \begin{vmatrix} 9 - r & -7 & 7 \\ 3 & -1 - r & 3 \\ -5 & 5 & -3 - r \end{vmatrix} = 0$$

Add the first column to the second, and then expand by cofactors of the second column to obtain

$$\begin{vmatrix} 9-r & 2-r & 7 \\ 3 & 2-r & 3 \\ -5 & 0 & -3-r \end{vmatrix}$$

$$= -(2-r)\begin{vmatrix} 3 & 3 \\ -5 & -3-r \end{vmatrix} + (2-r)\begin{vmatrix} 9-r & 7 \\ -5 & -3-r \end{vmatrix} = 0$$

Expanding the 2×2 determinants gives

$$(r-2)[-3(3+r)+15] - (r-2)[(r-9)(r+3)+35] = 0$$

Factoring and collecting terms leads to

$$(r-2)\{[-3r+6] - [r^2 - 6r + 8]\} = (r-2)\{(-r+2)(r-1)\} = 0$$

The distinct roots are $r = 1$ and $r = 2$.

Now the column vectors of P are the vectors $\mathbf{p}$ for which $(A - rI)\mathbf{p} = \mathbf{0}$. Substituting first $r = 1$, then $r = 2$ into (2), we must therefore solve

$$\begin{array}{cc}
\text{for } r = 1 & \text{for } r = 2 \\[4pt]
\begin{bmatrix} 8 & -7 & 7 \\ 3 & -2 & 3 \\ -5 & 5 & -4 \end{bmatrix}\begin{bmatrix} p_1 \\ p_2 \\ p_3 \end{bmatrix} = \begin{bmatrix} 0 \\ 0 \\ 0 \end{bmatrix}, & \begin{bmatrix} 7 & -7 & 7 \\ 3 & -3 & 3 \\ -5 & 5 & -5 \end{bmatrix}\begin{bmatrix} p_1 \\ p_2 \\ p_3 \end{bmatrix} = \begin{bmatrix} 0 \\ 0 \\ 0 \end{bmatrix}
\end{array}$$

Proceed in the now familiar way to find that:

for $r = 1$	for $r = 2$
the solution space has a basis of	the solution space has a basis of
$(7, 3, -5)$	$(1, 1, 0)$
	$(-1, 0, 1)$

Use these vectors as column vectors to form

$$P = \begin{bmatrix} 7 & 1 & -1 \\ 3 & 1 & 0 \\ -5 & 0 & 1 \end{bmatrix}$$

If all goes well, it should now be the case that $P^{-1}AP = D$. First we do the usual necessary work to find

$$P^{-1} = \begin{bmatrix} -1 & 1 & -1 \\ 3 & -2 & 3 \\ -5 & 5 & -4 \end{bmatrix}$$

and then multiply P^{-1}, A, and P to find

$$P^{-1}AP = \begin{bmatrix} -1 & 1 & -1 \\ 3 & -2 & 3 \\ -5 & 5 & -4 \end{bmatrix} \begin{bmatrix} 9 & -7 & 7 \\ 3 & -1 & 3 \\ -5 & 5 & -3 \end{bmatrix} \begin{bmatrix} 7 & 1 & -1 \\ 3 & 1 & 0 \\ -5 & 0 & 1 \end{bmatrix}$$

$$= \begin{bmatrix} 1 & 0 & 0 \\ 0 & 2 & 0 \\ 0 & 0 & 2 \end{bmatrix} \quad \square$$

All went well; in fact, things went better than we had any right to expect. In the first place, all values of r were integers; they could have been irrational, or worse yet, complex. Also, in the case of the root $r = 2$, the corresponding matrix $A - 2I$ had a null space with two basis vectors, giving us a total of three vectors with which to form the columns of P. As a final stroke of good luck, the three vectors used to form P turned out to be linearly independent so that P was nonsingular.

Our problem now is to discover how much of this was good luck and how much, if any of it, can be counted upon to happen every time. One might, for instance, suppose that the 2-dimensional null space of $A - 2I$ is related to the fact that $r = 2$ is a double root of equation (2). Unfortunately, this is not something we can rely upon, as we shall now see.

Example B   Suppose $T: R^3 \to R^3$ is represented with respect to the standard basis by

$$A = \begin{bmatrix} 6 & -3 & -1 \\ 11 & -6 & -3 \\ -10 & 7 & 5 \end{bmatrix}$$

Can we find a nonsingular matrix P such that $P^{-1}AP$ is a diagonal matrix?

We begin as in Example A by looking for choices of r for which

(3)
$$\det(A - rI) = \begin{vmatrix} 6 - r & -3 & -1 \\ 11 & -6 - r & -3 \\ -10 & 7 & 5 - r \end{vmatrix} = 0$$

Evaluating this determinant by any of the standard methods gives

$$r^3 - 5r^2 + 8r - 4 = 0$$

Using synthetic division (which may cause you to review your college algebra), we discover that this factors into

$$(r - 1)(r - 2)^2 = 0$$

The distinct roots are again, as in Example A, $r = 1$ and $r = 2$. And, as in Example A, $r = 2$ is a root of multiplicity 2. Proceeding with confidence, we find that:

$$\text{for } r = 1$$

$$\begin{bmatrix} 5 & -3 & -1 \\ 11 & -7 & -3 \\ -10 & 7 & 4 \end{bmatrix} \begin{bmatrix} p_1 \\ p_2 \\ p_3 \end{bmatrix} = \begin{bmatrix} 0 \\ 0 \\ 0 \end{bmatrix}$$

the solution space is spanned by

$$(1, 2, -1)$$

$$\text{for } r = 2$$

$$\begin{bmatrix} 4 & -3 & -1 \\ 11 & -8 & -3 \\ -10 & 7 & 3 \end{bmatrix} \begin{bmatrix} p_1 \\ p_2 \\ p_2 \end{bmatrix} = \begin{bmatrix} 0 \\ 0 \\ 0 \end{bmatrix}$$

the solution space is spanned by

$$(1, 1, 1)$$

Alas, our good luck has run out. We do not have enough vectors even to form P, much less to worry about whether P is nonsingular. Our efforts on this problem have ground to a halt. □

Is there some way to finish the problem started in Example B? If not, that is if matrix A is not similar to a diagonal matrix, how close can we get to a diagonal matrix? And are there conditions that will tell us ahead of time whether or not A is similar to a diagonal matrix? Some of these questions, together with some of the questions raised following Example A, will be answered in this chapter. Others are answered in more advanced courses. To get a start on any of them, it is necessary to learn some terminology.

TERMINOLOGY

Equations (2) and (3) are called the characteristic equations of the respective matrices. Such an equation is defined for any $n \times n$ matrix A.

The **characteristic equation** of matrix A is defined to be $\det(A - rI) = 0$. A root of this equation is called an **eigenvalue**. A nonzero vector $\mathbf{p}$ satisfying $A\mathbf{p} = r\mathbf{p}$ is said to be an **eigenvector** of A belonging to r.

In Example A, the eigenvalues are $r = 1$ and $r = 2$. The vector $(7, 3, -5)$ is an eigenvector belonging to $r = 1$ since

$$\begin{bmatrix} 9 & -7 & 7 \\ 3 & -1 & 3 \\ -5 & 5 & -3 \end{bmatrix} \begin{bmatrix} 7 \\ 3 \\ -5 \end{bmatrix} = \begin{bmatrix} 7 \\ 3 \\ -5 \end{bmatrix}$$

The vectors $(1, 1, 0)$ and $(-1, 0, 1)$ are eigenvectors belonging to $r = 2$, since

$$\begin{bmatrix} 9 & -7 & 7 \\ 3 & -1 & 3 \\ -5 & 5 & -3 \end{bmatrix} \begin{bmatrix} 1 \\ 1 \\ 0 \end{bmatrix} = 2 \begin{bmatrix} 1 \\ 1 \\ 0 \end{bmatrix}$$

and

$$\begin{bmatrix} 9 & -7 & 7 \\ 3 & -1 & 3 \\ -5 & 5 & -3 \end{bmatrix} \begin{bmatrix} -1 \\ 0 \\ 1 \end{bmatrix} = 2 \begin{bmatrix} -1 \\ 0 \\ 1 \end{bmatrix}$$

If **p** is an eigenvector belonging to r, so that $A\mathbf{p} = r\mathbf{p}$, then for any real number t, $A(t\mathbf{p}) = tA\mathbf{p} = r(t\mathbf{p})$. This shows that if **p** is an eigenvector belonging to r, then so is $t\mathbf{p}$ for any real number t. You might verify for the example just given that $(3, 3, 0)$ and $(5, 0, -5)$ are both eigenvectors belonging to 2.

Eigenvalues are sometimes called **characteristic values** or **proper values,** the latter being used by people who dislike the hybrid word *eigenvalue; eigen* is the German word for *proper*. Corresponding to alternate names for the eigenvalues, the eigenvectors are also known as **characteristic vectors** or **proper vectors.**

The characteristic equation of an $n \times n$ matrix is an nth degree polynomial equation $p(x) = 0$. If we allow the use of complex numbers, then the fundamental theorem of algebra implies that $p(x) = 0$ will have n roots $r_1, \ldots, r_n$. Hence, $p(x)$ can be written in the factored form

$$p(x) = (x - r_1)(x - r_2) \ldots (x - r_n)$$

If equal factors are grouped together,

$$p(x) = (x - r_1)^{m_1} \ldots (x - r_i)^{m_i} \ldots (x - r_k)^{m_k}, \qquad m_1 + \cdots + m_k = n$$

Then r_i is said to be an eigenvalue of **multiplicity** m_i.

Just before defining a vector space (Section 4-2), we said that there would come a time when we would want to allow scalars to be complex numbers. The time has come. To be able to say without exception that an $n \times n$ matrix has n eigenvalues, we must understand that some or all of them may be complex. This same understanding will greatly simplify our work throughout this chapter. The entries in our vectors and matrices, and the scalars by which they are multiplied, may be drawn from the set C of complex numbers. Consistency then requires that the space of n-tuples be C^n.

Such an understanding greatly simplifies theory, but it could make computations complex. In fact, it won't in this book because our examples and problems (except for a few starred ones that carry warnings) are contrived to avoid the appearance of complex numbers. This being the case, you should take care not to introduce them unnecessarily. For instance, if the entries of A and an eigenvalue r are real, there will surely be a real eigenvector **v** belonging to r (Problem 23). It is permissible, but contrary to the spirit of things, to observe that the product $c\mathbf{v}$ of **v** with a complex scalar c is also an eigenvector. The general rule is this. Confine attention to real numbers unless complex numbers are forced on you as eigenvalues. With such an understanding, the C^n in our theorems can be replaced with R^n whenever A and all its eigenvalues are real.

EIGENVALUES AND EIGENVECTORS

We have seen (in Example A for eigenvalue $r = 2$) that two linearly independent vectors may belong to the same eigenvalue. Suppose now that $\mathbf{v}_1$ and $\mathbf{v}_2$ are any two vectors that satisfy, for a fixed eigenvalue r of matrix A, the equation $A\mathbf{v} = r\mathbf{v}$. Then for any two scalars s_1 and s_2,

$$
\begin{aligned}
A(s_1\mathbf{v}_1 + s_2\mathbf{v}_2) &= s_1 A\mathbf{v}_1 + s_2 A\mathbf{v}_2 \\
&= s_1 r\mathbf{v}_1 + s_2 r\mathbf{v}_2 \\
&= r(s_1\mathbf{v}_1 + s_2\mathbf{v}_2)
\end{aligned}
$$

This proves that $s_1\mathbf{v}_1 + s_2\mathbf{v}_2$ also satisfies $A\mathbf{v} = r\mathbf{v}$, hence that the solutions to this equation form a subspace of C^n.

> *If r is an eigenvalue of the $n \times n$ matrix A, the set of all vectors $\mathbf{v}$ satisfying $A\mathbf{v} = r\mathbf{v}$ form a subspace of C^n.*   **THEOREM A**

This subspace contains all the eigenvectors belonging to r and the **0** vector (which by definition is not an eigenvector).

We have defined eigenvalues and eigenvectors for a given matrix A. Since this entire discussion began by asking for a simple matrix representation for a linear transformation, and since any two matrix representations of the same transformation are similar matrices, it is natural to wonder how the eigenvalues and eigenvectors of similar matrices are related.

> *If $A \sim B$, then A and B have the same characteristic equation, hence the same eigenvalues.*   **THEOREM B**

Proof: Since $A = P^{-1}BP$,

$$\det (A - rI) = \det (P^{-1}BP - rI)$$

Now observe that $P^{-1}BP - rI = P^{-1}(B - rI)P$ and write

$$
\begin{aligned}
\det (A - rI) = \det P^{-1}(B - rI)P &= \det (B - rI) \\
&= (\det P^{-1})(\det (B - rI))(\det P) \\
&= \det (B - rI) \quad \square
\end{aligned}
$$

Theorem B opens the door to defining eigenvalues and eigenvectors for a linear transformation $T\colon C^n \to C^n$. We shall say that r is an eigenvalue of T and that a nonzero vector $\mathbf{p}$ is an eigenvector of T belonging to r if $T(\mathbf{p}) = r\mathbf{p}$. The eigenvalues of T may be found by finding the eigenvalues of any matrix representation of A because all such representations are similar.

DIAGONALIZATION

If for a given matrix A we can find a diagonal matrix D that is similar to A, then A is said to be **diagonalizable**. As you have surely guessed by now, not every matrix is diagonalizable. Which ones are? More will be said about this in the next section. This much can be said now: The steps taken in our examples above are always appropriate first steps, and when (as in Example A) things seem to be working out, our procedure is also an efficient way to solve the problem. The outline of this procedure will be referred to often in the following sections.

Diagonalization of an $n \times n$ matrix A

Step 1: Solve the characteristic equation $|A - rI| = 0$ to determine n eigenvalues $r_1, \ldots, r_n$.

If these roots are complex, we will, without the benefit of more techniques than are presented here, be stopped in our tracks at this point. This difficulty is computational in nature, and is avoided in the problem sets by a judicious choice of problems.

Step 2: For each eigenvalue r_i, find a basis for the solution space of the system $(A - r_i I)\mathbf{p} = \mathbf{0}$.

Step 3: Use the eigenvectors determined in Step 2 as column vectors to form an $n \times n$ matrix P.

If less than n eigenvectors were determined in Step 2, we are again stopped, this time by difficulties intrinsic to the problem itself.

Step 4: Check to see that P is nonsingular.

If it is not, we are again stopped.

Step 5: Matrix A is similar to the diagonal matrix $D = P^{-1}AP$, and the entries on the diagonal are the eigenvalues of A.

If A is the matrix representation of a transformation $T: R^n \rightarrow R^n$ with respect to the basis of column vectors $\mathbf{u}_1, \ldots, \mathbf{u}_n$, then D represents T with respect to the basis of column vectors $\mathbf{v}_1, \ldots, \mathbf{v}_n$ where

$$[\mathbf{v}_1 \quad \cdots \quad \mathbf{v}_n] = [\mathbf{u}_1 \quad \cdots \quad \mathbf{u}_n]P$$

We have a good outline to follow. It remains to identify those matrices for which all the steps can be carried out.

A METHOD THAT ALLOWS MISTAKES

All of the methods we have used to solve a system of n equations in n variables have aimed at finding the exact solution to the system. They are therefore called **exact methods,** in contrast with **iterative methods** that are rules for operating on an approximate solution s_k in order to obtain an improved approximate solution s_{k+1}. The idea will be illustrated by using the **Jacobi method** for solving the system

$$\begin{aligned} 4x_1 + x_2 - 4x_3 &= -5 \\ x_1 + 2x_2 + x_3 &= 1 \\ - x_2 + 4x_3 &= 9 \end{aligned}$$

The first step is to solve the first equation for x_1, the second for x_2, and so on.

$$\begin{aligned} x_1 &= \phantom{-\tfrac{1}{2}x_1} -\tfrac{1}{4}x_2 + x_3 - \tfrac{5}{4} \\ x_2 &= -\tfrac{1}{2}x_1 \phantom{-\tfrac{1}{4}x_2} - \tfrac{1}{2}x_3 + \tfrac{1}{2} \\ x_3 &= \phantom{-\tfrac{1}{2}x_1} \tfrac{1}{4}x_2 \phantom{-\tfrac{1}{2}x_3} + \tfrac{9}{4} \end{aligned} \qquad (i)$$

Now, if you have some idea of what the solution might be, say $s_0 = (s_{01}, s_{02}, s_{03})$, set the x_i on the right-hand side of (i) equal to these values. The resulting values on the left-hand side become the next approximation. If you have no idea of the exact solution, begin with $s_0 = 0$. This leads to the successive approximations listed below.

$$\begin{aligned} s_0 &= (0,0,0) & s_3 &= (\tfrac{9}{8}, -\tfrac{18}{16}, \tfrac{18}{8}) \\ s_1 &= (-\tfrac{5}{4}, \tfrac{1}{2}, \tfrac{9}{4}) & s_4 &= (\tfrac{41}{32}, -\tfrac{19}{16}, \tfrac{63}{32}) \\ s_2 &= (\tfrac{7}{8}, 0, \tfrac{19}{8}) & s_5 &= (\tfrac{65}{64}, -\tfrac{36}{32}, \tfrac{125}{64}) \end{aligned}$$

You can see that these approximations are converging to the exact solution which is $(1, -1, 2)$.

Two questions will agitate the inquiring mind: Will this always work? Why not just use available exact methods? We can only hint at the deeper answers in the space available.

If we begin with a system $A\mathbf{x} = \mathbf{b}$, we arrive at (i) by writing A as the sum of a diagonal matrix D and a matrix T having only zero entries on the diagonal; $A = D + T$. Then the system becomes

$(D + T)\mathbf{x} = \mathbf{b}$ or $D\mathbf{x} = -T\mathbf{x} + \mathbf{b}$. Equations (i) are obtained by multiplying by D^{-1}:

$$\mathbf{x} = -D^{-1}T\mathbf{x} + D^{-1}\mathbf{b}$$

The recursion formula used is

$$\mathbf{s}_{k+1} = -D^{-1}T\mathbf{s}_k + D^{-1}\mathbf{b}$$

A theorem from numerical analysis then says that the method converges whenever all the eigenvalues r of $-D^{-1}T$ satisfy $|r| < 1$. (Eigenvalues in our example are $\frac{1}{2}, \frac{1}{4} \pm i\sqrt{3}/4$.)

Why use such a method? Well, sometimes one shouldn't use the Jacobi method, since a very slight modification (the Gauss-Seidel method) frequently converges faster. As to ever using iterative methods, we shall only say that they are often preferable for machine computation since they have more tolerance for round-off errors, and they require less storage space if the system is sparse (has many zero coefficients). (See Ralston, A. 1965. *A first course in numerical analysis*. New York: McGraw-Hill.)

PROBLEM
SET 6-2

For each of the following matrices, proceed as far as you can with the process of finding P such that $P^{-1}AP$ is diagonal.

1. $A = \begin{bmatrix} -5 & 4 \\ -8 & 7 \end{bmatrix}$

2. $A = \begin{bmatrix} -5 & 3 \\ -6 & 4 \end{bmatrix}$

3. $A = \begin{bmatrix} -7 & -6 \\ 15 & 12 \end{bmatrix}$

4. $A = \begin{bmatrix} 8 & 6 \\ -15 & -11 \end{bmatrix}$

5. $A = \begin{bmatrix} 4 & 0 & 2 \\ 2 & 3 & 2 \\ -3 & 0 & -1 \end{bmatrix}$

6. $A = \begin{bmatrix} 5 & 0 & 4 \\ 3 & 2 & 3 \\ -6 & 0 & -5 \end{bmatrix}$

7. $A = \begin{bmatrix} 0 & 0 & 2 \\ -2 & 1 & 4 \\ -1 & 0 & 3 \end{bmatrix}$

8. $A = \begin{bmatrix} 5 & 0 & -6 \\ -3 & -1 & 3 \\ 3 & 0 & -4 \end{bmatrix}$

9. $A = \begin{bmatrix} -5 & 0 & -4 \\ -5 & -1 & -4 \\ 9 & 0 & 7 \end{bmatrix}$

10. $A = \begin{bmatrix} 8 & 0 & 6 \\ 1 & -1 & 1 \\ -9 & 0 & -7 \end{bmatrix}$

11. $A = \begin{bmatrix} -1 & 0 & 1 & 0 \\ -4 & 3 & 3 & -4 \\ -6 & 0 & 4 & 0 \\ -2 & 2 & 2 & -3 \end{bmatrix}$

12. $A = \begin{bmatrix} -8 & 4 & 1 & -4 \\ -27 & 17 & 3 & -18 \\ -14 & 8 & 1 & -8 \\ -17 & 11 & 2 & -12 \end{bmatrix}$

In Problems 13–16 you are given the coordinates of a vector **w** *in* R^3, *and a matrix* A *that represents a transformation* $T: R^3 \to R^3$, *both expressed with respect to the standard basis. In each problem,*

(a) *Find a basis* B *for* R^3 *such that the matrix of* T *with respect to* B *is a diagonal matrix* D.

(b) *Find the coordinates of* **w** *with respect to the basis* B.

(c) *Find* $T(\mathbf{w})$ *using both* $A\mathbf{w}_S$ *and* $D\mathbf{w}_B$. *Verify your answers.*

13. $\mathbf{w} = \begin{bmatrix} -1 \\ 2 \\ 1 \end{bmatrix}$, $\quad A = \begin{bmatrix} 3 & 2 & -1 \\ -4 & -3 & 1 \\ -2 & -2 & 2 \end{bmatrix}$

14. $\mathbf{w} = \begin{bmatrix} 1 \\ -2 \\ 2 \end{bmatrix}$, $\quad A = \begin{bmatrix} 3 & 2 & -1 \\ -4 & -3 & 1 \\ -6 & -6 & 2 \end{bmatrix}$

15. $\mathbf{w} = \begin{bmatrix} 2 \\ -1 \\ -3 \end{bmatrix}$, $\quad A = \begin{bmatrix} 0 & -2 & 1 \\ 1 & 3 & -1 \\ 0 & 0 & 1 \end{bmatrix}$

16. $\mathbf{w} = \begin{bmatrix} -2 \\ 1 \\ 1 \end{bmatrix}$, $\quad A = \begin{bmatrix} 5 & 6 & -3 \\ -3 & -4 & 3 \\ 0 & 0 & 2 \end{bmatrix}$

17. Let $T: R^2 \to R^2$ be a reflection through the line $x_2 = x_1$. From geometric considerations, describe the eigenvectors of T. Then find the matrix representation of T and the eigenvectors of the matrix. Does everything check?

18. Let $T: R^2 \to R^2$ be a rotation about the origin. Follow the instructions for Problem 17.

19. Let M be a 5×5 matrix which has three distinct real eigenvalues r_1, r_2, and r_3 with corresponding eigenvectors $\mathbf{v}_1, \mathbf{v}_2$, and $\mathbf{v}_3$. Is it certain that we can find two vectors $\mathbf{u}_1$ and $\mathbf{u}_2$ in R^5 so that the matrix $[\mathbf{v}_1 \quad \mathbf{v}_2 \quad \mathbf{v}_3 \quad \mathbf{u}_1 \quad \mathbf{u}_2]$ will be nonsingular?

20. Using the matrix M and its eigenvectors $\mathbf{v}_1, \mathbf{v}_2$, and $\mathbf{v}_3$ as defined in Problem 19, let V be the subspace of R^5 spanned by $\mathbf{v}_1, \mathbf{v}_2$, and $\mathbf{v}_3$. Can we be sure that the linear transformation $T: R^5 \to R^5$ defined by $\mathbf{y} = M\mathbf{x}$ will map any vector of V into another vector of V?

21. Prove that if r is an eigenvalue of matrix A, then r^2 is an eigenvalue of A^2.

22. Prove that if the entries of an $n \times n$ matrix A are all real, then there will surely be a real eigenvector corresponding to any real eigenvalue. *Hint:* The coefficient matrix of the system $[A - rI]\mathbf{x}$ is singular.

6-3 DIAGONALIZATION OF SYMMETRIC MATRICES

We have set for ourselves the problem of deciding which $n \times n$ matrices are diagonalizable. When confronted with such a problem, it is often profitable to consider special cases. The idea is to choose special cases which arise naturally in applications and for which we can make more progress than can be made in the general case. Neither goal should be underestimated as a motivating force.

Our discussion of the importance of square matrices in Section 6-1 mentioned in equations (1) and (2) of that section the two matrices

$$\begin{bmatrix} a & \dfrac{b}{2} \\ \dfrac{b}{2} & a \end{bmatrix}, \quad \begin{bmatrix} a & \dfrac{d}{2} & \dfrac{f}{2} \\ \dfrac{d}{2} & b & \dfrac{e}{2} \\ \dfrac{f}{2} & \dfrac{e}{2} & c \end{bmatrix}$$

Both exhibit a symmetry about the main diagonal that can be described precisely in terms of transposes: A square matrix A is said to be **symmetric** if $A^t = A$. The outline at the conclusion of Section 6-2 emphasized that theoretical deficiencies or computational complexity may impede progress when we try to diagonalize a matrix. In this section we shall see that if the matrix is symmetric, all the theoretical and some of the important computational difficulties are removed, allowing a straightforward passage to our diagonal goal. Stated more concisely, our goal is to show that every real symmetric matrix can be diagonalized.

Since complex roots of the characteristic equation are the first possible block to further progress, we begin with the following theorem:

THEOREM A | *The eigenvalues of a real symmetric matrix A are all real.*

Proof: Let $r = a + bi$ be an eigenvalue of the symmetric matrix A. Our goal is to show that $b = 0$. We begin by noting that

$$A - rI = A - (a + bi)I$$

is singular. So, then, is its product with any other matrix. In particular, the matrix B is singular where

$$\begin{aligned} B &= [A - (a + bi)I][A - (a - bi)I] \\ &= A^2 - 2aA + a^2I + b^2I \\ &= (A - aI)^2 + b^2I \end{aligned}$$

This last sum involves real matrices only, so B is a real singular matrix. It follows that there is a nonzero real column vector $\mathbf{x}$ such that $B\mathbf{x} = \mathbf{0}$. Surely then,

$$0 = \mathbf{x}^t B \mathbf{x} = \mathbf{x}^t[(A - aI)^2 + b^2 I]\mathbf{x}$$

Since A is symmetric, $(A - aI)^t = A - aI$, enabling us to write this last expression in the form

$$0 = \mathbf{x}^t[(A - aI)^t(A - aI) + b^2 I]\mathbf{x}$$
$$0 = \mathbf{x}^t(A - aI)^t(A - aI)\mathbf{x} + b^2 \mathbf{x}^t I \mathbf{x}$$

Set $\mathbf{y} = (A - aI)\mathbf{x}$ and substitute to get

$$0 = \mathbf{y}^t \mathbf{y} + b^2 \mathbf{x}^t \mathbf{x} = |\mathbf{y}|^2 + b^2 |\mathbf{x}|^2$$

This can only happen if both $|\mathbf{y}| = 0$ and $b|\mathbf{x}| = 0$. And finally, since $\mathbf{x}$ is a nonzero vector, we have $b = 0$. □

The eigenvectors of a real symmetric matrix are vectors in R^n.   *COROLLARY*

The corollary simply observes that if the entries of the symmetric matrix A are real, then, since r must be real, the solutions to $(A - rI)\mathbf{x} = \mathbf{0}$ must be real.

With this taste of real progress, our attention shifts to obtaining a sufficient number of linearly independent eigenvectors.

If A is a real symmetric matrix, then eigenvectors belonging to distinct eigenvalues are orthogonal (and so surely linearly independent).   *THEOREM B*

Proof: Let $\mathbf{p}_1$ and $\mathbf{p}_2$ be eigenvectors belonging to distinct eigenvalues r_1 and r_2, respectively. We will think of $\mathbf{p}_1$ and $\mathbf{p}_2$ as column vectors so that we can write the dot product $\mathbf{p}_1 \cdot \mathbf{p}_2$ in the form $\mathbf{p}_1^t \mathbf{p}_2$. Now the 1×1 matrix $\mathbf{p}_1^t A \mathbf{p}_2$ is obviously symmetric, so

$$\mathbf{p}_1^t A \mathbf{p}_2 = (\mathbf{p}_1^t A \mathbf{p}_2)^t = \mathbf{p}_2^t A^t \mathbf{p}_1 = \mathbf{p}_2^t A \mathbf{p}_1 \qquad (1)$$

Also, since $\mathbf{p}_1$ and $\mathbf{p}_2$ are eigenvectors,

$$\mathbf{p}_1^t A \mathbf{p}_2 = \mathbf{p}_1^t r_2 \mathbf{p}_2 = r_2 \mathbf{p}_1^t \mathbf{p}_2$$
$$\mathbf{p}_2^t A \mathbf{p}_1 = \mathbf{p}_2^t r_1 \mathbf{p}_1 = r_1 \mathbf{p}_2^t \mathbf{p}_1$$

Substitution in (1) gives

$$r_2 \mathbf{p}_1^t \mathbf{p}_2 = r_1 \mathbf{p}_2^t \mathbf{p}_1$$

Since the dot product commutes, $\mathbf{p}_1^t \mathbf{p}_2 = \mathbf{p}_2^t \mathbf{p}_1$ and we conclude that

$$(r_2 - r_1)\mathbf{p}_1^t \mathbf{p}_2 = 0$$

By hypothesis, $r_2 \neq r_1$, so $r_2 - r_1 \neq 0$ and it follows that $\mathbf{p}_1^t \mathbf{p}_2 = 0$; $\mathbf{p}_1$ and $\mathbf{p}_2$ are orthogonal. The linear independence of an orthogonal set of vectors was discussed in Section 4-7.

□

So much for distinct eigenvalues. It is the eigenvalues that appear as multiple roots of the characteristic equation that have caused us trouble in the past. Symmetry changes all that.

THEOREM C

Let r be a root of multiplicity k of the characteristic equation of the real symmetric matrix A. Then there are k linearly independent eigenvectors belonging to r.

It will quickly be apparent that we can actually improve this theorem, so rather than prove it now, let us review what we now know about the prospects for diagonalizing a real symmetric matrix A.

Step 1: Solve $|A - rI| = 0$ to find the n eigenvalues $r_1, \ldots, r_n$. They are all real.

Step 2: For each characteristic root r_i, find a basis for the subspace spanned by the eigenvectors that belong to r_i.
(a) *Eigenvectors belonging to distinct eigenvalues will be orthogonal vectors in R^n.*
(b) *There will be k linearly independent eigenvectors belonging to an eigenvalue that occurs k times in the listing in Step 1.*

We are thus assured of finding n linearly independent eigenvectors $\mathbf{p}_1, \ldots, \mathbf{p}_n$ belonging to the respective eigenvalues.

Step 3: Form the nonsingular matrix $P = [\mathbf{p}_1 \quad \cdots \quad \mathbf{p}_n]$.

Step 4: The desired diagonal matrix is $D = P^{-1}AP$.

The column vectors of P that belong to distinct eigenvalues of A are orthogonal. Consider now the possibility that there are k linearly independent column vectors of P that are eigenvectors belonging to the same eigenvalue r_i (meaning of course that r_i is a root of multiplicity k of the characteristic equation). These k vectors span a subspace of R^n. Calling upon the Gram-Schmidt orthogonalization process, we can find an orthogonal basis for this subspace. If this is done for each eigenvalue with multiplicity greater than 1, then any two column vectors of P, whether belonging to the same or to distinct eigenvalues, will be orthogonal. Finally, each column vector $\mathbf{p}_i$ could be normalized so that $\mathbf{p}_i \cdot \mathbf{p}_i = 1$.

A real $n \times n$ matrix P is said to be **orthogonal** if the column vectors of P form an orthonormal set. Note that P is orthogonal if and only if $PP^t = P^tP = I$. Stated another way, P is orthogonal if and only if $P^{-1} = P^t$. The steps above can obviously be modified so that the matrix P formed in Step 3 is orthogonal.

Find an orthogonal matrix P that diagonalizes the symmetric matrix

Example A

$$A = \begin{bmatrix} 2 & -1 & -1 \\ -1 & 2 & -1 \\ -1 & -1 & 2 \end{bmatrix}$$

The characteristic equation is, after some algebra,

$$\begin{bmatrix} 2-r & -1 & -1 \\ -1 & 2-r & -1 \\ -1 & -1 & 2-r \end{bmatrix} = -r(r-3)^2 = 0$$

To get the eigenvectors corresponding to the eigenvalue $r = 3$, we first note that

$$\begin{bmatrix} -1 & -1 & -1 \\ -1 & -1 & -1 \\ -1 & -1 & -1 \end{bmatrix} \sim \begin{bmatrix} 1 & 1 & 1 \\ 0 & 0 & 0 \\ 0 & 0 & 0 \end{bmatrix}$$

A basis for the null space, found in the usual way, is given by

$$\mathbf{u}_1 = (-1, 1, 0), \qquad \mathbf{u}_2 = (1, 0, -1)$$

These are not orthogonal, but the Gram-Schmidt process (Section 4-7) gives

$$\mathbf{v}_1 = \mathbf{u}_1 = (-1, 1, 0)$$

$$\mathbf{v}_2 = \mathbf{u}_2 - \frac{\mathbf{v}_1 \bullet \mathbf{u}_2}{\mathbf{v}_1 \bullet \mathbf{v}_1}\mathbf{v}_1 = (1, 0, -1) - (-\tfrac{1}{2})(-1, 1, 0)$$

$$\mathbf{v}_2 = (\tfrac{1}{2}, \tfrac{1}{2}, -1)$$

Moving on to $r = 0$, we are left with finding the null space of A itself.

$$A = \begin{bmatrix} 2 & -1 & -1 \\ -1 & 2 & -1 \\ -1 & -1 & 2 \end{bmatrix} \sim \begin{bmatrix} 1 & 0 & -1 \\ 0 & 1 & -1 \\ 0 & 0 & 0 \end{bmatrix}$$

The null space is spanned by the vector

$$\mathbf{v}_3 = (1, 1, 1)$$

which is (just as the theory promised) orthogonal to both $\mathbf{v}_1$ and $\mathbf{v}_2$.

These three vectors, normalized, are then used as column vectors to form

$$P = \begin{bmatrix} -\dfrac{1}{\sqrt{2}} & \dfrac{1}{\sqrt{6}} & \dfrac{1}{\sqrt{3}} \\ \dfrac{1}{\sqrt{2}} & \dfrac{1}{\sqrt{6}} & \dfrac{1}{\sqrt{3}} \\ 0 & -\dfrac{2}{\sqrt{6}} & \dfrac{1}{\sqrt{3}} \end{bmatrix}$$

Since P is orthogonal, $P^{-1} = P^t$ and it is a matter of straightforward calculation to verify that

$$P^{-1}AP = \begin{bmatrix} -\dfrac{1}{\sqrt{2}} & \dfrac{1}{\sqrt{2}} & 0 \\ \dfrac{1}{\sqrt{6}} & \dfrac{1}{\sqrt{6}} & -\dfrac{2}{\sqrt{6}} \\ \dfrac{1}{\sqrt{3}} & \dfrac{1}{\sqrt{3}} & \dfrac{1}{\sqrt{3}} \end{bmatrix} \begin{bmatrix} 2 & -1 & -1 \\ -1 & 2 & -1 \\ -1 & -1 & 2 \end{bmatrix} \begin{bmatrix} -\dfrac{1}{\sqrt{2}} & \dfrac{1}{\sqrt{6}} & \dfrac{1}{\sqrt{3}} \\ \dfrac{1}{\sqrt{2}} & \dfrac{1}{\sqrt{6}} & \dfrac{1}{\sqrt{3}} \\ 0 & -\dfrac{2}{\sqrt{6}} & \dfrac{1}{\sqrt{3}} \end{bmatrix}$$

$$= \begin{bmatrix} 3 & 0 & 0 \\ 0 & 3 & 0 \\ 0 & 0 & 0 \end{bmatrix} \quad \square$$

The promised improvement of Theorem C is at hand. Its proof is given in Section 6-4.

THEOREM D

If A is a real symmetric matrix, there is an orthogonal matrix P such that $P^{-1}AP = D$ is diagonal.

*PROBLEM
SET 6-3*

1. Multiply $\quad [x_1\, x_2\, x_3] \begin{bmatrix} 2 & \frac{1}{2} & -1 \\ \frac{1}{2} & 0 & 1 \\ -1 & 1 & -1 \end{bmatrix} \begin{bmatrix} x_1 \\ x_2 \\ x_3 \end{bmatrix}$

2. Multiply $\quad [x_1\, x_2\, x_3] \begin{bmatrix} -1 & 1 & -\frac{1}{2} \\ 1 & 2 & 1 \\ -\frac{1}{2} & 1 & 3 \end{bmatrix} \begin{bmatrix} x_1 \\ x_2 \\ x_3 \end{bmatrix}$

Show that the matrices given in Problems 3–6 are orthogonal.

3. $\begin{bmatrix} \dfrac{1}{\sqrt{5}} & -\dfrac{2}{\sqrt{5}} \\ -\dfrac{2}{\sqrt{5}} & -\dfrac{1}{\sqrt{5}} \end{bmatrix}$

4. $\begin{bmatrix} \dfrac{3}{5} & \dfrac{4}{5} \\ -\dfrac{4}{5} & \dfrac{3}{5} \end{bmatrix}$

5. $\begin{bmatrix} \dfrac{2}{3} & \dfrac{1}{\sqrt{5}} & \dfrac{4}{3\sqrt{5}} \\[2ex] -\dfrac{2}{3} & 0 & \dfrac{5}{3\sqrt{5}} \\[2ex] \dfrac{1}{3} & -\dfrac{2}{\sqrt{5}} & \dfrac{2}{3\sqrt{5}} \end{bmatrix}$
   6. $\begin{bmatrix} \dfrac{1}{3\sqrt{2}} & \dfrac{2}{3} & \dfrac{1}{\sqrt{2}} \\[2ex] \dfrac{4}{3\sqrt{2}} & -\dfrac{1}{3} & 0 \\[2ex] \dfrac{1}{3\sqrt{2}} & \dfrac{2}{3} & -\dfrac{1}{\sqrt{2}} \end{bmatrix}$

(*Problem set continues on p. 254*)

SECOND DERIVATIVES AND SYMMETRIC MATRICES

In the vignette in Section 5-3, we pointed out that the derivative of a real-valued function $f(x_1, x_2)$ in two variables is a linear transformation from R^2 to R represented by the gradient matrix

$$f'(\mathbf{x}) = [f_1(\mathbf{x}) \quad f_2(\mathbf{x})]$$

This 1×2 matrix may, of course, be viewed as a vector in R^2. Then we have a (nonlinear) transformation $f': R^2 \to R^2$ defined by

$$y_1 = f_1(x_1, x_2)$$
$$y_2 = f_2(x_1, x_2)$$

It so happens that in the same note we pointed out in equation (iii) that the derivative of a transformation of this form is a linear transformation from R^2 to R^2 represented by the Jacobian matrix

$$f''(\mathbf{x}) = \begin{bmatrix} \dfrac{\partial y_1}{\partial x_1} & \dfrac{\partial y_1}{\partial x_2} \\[2ex] \dfrac{\partial y_2}{\partial x_1} & \dfrac{\partial y_2}{\partial x_2} \end{bmatrix} = \begin{bmatrix} f_{11}(\mathbf{x}) & f_{12}(\mathbf{x}) \\ f_{21}(\mathbf{x}) & f_{22}(\mathbf{x}) \end{bmatrix}$$

Now it is proved in calculus that if all the second partial derivatives of f are continuous, then $f_{12}(\mathbf{x}) = f_{21}(\mathbf{x})$. This equality actually holds under even more general conditions, meaning that for most functions occurring in applications, the matrix representing $f''(\mathbf{x})$ is symmetric.

To be specific, consider $f(x_1, x_2) = x_1^{3/2} + \sqrt{x_1 x_2} - x_2^{3/2}$. The nonlinear transformation $f': R^2 \to R^2$ is described by

$$y_1 = f_1(x_1, x_2) = \tfrac{3}{2}x_1^{1/2} + \tfrac{1}{2}x_1^{-1/2}x_2^{1/2}$$
$$y_2 = f_2(x_1, x_2) = -\tfrac{3}{2}x_2^{1/2} + \tfrac{1}{2}x_1^{1/2}x_2^{-1/2}$$

The Jacobian matrix representing $f''(\mathbf{x})$ is

$$f''(\mathbf{x}) = \begin{bmatrix} \tfrac{3}{4}x_1^{-1/2} - \tfrac{1}{4}x_1^{-3/2}x_2^{1/2} & \tfrac{1}{4}x_1^{-1/2}x_2^{-1/2} \\[2ex] \tfrac{1}{4}x_1^{-1/2}x_2^{-1/2} & -\tfrac{3}{4}x_2^{-1/2} - \tfrac{1}{4}x_1^{1/2}x_2^{-3/2} \end{bmatrix}$$

In Problems 7–14, find an orthogonal matrix P and a diagonal matrix D such that $P^{-1}AP = D$.

7. $A = \begin{bmatrix} 5 & 8 & 10 \\ 8 & 11 & -2 \\ 10 & -2 & 2 \end{bmatrix}$

8. $A = \begin{bmatrix} 11 & 2 & -8 \\ 2 & 2 & 10 \\ -8 & 10 & 5 \end{bmatrix}$

9. $A = \begin{bmatrix} -1 & -2 & -5 \\ -2 & 2 & 2 \\ -5 & 2 & -1 \end{bmatrix}$

10. $A = \begin{bmatrix} 1 & -1 & 2 \\ -1 & -2 & 1 \\ 2 & 1 & 1 \end{bmatrix}$

11. $A = \begin{bmatrix} 1 & -8 & -4 \\ -8 & 1 & -4 \\ -4 & -4 & 7 \end{bmatrix}$

12. $A = \begin{bmatrix} -1 & 8 & 4 \\ 8 & -1 & 4 \\ 4 & 4 & -7 \end{bmatrix}$

13. $A = \begin{bmatrix} 2 & -2 & 1 \\ -2 & -1 & 2 \\ 1 & 2 & 2 \end{bmatrix}$

14. $A = \begin{bmatrix} -5 & 2 & -1 \\ 2 & -2 & -2 \\ -1 & -2 & -5 \end{bmatrix}$

15. Let $T: R^3 \to R^3$ be described by the matrix of Problem 5. Verify that for $\mathbf{x} = (1, -1, 2)$, $|\mathbf{x}| = |T(\mathbf{x})|$.

16. Let $S: R^3 \to R^3$ be described by the matrix of Problem 6. Verify that for $\mathbf{x} = (1, -1, 2)$, $|\mathbf{x}| = |S(\mathbf{x})|$.

17. Show that if $T: R^n \to R^n$ is a linear transformation that preserves lengths of vectors, and if $\mathbf{p}$ and $\mathbf{q}$ are arbitrary vectors in R^n, then $|\mathbf{p} - \mathbf{q}| = |T(\mathbf{p}) - T(\mathbf{q})|$.

18. Show that if $T: R^3 \to R^3$ is a linear transformation that preserves the length of vectors, then, if $\mathbf{p}$ and $\mathbf{q}$ are orthogonal vectors, $T(\mathbf{p})$ and $T(\mathbf{q})$ are also orthogonal.

19. Show that if A is orthogonal, then $\det A = \pm 1$.

20. Show that if A is a 3×3 matrix with the property that the angle between $\mathbf{p}$ and $\mathbf{q}$ always equals the angle between $A\mathbf{p}$ and $A\mathbf{q}$, then A is orthogonal.

21. Write $\mathbf{x}^t B \mathbf{x}$ as a polynomial where

$$B = \begin{bmatrix} 0 & 1 & 2 \\ -1 & 0 & -3 \\ -2 & 3 & 0 \end{bmatrix}, \qquad \mathbf{x} = \begin{bmatrix} x_1 \\ x_2 \\ x_3 \end{bmatrix}$$

22. The matrix B of Problem 21 is said to be **skew symmetric** since $B^t = -B$. Show that if S is any 3×3 skew symmetric matrix, then for all $\mathbf{x}$, $\mathbf{x}^t S \mathbf{x} = 0$.

23. A function $B(\mathbf{u}, \mathbf{v})$ of two vectors $\mathbf{u}$ and $\mathbf{v}$ in R^3 is called **bilinear** if

$$B(\mathbf{u}_1 + \mathbf{u}_2, \mathbf{v}) = B(\mathbf{u}_1, \mathbf{v}) + B(\mathbf{u}_2, \mathbf{v})$$
$$B(r\mathbf{u}, \mathbf{v}) = rB(\mathbf{u}, \mathbf{v})$$
$$B(\mathbf{u}, \mathbf{v}_1 + \mathbf{v}_2) = B(\mathbf{u}, \mathbf{v}_1) + B(\mathbf{u}, \mathbf{v}_2)$$
$$B(\mathbf{u}, r\mathbf{v}) = rB(\mathbf{u}, \mathbf{v})$$

(Problem set continues on p. 256)

CHIPPING AWAY;
A METHOD WITH POWERS

In theory, we find the eigenvalues and eigenvectors of an $n \times n$ matrix by finding the characteristic equation, finding the roots of this nth degree equation, and then finding the eigenvalues. In practice, the difficulties are even greater than you already expect. Not only is it tedious to find roots of an nth degree equation, but we necessarily work with decimal approximations—with or without the aid of a computer. The problem with this is that small changes in the coefficients of a polynomial can lead to large changes in its roots. The inevitable rounding errors in a large problem therefore lead to highly questionable results.

When a problem seems too difficult to solve in general, mathematicians typically chip away at it by considering special cases. The calculation of eigenvalues and eigenvectors is one of the difficult problems of current interest in mathematical research. The flavor of such work can be seen from the following "chip" knocked off of the general problem.

Suppose that A is an $n \times n$ matrix with distinct eigenvalues $r_1, \ldots, r_n$, arranged so that $|r_1| > |r_2| > \cdots > |r_n|$. Let $v_1, \ldots, v_n$ be the corresponding eigenvectors, and suppose (as is the case in many applications) that we are primarily interested in finding the eigenvalue r_1 of largest magnitude, the so-called **dominant** eigenvalue.

Begin with an arbitrary vector x_0 in R^n. Since the eigenvectors belonging to distinct eigenvalues are linearly independent, $v_1, \ldots, v_n$ form a basis, and we can write $x_0 = c_1 v_1 + c_2 v_2 + \cdots + c_n v_n$. Using the linearity of A and the fact that $A v_i = r_i v_i$, we generate

$$x_1 = A x_0 \qquad = c_1 r_1 v_1 + c_2 r_2 v_2 + \cdots + c_n r_n v_n$$
$$x_2 = A^2 x_0 = A x_1 = c_1 r_1^2 v_1 + c_2 r_2^2 v_2 + \cdots + c_n r_n^2 v_n$$
$$\vdots \qquad\qquad \vdots \qquad\quad \vdots \qquad\qquad \vdots$$
$$x_k = A^k x_0 \qquad = c_1 r_1^k v_1 + c_2 r_2^k v_2 + \cdots + c_n r_n^k v_n$$

From the last expression,

$$\frac{1}{r_1^k} x_k = c_1 v_1 + c_2 \left(\frac{r_2}{r_1}\right)^k v_2 + \cdots + c_n \left(\frac{r_n}{r_1}\right)^k v_n$$

As k increases, each of the terms $(r_i/r_1)^k$ approaches 0; so for large

k, $(1/r_1^k)\mathbf{x}_k \approx c_1\mathbf{v}_1$, or $\mathbf{x}_k \approx c_1 r_1^k \mathbf{v}_1$. This enables us to write

$$A\mathbf{x}_k \approx A(A^k\mathbf{x}_0) = A^{k+1}\mathbf{x}_0 = \mathbf{x}_{k+1} \approx c_1 r_1^{k+1}\mathbf{v}_1$$
$$= r_1(c_1 r_1^k \mathbf{v}_1) \approx r_1\mathbf{x}_k$$

We have proved that for large k, $A\mathbf{x}_k \approx r_1\mathbf{x}_k$. Here, r_1 is the dominant eigenvalue that we sought, and $\mathbf{x}_k$ is its corresponding eigenvector.

In practice, it is important to normalize the vector $\mathbf{x}_i$ at each iteration to avoid computations with large numbers. Of course this will not alter the eigenvector, since a scalar multiple of an eigenvector is still an eigenvector.

As an example, let us calculate the dominant eigenvalue of the matrix

$$A = \begin{bmatrix} 2 & 4 \\ 3 & 13 \end{bmatrix},$$

beginning with $\mathbf{x}_0 = \begin{bmatrix} 1 \\ 0 \end{bmatrix}$.

$$\mathbf{x}_1 = A\mathbf{x}_0 = \begin{bmatrix} 2 & 4 \\ 3 & 13 \end{bmatrix}\begin{bmatrix} 1 \\ 0 \end{bmatrix} = \begin{bmatrix} 2 \\ 3 \end{bmatrix}$$

We normalize $\mathbf{x}_1$, getting

$$\mathbf{x}_1' = \frac{\mathbf{x}_1}{\|\mathbf{x}_1\|} = \frac{1}{\sqrt{13}}\begin{bmatrix} 2 \\ 3 \end{bmatrix} = \begin{bmatrix} .5547 \\ .8321 \end{bmatrix}$$

Then

$$x_2 = A\mathbf{x}_1' = \begin{bmatrix} 2 & 4 \\ 3 & 13 \end{bmatrix}\begin{bmatrix} .5547 \\ .8321 \end{bmatrix} = \begin{bmatrix} 4.4378 \\ 12.4814 \end{bmatrix}$$

You are asked in Problem 29 to carry this process one step further, and to compare the result after three steps with the dominant eigenvalue, which you can easily compute by other methods for the given matrix A.

Show that if A is a 3×3 matrix, then

$$B(\mathbf{u}, \mathbf{v}) = \mathbf{u}^t A\mathbf{v}$$

defines a bilinear transformation.

24. A bilinear transformation B is called **symmetric** if $B(\mathbf{u}, \mathbf{v}) = B(\mathbf{v}, \mathbf{u})$ for every $\mathbf{u}$ and $\mathbf{v}$. Prove that the bilinear transformation B defined in Problem 23 is symmetric if and only if A is a symmetric matrix.

25. Prove that if the matrix A is **orthogonally similar** to a diagonal matrix D, that is, if there is an orthogonal matrix P such that $P^{-1}AP = D$, then A is symmetric.

*26. Suppose A is orthogonal and $A \sim B$. Must B be orthogonal?

*27. Let A be a matrix having all real entries and the real eigenvalue r. Show that there must be an eigenvector $\mathbf{v}$ belonging to r which has only real entries. *Hint:* Suppose

$$[A - rI] \begin{bmatrix} a_1 + b_1 i \\ \vdots \\ a_n + b_n i \end{bmatrix} = \begin{bmatrix} 0 \\ \vdots \\ 0 \end{bmatrix}$$

Consider the vector $\mathbf{v} = (a_1, \ldots, a_n)$.

28. Write the Jacobian matrix representing $f''(x)$ if $f(x_1, x_2) = \dfrac{x_1^2}{x_2} - x_1 x_2^2$. ▣

29. Continue "Chipping Away" by finding $\mathbf{x}_3 = A\mathbf{x}_2$. Compare your results with the dominant eigenvalue of the given matrix A. ▣

QUADRATIC FORMS 6-4

When introducing the topic of changing bases in a vector space (Section 4-5), we discussed the technique of rotation of axes as an aid to graphing the general second degree equation

$$ax_1^2 + bx_1x_2 + cx_2^2 + dx_1 + ex_2 + f = 0 \qquad (1)$$

One thinks of rotating the x_1x_2 axes through an angle θ to obtain the $x_1'x_2'$ axes (Figure 6-1). This amounts to changing from the basis B to B' where

$$B = \left\{ \mathbf{u}_1 = \begin{bmatrix} 1 \\ 0 \end{bmatrix}, \mathbf{u}_2 = \begin{bmatrix} 0 \\ 1 \end{bmatrix} \right\}, \qquad B' = \left\{ \mathbf{v}_1 = \begin{bmatrix} \cos\theta \\ \sin\theta \end{bmatrix}, \mathbf{v}_2 = \begin{bmatrix} -\sin\theta \\ \cos\theta \end{bmatrix} \right\}$$

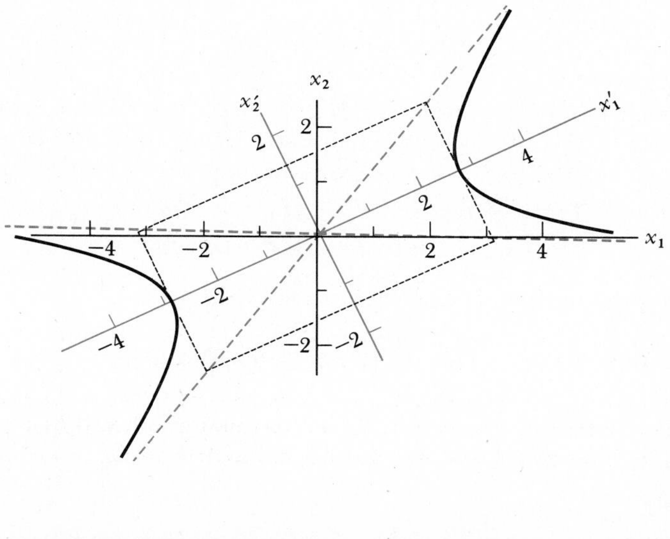

Figure 6-1

If the notation introduced in Section 4-5 is employed, the relationship between the two bases can be exhibited as a matrix product involving the two matrices $U = [\mathbf{e}_1 \quad \mathbf{e}_2]$ and $V = [\mathbf{v}_1 \quad \mathbf{v}_2]$. According to Theorem A of Section 4-5, there exists a matrix P such that $V = UP$. In our particular case, P is easily determined because $U = I$ and so $V = P$. This theorem also tells us that the coordinates with respect to the two bases are related by

(2)
$$\begin{bmatrix} x_1 \\ x_2 \end{bmatrix} = \begin{bmatrix} \cos\theta & -\sin\theta \\ \sin\theta & \cos\theta \end{bmatrix} \begin{bmatrix} x_1' \\ x_2' \end{bmatrix}$$

The idea is to substitute

$$x_1 = x_1'\cos\theta - x_2'\sin\theta, \qquad x_1 = x_1'\sin\theta + x_2'\cos\theta$$

into (1) and then choose θ so as to obtain an equation of the form

(3)
$$a'(x_1')^2 + c'(x_2')^2 + d'x_1' + e'x_2' + f' = 0$$

The actual choosing of θ leads to some involved trigonometric equations. We said in our discussion in Section 4-5 that one of the triumphs of linear algebra is that we can actually transform an equation of the form (1) into an equation of the form (3) with no trigonometry at all. We are now in a position to see how this is done.

Since the linear terms of (1) and (3) cause no difficulty, we focus attention on the quadratic terms which can be written (equation (1), Section 6-1) respectively as

(4)
$$\mathbf{x}^t A \mathbf{x} \quad \text{where} \quad A = \begin{bmatrix} a & \dfrac{b}{2} \\ \dfrac{b}{2} & c \end{bmatrix}$$

and

$$(\mathbf{x}')^t D \mathbf{x}' \quad \text{where} \quad D = \begin{bmatrix} a' & 0 \\ 0 & c' \end{bmatrix}$$

The matrix A is real and symmetric so, according to Theorem D of Section 6-3, there is an orthogonal matrix P such that $P^{-1}AP = D$, a diagonal matrix. Suppose that we substitute $\mathbf{x} = P\mathbf{x}'$ in (4), remembering that $\mathbf{x}^t = (P\mathbf{x}')^t = (\mathbf{x}')^t P^t = (\mathbf{x}')^t P^{-1}$. We get

$$\mathbf{x}^t A \mathbf{x} = (\mathbf{x}')^t P^{-1} A P \mathbf{x}' = (\mathbf{x}')^t D \mathbf{x}'$$

The right-hand side of this equation is of the form

$$\begin{bmatrix} x_1' & x_2' \end{bmatrix} \begin{bmatrix} a' & 0 \\ 0 & c' \end{bmatrix} \begin{bmatrix} x_1' \\ x_2' \end{bmatrix} = a'(x_1')^2 + c'(x_2')^2$$

which is the quadratic portion of (3). And though we had no reason to notice it before, we now notice that the matrix in (2) is orthogonal.

It is now evident that we have two methods for solving this problem. One requires trigonometry to determine θ in (2). The other uses the theory of similar matrices. It may be instructive to illustrate the new method by solving again the problem first solved with trigonometric methods in Section 4-5.

Graph $4x_1x_2 - 3x_2^2 = 8$.   *Example A*

The equation can be written in the form

$$[x_1 \quad x_2]\begin{bmatrix} 0 & 2 \\ 2 & -3 \end{bmatrix}\begin{bmatrix} x_1 \\ x_2 \end{bmatrix} = 8$$

The characteristic equation of the symmetric matrix is

$$\begin{vmatrix} 0 - r & 2 \\ 2 & -3 - r \end{vmatrix} = r(r + 3) - 4 = (r + 4)(r - 1) = 0$$

The eigenvectors are found in the usual way. For

$$r = 1 \qquad\qquad\qquad\qquad\qquad r = -4$$

$$\begin{bmatrix} -1 & 2 \\ 2 & -4 \end{bmatrix}\begin{bmatrix} p_1 \\ p_2 \end{bmatrix} = \begin{bmatrix} 0 \\ 0 \end{bmatrix} \qquad \begin{bmatrix} 4 & 2 \\ 2 & 1 \end{bmatrix}\begin{bmatrix} p_1 \\ p_2 \end{bmatrix} = \begin{bmatrix} 0 \\ 0 \end{bmatrix}$$

the solution space is spanned | the solution space is spanned
by $(2, 1)$ | by $(1, -2)$

As the theory guarantees, the two eigenvectors are orthogonal. Normalizing each of them, we form

$$P = \begin{bmatrix} \dfrac{2}{\sqrt{5}} & \dfrac{1}{\sqrt{5}} \\ \dfrac{1}{\sqrt{5}} & -\dfrac{2}{\sqrt{5}} \end{bmatrix}$$

The new basis vectors are the columns of $V = UP$, and since $U = I$ when we start with the standard basis, our new basis is

$$\mathbf{v}_1 = \begin{bmatrix} \dfrac{2}{\sqrt{5}} \\ \dfrac{1}{\sqrt{5}} \end{bmatrix}, \quad \mathbf{v}_2 = \begin{bmatrix} \dfrac{1}{\sqrt{5}} \\ -\dfrac{2}{\sqrt{5}} \end{bmatrix}$$

Doubters can multiply to verify that

$$P^tAP = \begin{bmatrix} \dfrac{2}{\sqrt{5}} & \dfrac{1}{\sqrt{5}} \\ \dfrac{1}{\sqrt{5}} & -\dfrac{2}{\sqrt{5}} \end{bmatrix}\begin{bmatrix} 0 & 2 \\ 2 & -3 \end{bmatrix}\begin{bmatrix} \dfrac{2}{\sqrt{5}} & \dfrac{1}{\sqrt{5}} \\ \dfrac{1}{\sqrt{5}} & -\dfrac{2}{\sqrt{5}} \end{bmatrix} = \begin{bmatrix} 1 & 0 \\ 0 & -4 \end{bmatrix}$$

The equation of the curve with respect to the new axes is

$$(x_1')^2 - 4(x_2')^2 = 8$$

See Figure 6-1.

Sometimes it is only required that the transformed equation be found. In this case, we would have been through as soon as the eigenvalues of $r = 1$ and $r = -4$ were determined. □

It might be difficult to argue that the use of a similarity transformation is a big improvement over the trigonometric substitution in the case of a quadratic form in 2 variables, but there is no doubt as to which method most easily generalizes to quadratic forms in 3 or more variables. Only one precaution is necessary to extend the method of similarity transformations to the case of n variables. Let us review the general procedure.

Suppose that with respect to some orthogonal basis $B = \{\mathbf{u}_1, \ldots, \mathbf{u}_n\}$, a quadratic form in n variables is expressed in the form $\mathbf{x}^t A \mathbf{x}$, A being a real $n \times n$ symmetric matrix. Theorem D of Section 6-3 guarantees an orthonormal matrix P such that $P^{-1}AP$ is diagonal. Let us define a new basis B' by

(5)
$$[\mathbf{v}_1 \quad \cdots \quad \mathbf{v}_n] = [\mathbf{u}_1 \quad \cdots \quad \mathbf{u}_n]P$$

Then the coordinates with respect to B and B' are related by $\mathbf{x} = P\mathbf{x}'$ and this substitution leads to the quadratic form

$$(\mathbf{x}')^t P^t A P \mathbf{x}' = (\mathbf{x}')^t D \mathbf{x}'$$

in which each term involves just one variable.

The one precaution mentioned above is this: Since B is an orthonormal basis, the matrix $U = [\mathbf{u}_1 \quad \cdots \quad \mathbf{u}_n]$ is by definition orthogonal. P is also orthogonal, so matrix $V = [\mathbf{v}_1 \quad \cdots \quad \mathbf{v}_n]$ satisfies

$$V^t = (UP)^t = P^t U^t = P^{-1} U^{-1}$$

which is the inverse of $V = UP$. Hence $V^t = V^{-1}$, so V is orthogonal and basis B' is orthonormal. That, however, is not enough in R^3 to call our change of basis a rotation. This is easily seen in Figure 6-2 where we

Figure 6-2

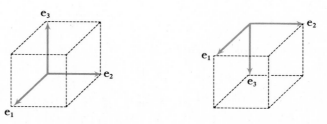

have two orthonormal bases for R^3. $B = \{\mathbf{u}_1 = \mathbf{e}_1, \mathbf{u}_2 = \mathbf{e}_2, \mathbf{u}_3 = \mathbf{e}_3\}$, and $B' = \{\mathbf{v}_1 = \mathbf{e}_1, \mathbf{v}_2 = \mathbf{e}_2, \mathbf{v}_3 = -\mathbf{e}_3\}$. No rotation of the coordinate axes determined by $\{\mathbf{u}_1, \mathbf{u}_2, \mathbf{u}_3\}$ will bring us to the axes determined by $\{\mathbf{v}_1, \mathbf{v}_2, \mathbf{v}_3\}$. The cube C' determined by the second basis is C turned "inside out", a geometric fact that will be seen analytically from the fact that the usual determinant formula for volume, without a correction for sign, would give a negative volume. The determinant of any orthogonal matrix is ± 1 (Problem 19, Problem Set 6-3), and (5) will be called a **rotation** if and only if the determinant of the orthogonal matrix P is 1.

In forming P, we have with each column vector a choice of using that vector or its negative. Thus, if we find that the P that we have formed has a determinant of -1, we may change the sign of any column vector to get the desired rotation.

We are now in a position to determine a rotation of axes that will change any quadratic equation in n variables to one in which the mixed-product terms are all missing.

Describe the graph of   *Example B*

$$2x_1^2 + 2x_2^2 + 2x_3^2 - 2x_1x_2 - 2x_1x_3 - 2x_2x_3 + 2x_1 + x_3 = 5$$

Ignoring first degree terms to begin with, write the quadratic terms in the form

$$[x_1 \quad x_2 \quad x_3] \begin{bmatrix} 2 & -1 & -1 \\ -1 & 2 & -1 \\ -1 & -1 & 2 \end{bmatrix} \begin{bmatrix} x_1 \\ x_2 \\ x_3 \end{bmatrix}$$

The symmetric matrix here is, by the best of luck, the one diagonalized in Example A of Section 6-3. A quick calculation shows, however, that the matrix P used in that example had a determinant of -1. To get a rotation, we arbitrarily change the sign of the middle column, getting

$$\begin{bmatrix} x_1 \\ x_2 \\ x_3 \end{bmatrix} = \begin{bmatrix} -\dfrac{1}{\sqrt{2}} & -\dfrac{1}{\sqrt{6}} & \dfrac{1}{\sqrt{3}} \\ \dfrac{1}{\sqrt{2}} & -\dfrac{1}{\sqrt{6}} & \dfrac{1}{\sqrt{3}} \\ 0 & \dfrac{2}{\sqrt{6}} & \dfrac{1}{\sqrt{3}} \end{bmatrix} \begin{bmatrix} x_1' \\ x_2' \\ x_3' \end{bmatrix} \qquad (6)$$

As shown in the example, this substitution leads to the quadratic form

$$[x_1' \quad x_2' \quad x_3'] \begin{bmatrix} 3 & 0 & 0 \\ 0 & 3 & 0 \\ 0 & 0 & 0 \end{bmatrix} \begin{bmatrix} x_1' \\ x_2' \\ x_3' \end{bmatrix}$$

Also, from (6),

$$x_1 = -\frac{1}{\sqrt{2}}x_1' - \frac{1}{\sqrt{6}}x_2' + \frac{1}{\sqrt{3}}x_3'$$

$$x_3 = \frac{2}{\sqrt{6}}x_2' + \frac{1}{\sqrt{3}}x_3'$$

so the linear part of the given equation gives us

$$2x_1 + x_3 = -\frac{2}{\sqrt{2}}x_1' + \frac{3}{\sqrt{3}}x_3'$$

The transformed equation is

$$3(x_1')^2 + 3(x_2')^2 - \frac{2}{\sqrt{2}}x_1' + \sqrt{3}x_3' = 5$$

Complete the square to obtain

$$3\left((x_1')^2 - \frac{2}{3\sqrt{2}}x_1' + \frac{1}{18}\right) + 3(x_2')^2 + \sqrt{3}x_3' = 5 + \frac{3}{18}$$

$$\left(x_1' - \frac{1}{3\sqrt{2}}\right)^2 + (x_2')^2 = -\frac{\sqrt{3}}{3}\left(x_3' - \frac{31}{6\sqrt{3}}\right)$$

The graph is a paraboloid. With respect to the rotated coordinate axes $x_1'x_2'x_3'$, its vertex is at $(1/3\sqrt{2}, 0, 31/6\sqrt{3})$. See Figure 6-3. □

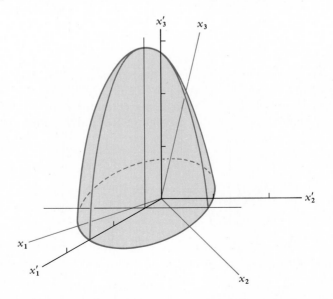

Figure 6-3

To bring matters to a tidy conclusion, it remains to prove Theorem D of the last section which we restate here for convenience.

If A is a real symmetric matrix, there is an orthogonal matrix P such that $P^{-1}AP = D$ is diagonal.

Proof: Our proof must be independent of the unproved Theorem of the same section. We use an induction argument. If A is 1×1, it is already diagonal, and elementary trigonometry can be used to spin out the required orthogonal matrix P for the case in which A is 2×2. That brings us to the 3×3 case which we will use to illustrate the induction argument.

The real 3×3 symmetric matrix A, according to Theorem A of the last section, has a real eigenvalue r_1. Let $\mathbf{u}_1$ be the corresponding (nonzero, of course) eigenvector. Together with some two vectors of the standard basis, we can obtain a basis for R^3, and the Gram-Schmidt orthogonalization process can then be used to obtain an orthonormal basis $\{ \mathbf{v}_1, \mathbf{v}_2, \mathbf{v}_3 \}$ in which the unit vector $\mathbf{v}_1$ is a multiple of $\mathbf{u}_1$, hence also an eigenvector of A belonging to r_1. Now form the matrix

$$Q = [\mathbf{v}_1 \quad \mathbf{v}_2 \quad \mathbf{v}_3]$$

Note that

$$AQ = [A\mathbf{v}_1 \quad A\mathbf{v}_2 \quad A\mathbf{v}_3] = [r_1\mathbf{v}_1 \quad A\mathbf{v}_2 \quad A\mathbf{v}_3]$$

Next form $M = Q^t AQ$ and note that

$$M = Q^t AQ = \begin{bmatrix} \mathbf{v}_1^t \\ \mathbf{v}_2^t \\ \mathbf{v}_3^t \end{bmatrix} [r_1\mathbf{v}_1 \quad A\mathbf{v}_2 \quad A\mathbf{v}_3]$$

Recall that the elements in the first column of the matrix M are obtained as the dot product of the row vectors $\mathbf{v}_1^t$, $\mathbf{v}_2^t$, and $\mathbf{v}_3^t$ with the column vector $r_1\mathbf{v}_1$. Since $\mathbf{v}_1$, $\mathbf{v}_2$, and $\mathbf{v}_3$ are orthonormal, it follows that M is of the form

$$M = \begin{bmatrix} r_1 & * & * \\ 0 & * & * \\ 0 & * & * \end{bmatrix}$$

We can do a little better by noting that

$$M^t = (Q^t AQ)^t = Q^t A^t (Q^t)^t = Q^t AQ = M$$

That is, M is symmetric. Hence, the first row is identical to the first column; M has the form

$$M = \begin{bmatrix} r_1 & 0 & 0 \\ 0 & & \\ 0 & & S \end{bmatrix}$$

where the 2×2 matrix S is symmetric. And we already know that any 2×2 symmetric matrix is orthogonally similar to a diagonal matrix, so there must be a 2×2 orthogonal matrix R such that

(7)
$$R^t S R = C$$

is a diagonal matrix. Finally, form

$$N = \begin{bmatrix} 1 & 0 & 0 \\ 0 & & \\ 0 & & R \end{bmatrix}$$

Note that N is orthogonal (the columns form an orthonormal set) and that

$$N^t M N = \begin{bmatrix} 1 & 0 & 0 \\ 0 & & R^t \\ 0 & & \end{bmatrix} \begin{bmatrix} r_1 & 0 & 0 \\ 0 & & S \\ 0 & & \end{bmatrix} \begin{bmatrix} 1 & 0 & 0 \\ 0 & & R \\ 0 & & \end{bmatrix}$$

$$= \begin{bmatrix} r_1 & 0 & 0 \\ 0 & & C \\ 0 & & \end{bmatrix} = D$$

where D, like C, is a diagonal matrix. Since $M = Q^t A Q$,

$$N^t Q^t A Q N = (QN)^t A (QN) = D$$

Finally, set $P = QN$. Since the product of two orthogonal matrices is orthogonal, we have exhibited the desired orthogonal matrix.

We could now prove the theorem for a 4×4 matrix, following the steps above and using at equation (7) what we just learned about 3×3 matrices, and so on. $\square$

In Problems 1–6, you are given equations that describe graphs drawn in R^2 using the standard basis. Represent these same graphs with respect to rotated axes by equations containing no mixed-product terms. Sketch the graphs, showing both sets of axes.

1. $5x_1^2 + 3x_1 x_2 + x_2^2 = 15$
2. $5x_1^2 + 6x_1 x_2 - 3x_2^2 = 9$
3. $9x_1^2 + 5x_1 x_2 - 3x_2^2 = 12$
4. $16x_1^2 + 24x_1 x_2 + 9x_2^2 + 10x_1 = 36$
5. $9x_1^2 + 24x_1 x_2 + 16x_2^2 + 10x_1 = 16$
6. $3x_1^2 - 10x_1 x_2 + 3x_2^2 = 6$

(*Problem set continues on p. 266*)

FACTORING SECOND DEGREE POLYNOMIALS

A second degree polynomial in n variables can sometimes be factored into two first degree polynomials using the following technique. Suppose we wish to factor

$$f(x_1, x_2) = x_1^2 - 5x_2^2 - 4x_1x_2 - 6x_2 - 1$$

Begin by forming a homogeneous polynomial

$$g(x_1, x_2, x_3) = x_3^2 f\left(\frac{x_1}{x_3}, \frac{x_2}{x_3}\right)$$

$$= x_1^2 - 5x_2^2 - x_3^2 - 4x_1x_2 - 6x_2x_3$$

$$= \mathbf{x}^t A \mathbf{x} \quad \text{where} \quad A = \begin{bmatrix} 1 & -2 & 0 \\ -2 & -5 & -3 \\ 0 & -3 & -1 \end{bmatrix} \quad \text{and} \quad \mathbf{x} = \begin{bmatrix} x_1 \\ x_2 \\ x_3 \end{bmatrix}$$

The matrix A has eigenvalues of 0, 2, and -7. Using the corresponding normalized eigenvectors to form column vectors, we get

$$P = \begin{bmatrix} \dfrac{2}{\sqrt{14}} & -\dfrac{2}{\sqrt{6}} & \dfrac{1}{\sqrt{21}} \\ \dfrac{1}{\sqrt{14}} & \dfrac{1}{\sqrt{6}} & \dfrac{4}{\sqrt{21}} \\ -\dfrac{3}{\sqrt{14}} & -\dfrac{1}{\sqrt{6}} & \dfrac{2}{\sqrt{21}} \end{bmatrix}$$

Substituting $\mathbf{x} = P\mathbf{x}'$,

$$\mathbf{x}^t A \mathbf{x} = (\mathbf{x}')^t P^t A P \mathbf{x}' = 2(x_2')^2 - 7(x_3')^2$$

The key that opens the door to this method is that A has just two nonzero eigenvalues, allowing us to factor this expression (possibly needing complex numbers) as the difference of two squares.

$$\mathbf{x}^t A \mathbf{x} = (\sqrt{2}x_2' - \sqrt{7}x_3')(\sqrt{2}x_2' + \sqrt{7}x_3') \qquad (i)$$

From $\mathbf{x}' = P^t\mathbf{x}$, we learn that

$$\sqrt{2}x_2' = -\frac{2}{\sqrt{3}}x_1 + \frac{1}{\sqrt{3}}x_2 - \frac{1}{\sqrt{3}}x_3$$

$$\sqrt{7}x_3' = \frac{1}{\sqrt{3}}x_1 + \frac{4}{\sqrt{3}}x_2 + \frac{2}{\sqrt{3}}x_3$$

Substitution into (i) gives

$$g(x_1, x_2, x_3)$$

$$= \left(-\frac{1}{\sqrt{3}}x_1 + \frac{5}{\sqrt{3}}x_2 + \frac{1}{\sqrt{3}}x_3\right)\left(-\frac{3}{\sqrt{3}}x_1 - \frac{3}{\sqrt{3}}x_2 - \frac{3}{\sqrt{3}}x_3\right)$$

$$x_3^2 f\left(\frac{x_1}{x_3}, \frac{x_2}{x_3}\right) = x_3^2\left(-\frac{x_1}{x_3} + 5\frac{x_2}{x_3} + 1\right)\left(-\frac{x_1}{x_3} - \frac{x_2}{x_3} - 1\right)$$

Therefore,

$$f(x_1, x_2) = (-x_1 + 5x_2 + 1)(-x_1 - x_2 - 1)$$

This is a great deal of work for so modest an accomplishment, but the same method will show that

$$2x_1^2 - 4x_2^2 - 6x_3^2 + 7x_1x_2 - x_1x_3$$
$$+ 14x_2x_3 + 2x_1 + 17x_2 - 11x_3 - 4$$
$$= (2x_1 - x_2 + 3x_3 + 4)(x_1 + 4x_2 - 2x_3 - 1)$$

In Problems 7–11 you are given equations that describe graphs drawn in R^3 using the standard basis. Represent these same graphs with respect to rotated axes by equations containing no mixed-product terms.

7. $5x_1^2 + 11x_2^2 + 2x_3^2 + 16x_1x_2 - 4x_2x_3 + 20x_1x_3 = 36$
   (See Problem 7, Problem Set 6-3.)

8. $11x_1^2 + 2x_2^2 + 5x_3^2 + 4x_1x_2 + 20x_2x_3 - 16x_1x_3 = 36$
   (See Problem 8, Problem Set 6-3.)

9. $-x_1^2 + 2x_2^2 - x_3^2 - 4x_1x_2 + 4x_2x_3 - 10x_1x_3 = 54$
   (See Problem 9, Problem Set 6-3.)

10. $x_1^2 - 2x_2^2 + x_3^2 - 2x_1x_2 + 2x_2x_3 + 4x_1x_3 = 27$
    (See Problem 10, Problem Set 6-3.)

*11. The matrix A of Problem 7, Problem Set 6-3, has the eigenvalue $r_1 = 9$ and the corresponding eigenvector is $\mathbf{u}_1^t = (1, -2, 2)$. Using these choices, follow the proof of Theorem D by determining in turn $\mathbf{v}_1, \mathbf{v}_2, \mathbf{v}_3$, then Q, M, S, R, N, and finally P.

12. Suppose that the substitution (2) transforms

$$Ax_1^2 + Bx_1x_2 + Cx_2^2 + Dx_1 + Ex_2 + F = 0$$

into the equation

$$A'(x_1')^2 + B'x_1'x_2' + C'(x_2')^2 + D'x_1' + E'x_2' + F' = 0$$

Prove that $B^2 - 4AC = (B')^2 - 4A'C'$.

POSITIVE MATRICES | 6-5

Consider the function

$$f(x_1, x_2) = \tfrac{34}{25}x_1^2 - \tfrac{24}{25}x_1 x_2 + \tfrac{41}{25}x_2^2$$

which can be written

$$f(x_1, x_2) = [x_1 \quad x_2] \begin{bmatrix} \tfrac{34}{25} & -\tfrac{12}{25} \\ -\tfrac{12}{25} & \tfrac{41}{25} \end{bmatrix} \begin{bmatrix} x_1 \\ x_2 \end{bmatrix} \tag{1}$$

The graph of $z = f(x_1, x_2)$ is easier to draw if we first make the change of variable (rotation of axes) that removes the $x_1 x_2$ term. Finding the eigenvalues and corresponding eigenvectors in the now familiar way, we are led to the substitution

$$\begin{bmatrix} x_1 \\ x_2 \end{bmatrix} = \begin{bmatrix} \tfrac{4}{5} & -\tfrac{3}{5} \\ \tfrac{3}{5} & \tfrac{4}{5} \end{bmatrix} \begin{bmatrix} x_1' \\ x_2' \end{bmatrix}$$

The equation to be graphed with respect to the new basis (with respect to the rotated axes) is

$$z = [x_1' \quad x_2'] \begin{bmatrix} 1 & 0 \\ 0 & 2 \end{bmatrix} \begin{bmatrix} x_1' \\ x_2' \end{bmatrix}$$

Several things are to be noted about the graph sketched in Figure 6-4. It holds water; more formally, it is **strictly convex.** There is a unique minimum point. For every choice of x_1 and x_2, $f(x_1, x_2) \geq 0$.

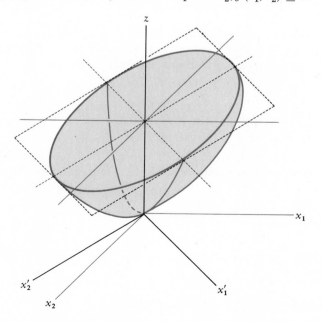

Figure 6-4

It is this last property that is generalized to $n \times n$ matrices. We call a real $n \times n$ symmetric matrix A **positive** if for every choice of $\mathbf{x} = (x_1, \ldots, x_n)$ the function

(2)
$$f(x_1, \ldots, x_n) = [x_1 \quad \cdots \quad x_n] A \begin{bmatrix} x_1 \\ \vdots \\ x_n \end{bmatrix} \geq 0$$

The matrix that defines f in (1) is an example of a positive matrix. If the inequality in (2) can be replaced by $>$ for all $\mathbf{x} \neq 0$, then the matrix A is said to be **positive definite. Negative** and **negative definite** matrices are similarly defined.

A positive or positive definite matrix is most easily recognized using the technique already illustrated. Diagonalize a given real symmetric matrix and examine the signs of the eigenvalues.

THEOREM A

A real symmetric matrix A is positive if its eigenvalues are all non-negative; it is positive definite if the eigenvalues are all positive.

Proof: Let P be the orthogonal matrix that diagonalizes A. Choose $\mathbf{x}$ arbitrarily and set $\mathbf{x}' = P^{-1}\mathbf{x}$. We are given that the eigenvalues $r_1, \ldots, r_n$ of A are all non-negative. Thus,

(3)
$$\mathbf{x}^t A \mathbf{x} = (\mathbf{x}')^t P^t A P \mathbf{x}' = r_1(\mathbf{x}_1')^2 + \cdots + r_n(\mathbf{x}_n')^2 \geq 0$$

Moreover, if $\mathbf{x} \neq 0$ and all the $r_i > 0$, then (3) must be strictly positive. $\square$

There is another characterization of positive matrices that we shall mention since it is often used in applications of matrix theory to economics. The statement of the result requires a definition. For the $n \times n$ matrix

$$A = \begin{bmatrix} a_{11} & a_{12} & \cdots & a_{1n} \\ a_{21} & a_{22} & \cdots & a_{2n} \\ \vdots & \vdots & & \vdots \\ a_{n1} & a_{n2} & \cdots & a_{nn} \end{bmatrix}$$

we define the **principal minors** to be the n determinants

$$a_{11}, \quad \begin{vmatrix} a_{11} & a_{12} \\ a_{21} & a_{22} \end{vmatrix}, \quad \ldots, \quad \begin{vmatrix} a_{11} & \cdots & a_{1k} \\ \vdots & & \vdots \\ a_{k1} & \cdots & a_{kk} \end{vmatrix}, \quad \ldots, \quad \begin{vmatrix} a_{11} & \cdots & a_{1n} \\ \vdots & & \vdots \\ a_{n1} & \cdots & a_{nn} \end{vmatrix}$$

THEOREM B

A real symmetric matrix A is positive definite if all of its principal minors are positive.

We omit the proof. (See Hoffman, K., and Kunze, R. 1971. *Linear algebra.* 2d ed. Englewood Cliffs, N.J.: Prentice-Hall.)

MINIMUMS FOR FUNCTIONS OF SEVERAL VARIABLES

Taylor's theorem says, for a function in one variable, that if f has two continuous derivatives in a neighborhood of a point x_0, then for sufficiently small h, there exists a point t between x_0 and $x_0 + h$ such that

$$f(x_0 + h) = f(x_0) + f'(x_0)h + \tfrac{1}{2}f''(t)h^2$$

If x_0 is a critical point of f, meaning that $f'(x_0) = 0$, then

$$f(x_0 + h) - f(x_0) = \tfrac{1}{2}f''(t)h^2 \qquad (i)$$

It follows that if $f''(t)$ is positive in a neighborhood of x_0, then $f(x_0 + h) - f(x_0) > 0$ for sufficiently small h, meaning that f has a relative minimum at x_0. This gives us the familiar second-derivative test for a minimum of f at a critical point x_0.

For a function of two variables, Taylor's theorem says that if f has two continuous derivatives in a neighborhood of $\mathbf{x}_0$, then for sufficiently small $\mathbf{h}$, there is a point $\mathbf{t}$ between $\mathbf{x}_0$ and $\mathbf{x}_0 + \mathbf{h}$ such that

$$f(\mathbf{x}_0 + \mathbf{h}) = f(\mathbf{x}_0) + f'(\mathbf{x}_0)\mathbf{h} + \tfrac{1}{2}\mathbf{h}^t f''(\mathbf{t})\mathbf{h}$$

Now if f has a critical point at $\mathbf{x}_0$, then

$$f'(\mathbf{x}_0) = [f_1(\mathbf{x}_0) \quad f_2(\mathbf{x}_0)] = [0 \quad 0],$$

$$\text{so} \quad f'(\mathbf{x}_0)\mathbf{h} = [0 \quad 0]\begin{bmatrix} h_1 \\ h_2 \end{bmatrix} = 0$$

What condition on the second derivative will show $\mathbf{x}_0$ to be a relative minimum? As in (i) above, we want

$$f(\mathbf{x}_0 + \mathbf{h}) - f(\mathbf{x}_0) = \tfrac{1}{2}[h_1 \quad h_2][f''(\mathbf{t})]\begin{bmatrix} h_1 \\ h_2 \end{bmatrix}$$

to be positive for all sufficiently small $\mathbf{h}$. That is, we want (see *Second Derivatives and Symmetric Matrices,* Section 6-3)

$$[f''(\mathbf{t})] = \begin{bmatrix} f_{11}(\mathbf{t}) & f_{12}(\mathbf{t}) \\ f_{21}(\mathbf{t}) & f_{22}(\mathbf{t}) \end{bmatrix} \qquad (ii)$$

to be positive definite.

The same reasoning for a function in n variables will give the same result. A critical point $\mathbf{x}_0$ of a function f will be a point at which f assumes a minimum if the $n \times n$ matrix defining $f''(\mathbf{t})$ is positive definite.

To determine whether a function $f(x_1, x_2)$ with a critical point at $\mathbf{x}_0$ has a maximum or minimum there, students in elementary calculus are advised (see Apostol, T. M. 1962. *Calculus II*. New York: Blaisdell) to calculate $f_{11}(\mathbf{x}_0), f_{12}(\mathbf{x}_0), f_{21}(\mathbf{x}_0),$ and $f_{22}(\mathbf{x}_0)$, and then to calculate $D = f_{11}(\mathbf{x}_0) f_{22}(\mathbf{x}_0) - f_{12}(\mathbf{x}_0) f_{21}(\mathbf{x}_0)$. Finally, the rule is that

If $D > 0$ and $f_{11}(\mathbf{x}_0) > 0$, f has a relative minimum at $\mathbf{x}_0$.
If $D > 0$ and $f_{11}(\mathbf{x}_0) < 0$, f has a relative maximum at $\mathbf{x}_0$.

Compare this advice with the conditions given in Theorem B to show that (ii) is positive definite.

PROBLEM SET 6-5

The matrices of Problems 1–6 are all positive. In each case, establish the fact using Theorem A, then again using Theorem B.

1. $\begin{bmatrix} 7 & \sqrt{3} \\ \sqrt{3} & 5 \end{bmatrix}$
2. $\begin{bmatrix} 11 & \sqrt{3} \\ \sqrt{3} & 9 \end{bmatrix}$

3. $\begin{bmatrix} 9 & 0 & -\sqrt{3} \\ 0 & 4 & 0 \\ -\sqrt{3} & 0 & 11 \end{bmatrix}$
4. $\begin{bmatrix} 7 & 0 & -3\sqrt{3} \\ 0 & 12 & 0 \\ -3\sqrt{3} & 0 & 13 \end{bmatrix}$

5. $\begin{bmatrix} 4 & 1 & 1 \\ 1 & 4 & 1 \\ 1 & 1 & 4 \end{bmatrix}$
6. $\begin{bmatrix} 5 & -1 & -1 \\ -1 & 5 & -1 \\ -1 & -1 & 5 \end{bmatrix}$

7. Prove that if A is positive definite and Q is nonsingular, then $B = Q^t A Q$ is also positive definite.

8. Prove that if A is positive definite, then there is a symmetric matrix B such that $B^2 = A$.

9. Prove that for any $m \times n$ matrix Q, the matrix $Q^t Q$ is positive.

10. Find the critical point of

$$f(x_1, x_2, x_3) = 2x_1^2 + 5x_2^2 + 2x_3^2 - 4x_1 x_2 - 2x_1 x_3 + 2x_2 x_3 - 2x_1 - 2x_3$$

and use the results of this section to prove it is a minimum.

11. Find the minimum of $f(x_1, x_2) = 2x_1^3 + (x_1 - x_2)^2 - 6x_2$.

6-6 NONSYMMETRIC MATRICES

The general question raised in Section 6-2 asked which matrices are diagonalizable. One answer, the symmetric matrices, was given in Section 6-3, and Sections 6-4 and 6-5 were digressions to topics in which symmetric matrices play an important role. We now return to the main question.

The $n \times n$ matrix A is similar to a diagonal matrix D if and only if there exist n eigenvectors of A that form a basis for C^n.

Proof: If A is similar to a diagonal matrix D, then there is a nonsingular matrix P such that $P^{-1}AP = D$; it follows that $AP = PD$. As was illustrated in some detail for the case $n = 2$ in Section 6-2, this says

$$A[\mathbf{P}_{\cdot 1} \quad \cdots \quad \mathbf{P}_{\cdot n}] = [r_1\mathbf{P}_{\cdot 1} \quad \cdots \quad r_n\mathbf{P}_{\cdot n}]$$

where $r_1, \ldots, r_n$ are the entries on the diagonal of D and $\mathbf{P}_{\cdot 1}, \ldots, \mathbf{P}_{\cdot n}$ are the column vectors of P. Hence,

$$A\mathbf{P}_{\cdot 1} = r_1\mathbf{P}_{\cdot 1}, \ldots, A\mathbf{P}_{\cdot n} = r_n\mathbf{P}_{\cdot n}$$

The column vectors of P are thus seen to be eigenvectors of A, and since P is nonsingular, these column vectors must be linearly independent, enough to guarantee that they form a basis for C^n (Theorem A, Section 2-6).

Now suppose that $\mathbf{P}_1, \ldots, \mathbf{P}_n$ are eigenvectors of A that form a basis for C^n. Then the matrix P formed by using these vectors as column vectors will be nonsingular.

$$\begin{aligned} AP &= A[\mathbf{P}_1 \quad \cdots \quad \mathbf{P}_n] = [A\mathbf{P}_1 \quad \cdots \quad A\mathbf{P}_n] \\ &= [r_1\mathbf{P}_1 \quad \cdots \quad r_n\mathbf{P}_n] \\ &= [\mathbf{P}_1 \quad \cdots \quad \mathbf{P}_n]\begin{bmatrix} r_1 & & 0 \\ & \ddots & \\ 0 & & r_n \end{bmatrix} = PD \end{aligned}$$

Multiplication on the left by P^{-1} gives $P^{-1}AP = D$. $\square$

We now have a condition which, if satisfied, guarantees that A is diagonalizable. In terms of the steps outlined in Section 6-2, however, little has been gained, since not until Step 4 is it known whether we can find eigenvectors that are linearly independent. The theorem tells us whether or not A can be diagonalized after we have either done it or wasted a lot of time trying.

It would clearly be better if we had a theorem that could tell us whether A is diagonalizable before all the work is done. Our next theorem moves us in that direction.

Eigenvectors of A that belong to distinct eigenvalues are linearly independent.

Proof: Let $\mathbf{p}_1, \ldots, \mathbf{p}_k$ be eigenvectors belonging to the distinct eigenvalues $r_1, \ldots, r_k$. Suppose that the set $S = \{\mathbf{p}_1, \ldots, \mathbf{p}_k\}$ is linearly dependent. Then the subspace spanned by the vectors of S has a basis that is a proper subset of S. If we choose our notation so that $\mathbf{p}_1, \ldots, \mathbf{p}_m$, $m < k$, is such a basis, it follows that

(1)
$$\mathbf{p}_{m+1} = s_1\mathbf{p}_1 + \cdots s_m\mathbf{p}_m$$

From the linearity properties of matrix multiplication,

$$A\mathbf{p}_{m+1} = s_1 A\mathbf{p}_1 + \cdots + s_m A\mathbf{p}_m$$

Since $A\mathbf{p}_i = r_i\mathbf{p}_i$ for $i = 1, \ldots, k$, this gives

(2)
$$r_{m+1}\mathbf{p}_{m+1} = s_1 r_1 \mathbf{p}_1 + \cdots + s_m r_m \mathbf{p}_m$$

Multiplying (1) by $-r_{m+1}$ and adding it to (2) gives

(3)
$$\mathbf{0} = (-r_{m+1}s_1 + s_1 r_1)\mathbf{p}_1 + \cdots + (-r_{m+1}s_m + s_m r_m)\mathbf{p}_m$$

But the set $\{\mathbf{p}_1, \ldots, \mathbf{p}_m\}$, being a basis for the span of S, must be linearly independent. Hence, all the coefficients in (3) are zero;

$$s_1(r_1 - r_{m+1}) = 0, \ldots, s_m(r_m - r_{m+1}) = 0$$

Since all the eigenvalues $r_1, \ldots, r_k$ are distinct, it follows that $s_1 = \cdots = s_m = 0$. This, combined with (1), says that the eigenvector $\mathbf{p}_{m+1} = \mathbf{0}$, contradicting the definition of eigenvector. We conclude that the set S is linearly independent. $\square$

While this theorem does not give necessary conditions for a matrix to be diagonalizable, it does give sufficient conditions that can be useful. Suppose that for an $n \times n$ matrix A, it turns out that all of the eigenvalues are distinct, a fact that will emerge in Step 1 of the outline we are following. Corresponding to each of these n eigenvalues will be an eigenvector, of course, and according to the theorem just proved, these n eigenvectors will be linearly independent. Theorem A then assures us that A is diagonalizable. We now have a result of the kind for which we have been looking.

THEOREM C

If an $n \times n$ matrix A has distinct eigenvalues $r_1, \ldots, r_n$, then A is similar to the diagonal matrix D having $r_1, \ldots, r_n$ as the entries on the diagonal.

Example A

Is there a diagonal matrix similar to the given matrix A?

$$A = \begin{bmatrix} 5 & -2 & 0 \\ -8 & 7 & 4 \\ 34 & -22 & -9 \end{bmatrix}$$

We begin as usual by finding the characteristic equation.

$$\begin{vmatrix} 5 - r & -2 & 0 \\ -8 & 7 - r & 4 \\ 34 & -22 & -9 - r \end{vmatrix} = 0$$

Expanding the determinant and simplifying leads to the equation

$$r^3 - 3r^2 - r + 3 = 0$$

which has the distinct roots $r_1 = 1, r_2 = -1, r_3 = 3$. According to Theorem C, A is similar to the diagonal matrix

$$D = \begin{bmatrix} 1 & 0 & 0 \\ 0 & -1 & 0 \\ 0 & 0 & 3 \end{bmatrix} \quad \square$$

Note that we have not found the matrix P for which $P^{-1}AP = D$. This could have been done, of course. We omitted finding P because it was not required to answer the question asked. This, it turns out, is often the case in applications. It is not unusual for interest to center on two questions:

1. Can matrix A be diagonalized?
2. If so, what is the diagonal matrix?

Theorem C enables us to answer both questions if it happens that the eigenvalues are distinct. We readily admit that Theorem C would not have been any help with Examples A and B of Section 6-2 since the eigenvalues in those cases were not distinct.

When the eigenvalues of matrix A are not distinct, this much can be said: If there are k distinct eigenvalues r_i, each having multiplicity m_i, $i = 1, \ldots, k$, and if there are m_i linearly independent eigenvectors that belong to r_i, then we will have $m_1 + \cdots + m_k = n$ linearly independent eigenvectors (for it will still be true that eigenvectors belonging to distinct eigenvalues will be linearly independent). It will follow from Theorem A that matrix A is again diagonalizable. This was the situation in Example A of Section 6-2.

There are applications of matrix algebra, some to appear in Chapter 8, in which we wish to find powers of a given matrix A. This computation is greatly simplified if A is diagonalizable.

Suppose A is diagonalizable so that

THEOREM D

$$P^{-1}AP = \begin{bmatrix} r_1 & & 0 \\ & \ddots & \\ 0 & & r_n \end{bmatrix} = D$$

Then

$$A^k = P \begin{bmatrix} r_1^k & & 0 \\ & \ddots & \\ 0 & & r_n^k \end{bmatrix} P^{-1} = PD^k P^{-1}$$

Proof: Since $A = PDP^{-1}$,

$$A^2 = (PDP^{-1})(PDP^{-1}) = PD^2 P^{-1}$$

and so on. $\square$

NONDIAGONALIZABLE MATRICES (OPTIONAL)

If $T: C^n \to C^n$ is represented by a nondiagonalizable matrix A, we are forced back to what we described as a deliberately vague question at the beginning of Section 6-2. Can we choose a basis for C^n so that T will be represented by a simple matrix of some sort? Stated another way, if A is not similar to a diagonal matrix, is there some other simple matrix to which it is similar? Though a full answer to this question is beyond the scope of this text, we will briefly sketch two answers.

Suppose that in addition to allowing nonzero entries on the diagonal, some 1s are also allowed on the diagonal just above the main diagonal. That is still a rather simple matrix, and any matrix A can be shown to be similar to such a matrix, the so-called **Jordan canonical form** of the matrix A. The idea may be illustrated by pointing out that the nondiagonalizable matrix of Example B of Section 6-2 is similar to the Jordan canonical form

$$\begin{bmatrix} 1 & 0 & 0 \\ 0 & 2 & 1 \\ 0 & 0 & 2 \end{bmatrix}$$

The entries on the main diagonal are still the eigenvalues. The 1s on the next higher diagonal "link" some (but not always all) of the repeated eigenvalues. Problems 9–12 at the end of this section help you anticipate the possibilities, but the placement of the 1s is a tricky business that is appropriately left as an advanced topic.

Another form of a matrix that might be considered simple is the upper-triangular matrix that has all entries *below* the main diagonal equal to 0. Every matrix A is similar to an upper-triangular matrix.

THEOREM E

Let A be an $n \times n$ matrix having eigenvalues $r_1, \ldots, r_n$. Then there exists a nonsingular matrix P such that

$$P^{-1}AP = \begin{bmatrix} r_1 & b_{12} & b_{13} & \cdots & b_{1n} \\ 0 & r_2 & b_{23} & \cdots & b_{2n} \\ & & \ddots & & \\ 0 & 0 & \cdots & & r_n \end{bmatrix}$$

Proof: Let $\mathbf{p}_1$ be an eigenvector belonging to r_1. There is a set of vectors $\mathbf{u}_i$, $i = 2, \ldots, n$, which together with $\mathbf{p}_1$ form a basis for C^n. Use these vectors as columns to form the nonsingular matrix $S = [\mathbf{p}_1 \quad \mathbf{u}_2 \quad \cdots \quad \mathbf{u}_n]$. Note immediately that

$$S^{-1}S = [S^{-1}\mathbf{p}_1 \quad S^{-1}\mathbf{u}_2 \quad \cdots \quad S^{-1}\mathbf{u}_n] = I$$

so, in particular, the column vector $S^{-1}\mathbf{p}_1$ is the standard basis vector $\mathbf{e}_1$. Since

$$AS = [A\mathbf{p}_1 \quad A\mathbf{u}_2 \quad \cdots \quad A\mathbf{u}_n] = [r_1\mathbf{p}_1 \quad A\mathbf{u}_2 \quad \cdots \quad A\mathbf{u}_n]$$

it follows that

$$S^{-1}AS = [r_1\mathbf{e}_1 \quad S^{-1}A\mathbf{u}_2 \quad \cdots \quad S^{-1}A\mathbf{u}_n]$$

This matrix has the form

$$S^{-1}AS = \begin{bmatrix} r_1 & b_{12} & b_{13} & \cdots & b_{1n} \\ 0 & & & & \\ \vdots & & & A_1 & \\ 0 & & & & \end{bmatrix}$$

where A_1 is an $(n-1) \times (n-1)$ submatrix. Theorem B of Section 6-2 says that $S^{-1}AS$ and A have the same characteristic equation, so

$$\det(A - rI) = \det(S^{-1}AS - rI)$$

$$= \det \begin{bmatrix} r_1 - r & b_{12} & b_{13} & \cdots & b_{1n} \\ 0 & & & & \\ \vdots & & & A_1 - rI & \\ 0 & & & & \end{bmatrix}$$

Expanding the determinant on the right by minors of the first column gives us

$$\det(A - rI) = (r_1 - r)\det(A_1 - rI)$$

Therefore, the eigenvalues of A_1 are $r_2, \ldots, r_n$.

The same calculations carried out for the matrix A_1 will lead us to an $(n-1) \times (n-1)$ nonsingular matrix S_1 such that

$$S_1^{-1}A_1S_1 = \begin{bmatrix} r_2 & b_{23} & \cdots & b_{2n} \\ 0 & & & \\ \vdots & & A_2 & \\ 0 & & & \end{bmatrix}$$

where A_2 has the eigenvalues $r_3, \ldots, r_n$. Now notice that the $n \times n$ matrix R defined by

$$R = \begin{bmatrix} 1 & 0 & \cdots & 0 \\ 0 & & & \\ \vdots & & S_1 & \\ 0 & & & \end{bmatrix} \qquad \text{has inverse} \qquad R^{-1} = \begin{bmatrix} 1 & 0 & \cdots & 0 \\ 0 & & & \\ \vdots & & S_1^{-1} & \\ 0 & & & \end{bmatrix}$$

Then $(SR)^{-1}ASR = R^{-1}(S^{-1}AS)R$ will be the matrix

$$\begin{bmatrix} r_1 & b_{12} & b_{13} & \cdots & b_{1n} \\ 0 & r_2 & b_{23} & \cdots & b_{2n} \\ \vdots & \vdots & & \begin{bmatrix} & & \\ & A_2 & \\ & & \end{bmatrix} & \\ 0 & 0 & & & \end{bmatrix}$$

The complete proof proceeds by induction. □

The theorem just proved can be used to establish a result that is absolutely fundamental to advanced work in matrix theory. For this purpose we define a polynomial function of a matrix as follows: If p is a polynomial defined by

$$p(r) = a_n r^n + a_{n-1} r^{n-1} + \cdots + a_1 r + a_0,$$

then for an $n \times n$ matrix A, we define

$$p(A) = a_n A^n + a_{n-1} A^{n-1} + \cdots + a_1 A + a_0 I$$

Since, with care about the order of multiplication, matrix algebra follows the familiar rules (associativity, the distributive law), any factorization of $p(x)$ leads to a corresponding factorization of $p(A)$. We are particularly interested in the characteristic polynomial $\det(A - Ir)$ of the $n \times n$ matrix A. If $r_1, \ldots, r_n$ are the roots of this polynomial (the eigenvalues of A), then this nth degree polynomial can be written

$$p(x) = (r - r_1)(r - r_2) \cdots (r - r_n)$$

It follows that

(4) $$p(A) = (A - r_1 I)(A - r_2 I) \cdots (A - r_n I)$$

We are now prepared to state the **Cayley-Hamilton theorem.**

THEOREM F   *If $p(r)$ is the characteristic polynomial of the $n \times n$ matrix A, then $p(A) = \mathcal{O}$, the zero matrix.*

Proof: The proof for an $n \times n$ matrix follows the same pattern as the proof given here for a 3×3 matrix A.

According to Theorem E, there exists a nonsingular matrix P such that

$$P^{-1}AP = T = \begin{bmatrix} r_1 & t_{12} & t_{13} \\ 0 & r_2 & t_{23} \\ 0 & 0 & r_3 \end{bmatrix}$$

Form the three matrices $M_i = T - r_i I$, $i = 1, 2, 3$. They have the following forms in which, for our purposes, it is enough to note where there are zero entries and where there might be

THE FIBONACCI SEQUENCE

The sequence 1, 1, 2, 3, 5, 8, 13, ... turns up in so many unexpected places, both in nature (count the scales on any one of the spiral-like paths that wind around a pine cone) and in pure mathematics, that it has for many people a haunting if not mystical quality. It is easy to define the sequence by a recursion formula:

$$x_n = x_{n-1} + x_{n-2}, \qquad x_0 = x_1 = 1$$

The trouble with recursion formulas is, of course, that to find the 100th term, you must first find the preceeding 99 terms. The challenge, therefore, is to find an explicit formula for x_n.

Begin by noting that

$$\begin{bmatrix} x_n \\ x_{n-1} \end{bmatrix} = \begin{bmatrix} 1 & 1 \\ 1 & 0 \end{bmatrix} \begin{bmatrix} x_{n-1} \\ x_{n-2} \end{bmatrix}$$

from which it follows that

$$\begin{bmatrix} x_2 \\ x_1 \end{bmatrix} = \begin{bmatrix} 1 & 1 \\ 1 & 0 \end{bmatrix} \begin{bmatrix} x_1 \\ x_0 \end{bmatrix}$$

$$\begin{bmatrix} x_3 \\ x_2 \end{bmatrix} = \begin{bmatrix} 1 & 1 \\ 1 & 0 \end{bmatrix} \begin{bmatrix} x_2 \\ x_1 \end{bmatrix} = \begin{bmatrix} 1 & 1 \\ 1 & 0 \end{bmatrix}^2 \begin{bmatrix} 1 \\ 1 \end{bmatrix}$$

and, in general,

$$\begin{bmatrix} x_n \\ x_{n-1} \end{bmatrix} = A^{n-1} \begin{bmatrix} 1 \\ 1 \end{bmatrix} \quad \text{where} \quad A = \begin{bmatrix} 1 & 1 \\ 1 & 0 \end{bmatrix}$$

It is easy to find that the eigenvalues of A are $\frac{1}{2}(1 \pm \sqrt{5})$, and that $A = P^{-1}DP$ where

$$D = \begin{bmatrix} \dfrac{1 + \sqrt{5}}{2} & 0 \\ 0 & \dfrac{1 - \sqrt{5}}{2} \end{bmatrix}$$

and

$$P = \begin{bmatrix} 1 + \sqrt{5} & 2 \\ 1 - \sqrt{5} & 2 \end{bmatrix}, \qquad P^{-1} = \begin{bmatrix} \dfrac{1}{2\sqrt{5}} & -\dfrac{1}{2\sqrt{5}} \\ \dfrac{-1 + \sqrt{5}}{4\sqrt{5}} & \dfrac{1 + \sqrt{5}}{4\sqrt{5}} \end{bmatrix}$$

Now is the time to call on Theorem D to help with the computation of A^{n-1}. We get

$$\begin{bmatrix} x_n \\ x_{n-1} \end{bmatrix} = A^{n-1} \begin{bmatrix} 1 \\ 1 \end{bmatrix} = P^{-1}D^{n-1}P\begin{bmatrix} 1 \\ 1 \end{bmatrix}$$

$$= \begin{bmatrix} \dfrac{1}{2\sqrt{5}} & -\dfrac{1}{2\sqrt{5}} \\ \dfrac{-1+\sqrt{5}}{4\sqrt{5}} & \dfrac{1+\sqrt{5}}{4\sqrt{5}} \end{bmatrix} \begin{bmatrix} \left(\dfrac{1+\sqrt{5}}{2}\right)^{n-1} & 0 \\ 0 & \left(\dfrac{1-\sqrt{5}}{2}\right)^{n-1} \end{bmatrix}$$

$$\times \begin{bmatrix} 1+\sqrt{5} & 2 \\ 1-\sqrt{5} & 2 \end{bmatrix}\begin{bmatrix} 1 \\ 1 \end{bmatrix}$$

Multiplication leads us to the conclusion,

$$x_n = \frac{1}{\sqrt{5}}\left(\frac{1+\sqrt{5}}{2}\right)^{n+1} - \frac{1}{\sqrt{5}}\left(\frac{1-\sqrt{5}}{2}\right)^{n+1}$$

nonzero entries, which we shall denote by *.

$$M_1 = \begin{bmatrix} 0 & * & * \\ 0 & * & * \\ 0 & 0 & * \end{bmatrix}, \qquad M_2 = \begin{bmatrix} * & * & * \\ 0 & 0 & * \\ 0 & 0 & * \end{bmatrix}, \qquad M_3 = \begin{bmatrix} * & * & * \\ 0 & * & * \\ 0 & 0 & 0 \end{bmatrix}$$

You should now convince yourself that if (x_1, x_2, x_3) is any vector in C^3 and M_i are matrices of the form just indicated, then

$$[x_1 \quad x_2 \quad x_3]M_1M_2M_3 = [0 \quad * \quad *]M_2M_3$$

$$= [0 \quad 0 \quad *]M_3 = [0 \quad 0 \quad 0]$$

Thus, the 3×3 matrix $M = M_1M_2M_3$ has the property that for *any* row matrix $\mathbf{x}$, $\mathbf{x}M = \mathbf{0}$. This means that M itself is the $\mathcal{O}$ matrix; $M_1M_2M_3 = \mathcal{O}$. Finally, since

$$M_i = T - r_iI = P^{-1}AP - r_iI = P^{-1}(A - r_iI)P$$

then

$$M_1M_2M_3 = P^{-1}[(A - r_1I)(A - r_2I)(A - r_3I)]P = \mathcal{O}$$

Setting $n = 3$ in (4), we see that we have proved $P^{-1}p(A)P = \mathcal{O}$. Multiplication on the left by P and on the right by P^{-1} gives us $p(A) = \mathcal{O}$. $\square$

The matrices given in Problems 1–4 are not diagonalizable. Verify that the eigenvectors belonging to distinct eigenvalues are linearly independent.

1. $\begin{bmatrix} -5 & 0 & -4 \\ -5 & -1 & -4 \\ 9 & 0 & 7 \end{bmatrix}$

2. $\begin{bmatrix} 8 & 0 & 6 \\ 1 & -1 & 1 \\ -9 & 0 & 7 \end{bmatrix}$

3. $\begin{bmatrix} 5 & 0 & 3 \\ 17 & 2 & 16 \\ -6 & 0 & -4 \end{bmatrix}$

4. $\begin{bmatrix} 3 & -1 & 2 \\ 9 & -7 & 10 \\ -3 & 2 & -4 \end{bmatrix}$

Show that the matrices given in Problems 5–8 are diagonalizable, and find a diagonal matrix to which the given matrix is similar.

5. $\begin{bmatrix} 0 & 0 & 2 \\ -8 & -1 & 14 \\ -1 & 0 & 3 \end{bmatrix}$

6. $\begin{bmatrix} -1 & 0 & -2 \\ 1 & 3 & 1 \\ 3 & 0 & 4 \end{bmatrix}$

7. $\begin{bmatrix} 7 & -1 & 4 \\ 0 & 2 & 0 \\ -8 & 1 & -5 \end{bmatrix}$

8. $\begin{bmatrix} 0 & -2 & 1 \\ -2 & 0 & -1 \\ -6 & -6 & 1 \end{bmatrix}$

Find the eigenvalues and the corresponding eigenvectors of the matrices given in Problems 9–12.

9. $\begin{bmatrix} 2 & 0 & 0 \\ 1 & 2 & 0 \\ 0 & 0 & -3 \end{bmatrix}$

10. $\begin{bmatrix} -1 & 0 & 0 \\ 1 & -1 & 0 \\ 0 & 0 & 2 \end{bmatrix}$

11. $\begin{bmatrix} 2 & 0 & 0 & 0 \\ 0 & 2 & 0 & 0 \\ 0 & 1 & 2 & 0 \\ 0 & 0 & 0 & 1 \end{bmatrix}$

12. $\begin{bmatrix} -1 & 0 & 0 & 0 \\ 1 & -1 & 0 & 0 \\ 0 & 0 & -1 & 0 \\ 0 & 0 & 0 & 2 \end{bmatrix}$

Problems 13–16 refer to matrices in Problem Set 6-2.

13. For the matrix of Problem 5, find A^4 in two ways.

14. For the matrix of Problem 6, find A^4 in two ways.

15. For the matrix of Problem 7, find A^4 in two ways.

16. For the matrix of Problem 8, find A^4 in two ways.

17. Prove that the characteristic equation of

$$\begin{bmatrix} 0 & 0 & -c \\ 1 & 0 & -b \\ 0 & 1 & -a \end{bmatrix}$$

is $x^3 + ax^2 + bx + c = 0$.

18. Verify that the matrix of Problem 2 satisfies its characteristic equation.

19. Verify that the matrix of Problem 3 satisfies its characteristic equation.

20. Verify that the matrix of Problem 17 satisfies its characteristic equation.

21. Prove that an $n \times n$ matrix is singular if and only if it has an eigenvalue of 0.

22. According to the Cayley-Hamilton theorem, any 3×3 matrix A must satisfy an equation of the form

$$a_3 A^3 + a_2 A^2 + a_1 A + a_0 I = 0$$

from which it follows that

$$A(a_3 A^2 + a_2 A + a_1 I) = -a_0 I$$

Use this expression to find an inverse for the matrix of Problem 3.

23. When will the method of Problem 22 fail? Relate this to the assertion proved in Problem 21.

24. Prove that an upper-triangular matrix is diagonalizable if the entries on the diagonal are distinct. Is the converse true?

CHAPTER 6
SELF-TEST

1. Prove that $r^3 + a_2 r^2 + a_1 r + a_0$ is the characteristic equation of

$$\begin{bmatrix} 0 & 0 & -a_0 \\ 1 & 0 & -a_1 \\ 0 & 1 & -a_2 \end{bmatrix}$$

2. The matrix M, given below, has an eigenvalue of 1. Find a basis for the vector space spanned by the eigenvectors that belong to 1.

$$M = \begin{bmatrix} 7 & -2 & -4 \\ 3 & 0 & -2 \\ 6 & -2 & -3 \end{bmatrix}$$

3. A linear transformation $T: R^2 \to R^2$ is described with respect to the standard basis by $\mathbf{y} = A\mathbf{x}$. Find which basis for R^2 should be used so that T will be represented by a diagonal matrix if

$$A = \begin{bmatrix} 2 & 2 \\ 5 & -1 \end{bmatrix}$$

4. Show how to find by inspection the eigenvalues of a matrix of the form

$$\begin{bmatrix} a & -b \\ b & a \end{bmatrix}$$

5. Can you diagonalize $\begin{bmatrix} 1 & 1 & 0 \\ 0 & 1 & 0 \\ 0 & 0 & 2 \end{bmatrix}$?

6. Describe the graph of the quadratic equation

$$3x_1^2 + 2x_2^2 + 4x_3^2 + 4x_1 x_2 + 4x_1 x_3 = 9$$

with an equation in which each term involves just one variable. Name the graph of this function.

7. Find A^5 by first diagonalizing A, given that

$$A = \begin{bmatrix} \frac{1}{3} & \frac{2}{3} \\ \frac{1}{2} & \frac{1}{2} \end{bmatrix}$$

8. Suppose that the symmetric $n \times n$ matrix B has n distinct eigenvalues r_1, $r_2, \ldots, r_n$. Prove that $\det B = r_1 \cdot r_2 \cdot \cdots r_n$.

9. Suppose that the $n \times n$ matrix A has the eigenvalue r_0.
   (a) Prove that A^t has the eigenvalue r_0.
   (b) Prove that A^{-1} has the eigenvalue $1/r_0$. Consider separately the case $r_0 = 0$.

10. The 3×3 matrix A is skew symmetric, meaning that $A^t = -A$. Show that the eigenvalues of A are all pure imaginary numbers, that is, numbers of the form ai, or 0. Can you improve on this result?

7 LINEAR PROGRAMMING

7-1 INTRODUCTION

Suppose the owner of a small machine shop has three machines A, B, and C which are used to produce three items designated here as I, II, and III. The time in hours required by each item on each machine is indicated in the following table:

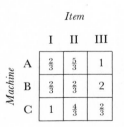

Item

		I	II	III
	A	$\frac{2}{3}$	$\frac{5}{3}$	1
Machine	B	$\frac{2}{3}$	$\frac{2}{3}$	2
	C	1	$\frac{4}{3}$	$\frac{2}{3}$

Suppose further that the profit to the owner on item I is $40; on items II and III it is $60 and $70, respectively. If the shop operates on a 40-hour work week, how many units of each item should be produced per week to maximize the owner's profit?

Let x_1, x_2, and x_3 be the number of each of the units to be produced per week. The profit to be maximized is

$$P = 40x_1 + 60x_2 + 70x_3$$

The hours required on machine A will then be $\frac{2}{3}x_1 + \frac{5}{3}x_2 + x_3$, and this total must not exceed 40 hours. Similar considerations for machines B and C give us two more inequalities, called the constraints to our problem. That is, we wish to choose x_1, x_2, and x_3 to maximize P, subject to the constraints

$$\frac{2}{3}x_1 + \frac{5}{3}x_2 + x_3 \leq 40$$
$$\frac{2}{3}x_1 + \frac{2}{3}x_2 + 2x_3 \leq 40$$
$$x_1 + \frac{4}{3}x_2 + \frac{2}{3}x_3 \leq 40$$
$$x_1 \geq 0, \qquad x_2 \geq 0, \qquad x_3 \geq 0$$

(1)

The problem just described is a simple example of what is called a **linear programming problem.** A useful and important form of a linear programming problem can be found in the transportation problem discussed in Section 2-8. The essential features of a linear programming problem are:

1. There is a function of several variables, called the **objective function,** which is to be optimized (maximized or minimized). This function is either linear or is the sum of a linear function and a constant.
2. The variables involved in the objective function must all satisfy a given set of linear inequalities and/or equations called **constraints.**
3. The nature of the problem is such that we always understand the constraints to include the restriction that all variables are nonnegative.

A practical technique for solving linear programming problems was developed by George Dantzig during World War II for applications in military logistics. In his book on the subject, Dantzig wrote,

In the summer of 1949 at the University of Chicago, a conference was held under the sponsorship of the Cowles Commission for Research in Economics; mathematicians, economists, statisticians from academic institutions and various government agencies presented research using the linear programming tool. The problems considered ranged from planning crop rotation to planning large scale military actions, from the routing of ships between harbors to the assessment of flow of commodities between industries of the economy. What was most surprising was that the research had taken place during the preceding two years. [Dantzig, G. B. 1963. Linear programming and extensions. Princeton: Princeton University Press.]

An annotated bibliography published in 1958 [Riley, Vera, and Gass, Saul I. 1958. *Bibliography of linear programming and related techniques.* Baltimore: Johns Hopkins Press.] listed applications in agriculture, finance, economics, personnel assignment, production scheduling, inventory control, traffic analysis, and industrial applications that were further divided into categories: chemical, coal, commercial airlines, communications, iron and steel, paper, petroleum (where the method has been particularly important), railroads, and others. Since 1958, many more applications have emerged.

In our brief introduction to this large and practical area of mathematics, we will standardize the problem in several ways. First, note the relationship between the functions $g(x_1, x_2) = d_1 x_1 + d_2 x_2$ and $f(x_1, x_2) = c_1 x_1 + c_2 x_2$ in which $c_1 = -d_1$, $c_2 = -d_2$. A point (r_1, r_2) at which $g(x_1, x_2)$ assumes a minimum will be a point at which $f(x_1, x_2)$ assumes a maximum. Since a similar comment would be true if more variables were involved, we will always assume that such a transformation has been made if necessary so that our objective function is to be maximized.

Secondly, it will be convenient if we take all the constraints to be inequalities. While it is hardly efficient to do so, there is no conceptual problem in replacing the equality

$$a_1 x_1 + a_2 x_2 = b_1$$

by the two inequalities

$$a_1 x_1 + a_2 x_2 \leq b_1$$
$$a_1 x_1 + a_2 x_2 \geq b_1$$

Finally, since the direction of an inequality can be reversed if both sides are multiplied by -1, we may specify the directions of the inequalities.

CANONICAL FORM

With the understandings just described, we may state the linear programming problem in the following canonical form:

Maximize the objective function

$$f(\mathbf{x}) = c_1 x_1 + \cdots + c_n x_n + d$$

subject to the constraints

$$a_{11} x_1 + \cdots + a_{1n} x_n \leq b_1$$
$$\vdots \qquad\qquad \vdots \qquad \vdots$$
$$a_{m1} x_1 + \cdots + a_{mn} x_n \leq b_m$$
$$x_1 \geq 0, \ldots, x_n \geq 0$$

With obvious choices for the column matrices $\mathbf{c}$, $\mathbf{x}$, and $\mathbf{b}$, and for the matrix A, the canonical form can then be written more compactly,

$$\text{Maximize} \quad f(\mathbf{x}) = \mathbf{c}^t \mathbf{x} + d$$
$$\text{subject to} \quad A\mathbf{x} \leq \mathbf{b}$$
$$\mathbf{x} \geq 0$$

A GEOMETRIC APPROACH | 7-2

We can get an intuitive feeling for the nature of our problem, introduce some of the common terminology, and illustrate some basic theorems by solving a few modest problems by geometric methods. Let's begin with an example in two dimensions.

Maximize

$$f(\mathbf{x}) = \tfrac{2}{5}x_1 + x_2 + 1$$

subject to

$$-2x_1 + x_2 \le 2$$
$$x_1 - 2x_2 \le 3$$
$$x_1 + 2x_2 \le 11$$
$$x_1 \ge 0, \qquad x_2 \ge 0$$

Example A

The inequality $-2x_1 + x_2 \le 2$ describes a *half-space* (see page 288) bounded by and including the line $-2x_1 + x_2 = 2$. (This half-space contains the origin, since $(0, 0)$ obviously satisfies the inequality.) Each of the other inequalities similarly defines a half-space, and the intersection of these half-spaces with the first quadrant (where $x_1 \ge 0$, $x_2 \ge 0$) gives us the set of points that satisfy all the constraints (Figure 7-1). This

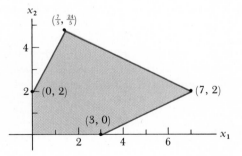

Figure 7-1

set is called the **feasible set** for the problem; its members, the **feasible solutions,** are candidates to be considered in our search for the point or points where f assumes its maximum value.

A 3-dimensional picture of the problem shows the graph of the objective function

$$z = f(x_1, x_2) = \tfrac{2}{5}x_1 + x_2 + 1$$

as a plane hovering over the feasible set $\mathcal{F}$ drawn in the x_1x_2 plane (Figure 7-2). Consider any point (r_1, r_2) interior to $\mathcal{F}$. Since we are free to move in any direction from (r_1, r_2), the tilt of the graph of the objec-

Figure 7-2

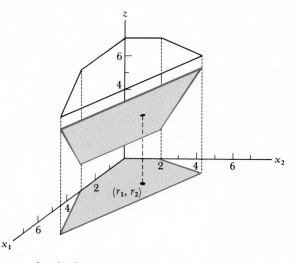

tive function makes it clear that $f(r_1, r_2)$ is *not* the maximum value that can be attained in $\mathcal{F}$. This argument, valid for any point interior to $\mathcal{F}$ so long as there is a "tilt" to the graph of the objective function, makes it clear that the maximum will always occur somewhere on the boundary. (If the graph has no tilt, then f is constant and its maximum occurs everywhere, including the boundary, of course.)

There is another argument that leads (or slides) to the same conclusion. The *level curves* of $z = f(x_1, x_2)$ are defined to be the graphs of

$$f(x_1, x_2) = k$$

where k is a constant. Drawn in the $x_1 x_2$ plane, they show all the points in the domain of f that produce values at the $z = k$ level of the graph of f. In Figure 7-3 we have shown the level curve

$$\tfrac{2}{5} x_1 + x_2 + 1 = 9$$

Figure 7-3

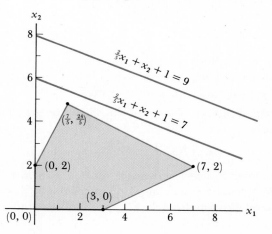

for our objective function. None of the points on this level curve are in the feasible set, which is another way of saying that we cannot achieve the lofty value of 9 with points that satisfy our constraints. Well, can we achieve a value of 7? We note that this level curve (Figure 7-3) doesn't intersect the feasible set either, but the two level curves now drawn make this much clear: as k is decreased toward a value that is attainable, the level curves slide along toward the feasible set in such a way as to make it obvious that the first intersection of a level curve and the feasible set will occur at $(\frac{7}{5}, \frac{24}{5})$. The maximum to be achieved is

$$f(\tfrac{7}{5}, \tfrac{24}{5}) = \tfrac{2}{5} \cdot \tfrac{7}{5} + (\tfrac{24}{5}) + 1 = \tfrac{159}{25} \quad \square$$

We next turn to the problem in 3 variables that was used in Section 7-1 to introduce the concept of linear programming. It is convenient to rewrite equations (1) without fractions; and at the same time the size of the numbers can be reduced if we multiply the given objective function by $\frac{1}{10}$.

Maximize

Example B

$$p = 4x_1 + 6x_2 + 7x_3$$

subject to

$$2x_1 + 5x_2 + 3x_3 \leq 120$$
$$x_1 + x_2 + 3x_3 \leq 60$$
$$3x_1 + 4x_2 + 2x_3 \leq 120$$
$$x_1 \geq 0, \qquad x_2 \geq 0, \qquad x_3 \geq 0$$

The feasible set is now a 3-dimensional polyhedron (Figure 7-4). The arguments used in Example A apply again. A maximum of a nonconstant function cannot occur at a point interior to the polyhedron. The level curves $4x_1 + 6x_2 + 7x_3 = k$ are planes that, for very large k, obviously miss the polyhedron. As k is decreased, we get a family of planes, all parallel, which may be thought of as sliding toward the polyhedron as k continuously decreases.

All of this is evident. The point at which one of these planes will first touch the polyhedron is not so evident, however, as it was in the 2-dimensional Example A. (It may be evident to readers with good conceptualization of figures in space; the author recalls his teacher of descriptive geometry who used to refer to "zero-visibility Roberts.") But this much should be evident to everyone (since it is evident to the author). The point (or one of the points) where these sliding planes first touch the polyhedron is a corner, more formally called a **vertex** of the polyhedron. It could happen, of course, that when a plane first touches the polyhedron, it will do so on an entire edge (like V_3V_5) or an entire face (like V_3, V_5, V_6, V_7). But even in such cases, a vertex is certain to be included among the points where a level surface first comes in contact with the polyhedron.

Therefore, one way to find the maximum value that the objective function assumes is to evaluate it at each vertex—or if you prefer, evaluate it at each vertex that cannot be visually eliminated. For completeness, we have identified with coordinates all the vertices of the polyhedron shown in Figure 7-4, and have evaluated the objective function at these points.

Vertex	$p = 4x_1 + 6x_2 + 7x_3$
$V_0(0, 0, 0)$	0
$V_1(40, 0, 0)$	160
$V_2(\frac{240}{7}, 0, \frac{60}{7})$	$197\frac{1}{7}$
$V_3(20, 10, 10)$	210
$V_4(0, 0, 20)$	140
$V_5(0, 15, 15)$	195
$V_6(0, 24, 0)$	144
$V_7(\frac{120}{7}, \frac{120}{7}, 0)$	$171\frac{3}{7}$

Obviously, the maximum of the objective function is 210, and this occurs at $(20, 10, 10)$.   □

FEASIBLE SETS

Some general observations about feasible sets are in order. The set of points in R^n that satisfy the inequality

$$a_1x_1 + \cdots + a_nx_n \leq b$$

is called a **half-space,** even when $n > 3$ and the geometric "feeling" has been left behind. Thus, the feasible set is always the intersection of a finite number of half-spaces. Such a set, called a **polyhedral set,** is convex. A set is **convex** if, whenever both x and y are in $\mathcal{F}$, then

$$z = \lambda x + (1 - \lambda)y, \qquad 0 \leq \lambda \leq 1$$

is also in $\mathcal{F}$; that is, $\mathcal{F}$ contains all points on the line segment joining x and y (Figure 7-5). It is easy to show that if both x and y satisfy a linear inequality, then z does also (see Problem 12). This is enough to prove the following theorem.

THEOREM A

The feasible set of a linear programming problem is a convex polyhedral set.

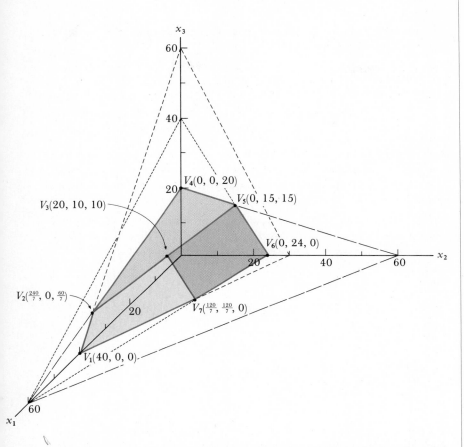

Figure 7-4

It is not so easy to prove that the maximum of the objective function will always occur at a vertex. Our first step in this direction, in fact, must be to sharpen up our concept of a vertex. We will say that **z** is an **extreme point** of the convex set $\mathscr{F}$ if **z** cannot be expressed as the midpoint of two other members of $\mathscr{F}$. The extreme points of a convex polyhedral set in two or three dimensions are exactly the vertices, but our definition has the advantage of making sense in higher dimensions where we all have visibility zero.

With the terminology available, we can now state the result which experience has already conditioned us to believe. (For a proof of this

Convex

Nonconvex

Figure 7-5

theorem, see Roberts, A. Wayne, and Varberg, Dale E. 1973. *Convex functions*. New York: Academic Press.)

THEOREM B

If the objective function of a linear programming problem is bounded above on the feasible set $\mathcal{F}$, then it attains a maximum on the feasible set, and this maximum is attained at an extreme point of $\mathcal{F}$.

Read the theorem carefully. It does not say that the maximum value won't also be assumed at some points that are not extreme; recall that a constant function assumes its maximum everywhere. It does say, however, that if a maximum is assumed anywhere, then there will be an extreme point where it is assumed.

We close this section with a series of linear programming problems in 2 variables. Though they all border on the trivial, they do illustrate things that may occur in much more complex situations.

Example C

Maximize

$$f(x_1, x_2) = x_1 + x_2$$

subject to

$$3x_1 + x_2 \leq 3$$
$$-x_1 + 2x_2 \leq -1$$
$$x_1 - x_2 \leq -2$$
$$x_1 \geq 0, \qquad x_2 \geq 0$$

As soon as you try to draw the feasible set $\mathcal{F}$, you see the trouble (Figure 7-6). There are no feasible points; $\mathcal{F} = \varnothing$.   □

Figure 7-6

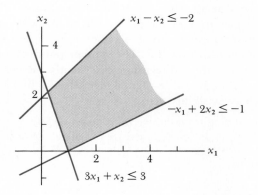

Example D

Maximize

$$f(x_1, x_2) = x_1 + x_2$$

subject to

$$x_1 - 2x_2 \leq 2$$
$$-3x_1 + 2x_2 \leq 3$$
$$x_1 \geq 0, \qquad x_2 \geq 0$$

In this example, the feasible set is unbounded (Figure 7-7). It contains every point, for example, of the form (t, t) where $t \geq 0$. And since $f(t, t) = t + t = 2t$, the objective function obviously does not attain a maximum value on $\mathcal{F}$. □

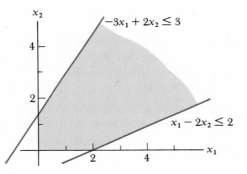

Figure 7-7

The difficulty in Example D is due to the fact that the feasible set is unbounded. Will an unbounded feasible set always lead to trouble in a linear programming problem? See Problem 13.

Maximize

$$f(x_1, x_2) = x_1 + x_2$$

Example E

subject to

$$x_1 - 2x_2 \leq 2$$
$$-3x_1 + 2x_2 \leq 3$$
$$x_1 + x_2 \leq 4$$
$$x_1 \geq 0, \qquad x_2 \geq 0$$

Since part of the boundary of the feasible set is parallel to the level curves of the objective function (Figure 7-8), the maximum will be attained at infinitely many points. □

The various possibilities that we have seen illustrated in this section cover all possibilities in the solving of linear programming problems:

1. There may be a unique point at which the objective function assumes a maximum.

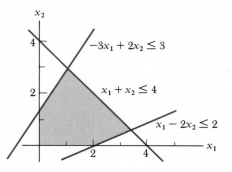

Figure 7-8

2. The feasible set may be empty, meaning of course that there is no feasible solution.
3. If the feasible set is unbounded, the objective function may grow without bound, meaning that there is no solution.
4. There may be infinitely many points at which the objective function assumes its maximum.

Whenever the objective function does assume a maximum on the feasible set, this maximum will always be assumed at an extreme point. For this reason, we shall have great interest in identifying the extreme points of convex polyhedral sets.

PROBLEM
SET 7-2

1. (a) Sketch the level curves $g(\mathbf{x}) = k$, where $k = 75, 60, 50, 40$, for the function $g(\mathbf{x}) = 5x_1 + 8x_2$.
   (b) On the same graph, draw the feasible set for Example A of this section.
   (c) At what point of the feasible set does the function g assume its maximum? What is this maximum value?

2. (a) Sketch the level curves $h(\mathbf{x}) = k$, where $k = 30, 20, 15, 10$, for the function $h(\mathbf{x}) = 3x_1 - 5x_2$.
   (b) On the same graph, draw the feasible set for Example A of this section.
   (c) At what point of the feasible set does the function h assume its maximum? What is this maximum value?

3. Maximize $z = 20x_1 + 24x_2$ subject to the constraints

$$3x_1 + 6x_2 \le 60$$
$$4x_1 + 2x_2 \le 32$$
$$x_1 \ge 0, \qquad x_2 \ge 0$$

by sketching the feasible set and then observing the level curves $20x_1 + 24x_2 = k$ as k decreases from 500.

4. Maximize $z = -x_1 + x_2$ subject to the constraints

$$2x_1 - 3x_2 \le 5$$
$$2x_1 + x_2 \le 7$$
$$x_1 \ge 0, \qquad x_2 \ge 0$$

by sketching the feasible set and then observing the level curves $-x_1 + 2x_2 = k$ as k decreases from 10.

5. Maximize $z = 3x_1 + 4x_2$ subject to the constraints

$$4x_1 + 9x_2 \geq 18$$
$$6x_1 - 3x_2 \leq 16$$
$$2x_1 + 3x_2 \leq 16$$
$$x_1 \geq 0, \qquad x_2 \geq 0$$

6. Maximize $z = 2x_1 + 3x_2$ subject to the constraints

$$x_1 + 2x_2 \leq 18$$
$$3x_1 + 2x_2 \leq 26$$
$$x_1 + 4x_2 \geq 3$$
$$x_1 \geq 0, \qquad x_2 \geq 0$$

7. Maximize $z = -x_1 + x_2 + x_3$ subject to the constraints

$$x_1 + 2x_2 + x_3 \leq 12$$
$$2x_1 + 4x_2 - x_3 \geq 8$$
$$x_1 - x_2 + 2x_3 \geq 4$$
$$x_1 \geq 0, \qquad x_2 \geq 0, \qquad x_3 \geq 0$$

8. Maximize $z = 2x_1 + 4x_2 - x_3$ subject to the constraints

$$2x_1 + 4x_2 + 3x_3 \leq 24$$
$$x_1 - 3x_2 + 9x_3 \geq 27$$
$$x_1 + x_2 \qquad \leq 6$$
$$x_1 \geq 0, \qquad x_2 \geq 0, \qquad x_3 \geq 0$$

9. Maximize $z = 2x_1 - x_2 + x_3$ subject to the constraints of Problem 8.

10. Maximize $z = x_1 + 2x_2 + 4x_3$ subject to the constraints of Problem 7.

11. Let $X(x_1, x_2)$ and $Y(y_1, y_2)$ be points in the plane, and let $Z(z_1, z_2)$ be on the line segments joining them. From the obvious vector equation (Figure 7-9)

$$\overrightarrow{OZ} = \overrightarrow{OY} + \lambda(\overrightarrow{YX}), \qquad 0 \leq \lambda \leq 1$$

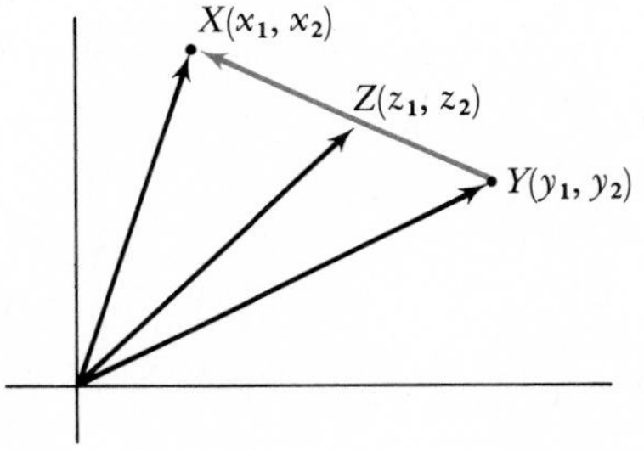

$X(x_1, x_2)$

$Z(z_1, z_2)$

$Y(y_1, y_2)$

Figure 7-9

deduce that

$$z_1 = \lambda x_1 + (1 - \lambda) y_1$$
$$z_2 = \lambda x_2 + (1 - \lambda) y_2$$

Compare these equations with those in the definition of convexity.

12. Suppose that the linear inequality

$$a_1 x_1 + \cdots + a_n x_n \leq b$$

is satisfied by both

$$\bar{\mathbf{x}} = \begin{bmatrix} \bar{x}_1 \\ \vdots \\ \bar{x}_n \end{bmatrix} \quad \text{and} \quad \bar{\mathbf{y}} = \begin{bmatrix} \bar{y}_1 \\ \vdots \\ \bar{y}_n \end{bmatrix}$$

Prove that if λ satisfies $0 \leq \lambda \leq 1$, then $\bar{\mathbf{z}} = \lambda \bar{\mathbf{x}} + (1 - \lambda) \bar{\mathbf{y}}$ also satisfies the inequality.

13. Maximize $z = 2x_1 - 3x_2 + 10$ subject to the constraints

$$3x_1 + 2x_2 \geq 12$$
$$-x_1 + x_2 \leq 3$$
$$x_1 - 2x_2 \leq 6$$
$$x_1 \geq 0, \qquad x_2 \geq 0$$

What does this example illustrate?

The last four problems hint at classic applications of linear programming. In each case, define appropriate unknowns, give the objective functions, and write down the constraints (as either equations or inequalities). Do not try to solve these problems; they are quite beyond the methods discussed in this section.

14. A student plans to build oak bookshelves in her study. She can buy oak boards in either 10 ft. or 12 ft. lengths. She wants 6 pieces that are 7 ft. long for uprights, 8 pieces that are 2.5 ft. long, and 16 pieces that are 3.5 ft. long. She is charged by the foot for the lumber. How many boards of each length should she buy? *Hint:* 12-ft. boards may be used as follows:

7 ft. uprights	3.5 ft. shelves	2.5 ft. shelves
1	1	0
1	0	2
0	3	0
0	2	2
0	1	3
0	0	4

Let x_1 through x_6 represent the number of 12-ft. boards used in each way. Similarly, let x_7 through x_{10} represent the number of 10-ft. boards used in each of the possible ways.

15. A business officer for a state university system having three campuses must buy fuel oil from four suppliers bidding for the business. The delivered cost per gallon varies with the relative location of the campuses and suppliers, these costs being given in the table below.

$$
\begin{array}{c}
\text{Campuses} \\
\begin{array}{cccc}
 & 1 & 2 & 3 \\
\end{array} \\
\text{Suppliers} \; \begin{array}{c} A \\ B \\ C \\ D \end{array}
\begin{bmatrix}
0.85 & 0.90 & 0.95 \\
0.85 & 0.92 & 1.00 \\
0.88 & 0.88 & 0.92 \\
0.90 & 0.87 & 0.85
\end{bmatrix}
\end{array}
$$

The monthly needs on Campuses 1, 2, and 3 during winter months are respectively 20,000, 12,500, and 15,000 gallons. Suppliers A, B, C, and D are respectively able to guarantee to supply totals of 25,000, 15,000, 15,000, and 10,000 gallons per month. How much should be contracted for from each supplier if the cost per month is to be minimized? *Hint:* Use 12 unknowns, each indicating the gallons delivered by a particular supplier to a particular campus.

16. A manufacturer has 3 plants that supply 4 warehouses with an identical product. Plants 1, 2, and 3 are capable of providing 500, 300, and 200 units respectively each month. Warehouses A, B, C, and D require 250, 300, 200, and 150 units per month. The costs of manufacturing and delivering a unit from a particular factory to a particular warehouse are indicated in the table. How should the warehouses be supplied to minimize costs?

| | | Warehouse | | |
	A	B	C	D
Plant 1	42	40	45	48
Plant 2	41	45	40	46
Plant 3	48	40	40	44

Hint: Use 12 unknowns, each representing the units delivered from a particular plant to a particular warehouse.

17. A diet-conscious student notices the following information on boxes of four cereals in the family kitchen:

	Calories	For a 1-ounce Serving Sugar	Vitamin C	Carbohydrate	Fiber	Cost
Cereal A	70	6 g	25%	22 g	8 g	5.6
Cereal B	140	15 g	—	30 g	—	6.2
Cereal C	80	11 g	—	28 g	4 g	5.0
Cereal D	110	3 g	100%	20 g	—	8.6

Having recently been impressed by what he has read about carbohydrate packing, the value of a high-fiber diet, and the value of holding down his intake of calories, he decides that his usual 3-ounce serving of cereal should be a mixture which (not counting the milk) meets the following conditions:

$$
\begin{aligned}
\text{total calories} &\quad \leq 300 \\
\text{total sugar} &\quad \leq 20 \text{ g} \\
\text{total carbohydrate} &\geq 70 \text{ g} \\
\text{total fiber} &\quad \geq 10 \text{ g}
\end{aligned}
$$

provides at least 75% of the daily requirement of Vitamin C

With these goals determined, he now wishes to minimize the cost of his breakfast.

7-3 EXTREME POINTS

We are now ready to confront what we refer to as our primal problem.

Primal Problem **P**

Maximize
$$f(\mathbf{x}) = c_1 x_1 + \cdots + c_n x_n + d$$

subject to
$$
\begin{aligned}
a_{11} x_1 + \cdots + a_{1n} x_n &\leq b_1 \\
\vdots \qquad\qquad \vdots &\quad \vdots \\
a_{m1} x_1 + \cdots + a_{mn} x_n &\leq b_m \\
x_1 \geq 0, \ldots, x_n &\geq 0
\end{aligned}
$$

According to Theorem B of Section 7-2, we should immediately go to extremes, specifically to the extreme points of the feasible set $\mathcal{F}$, in looking for a solution. However, most of us have difficulty in identifying these extreme points even if the feasible set is in R^3; and we all need help if it is in R^n, $n > 3$.

Suppose that to each of the first m constraint inequalities we add a variable that enables us to rewrite the inequality as an equality. These variables, which necessarily assume nonnegative values, are called **slack variables.** Their introduction leads us to consider a new problem (which is obviously related to the primal problem).

Related Problem **P**$_R$

Maximize

$$F(\mathbf{x}_R) = c_1 x_1 + \cdots + c_n x_n + 0 x_{n+1} + \cdots + 0 x_{n+m} + d$$

subject to
$$
\begin{aligned}
a_{11} x_1 + \cdots + a_{1n} x_n + x_{n+1} \qquad\qquad &= b_1 \\
\vdots \qquad\qquad \vdots \qquad\qquad &\quad \vdots \\
a_{m1} x_1 + \cdots + a_{mn} x_n \qquad\quad + x_{n+m} &= b_m \\
x_1 \geq 0, \ldots, x_{n+m} \geq 0
\end{aligned}
$$

The equality constraints of the related problem can be written in the form

$$
\begin{aligned}
x_{n+1} &= b_1 - (a_{11}x_1 + \cdots + a_{1n}x_n) \\
\vdots \quad &\quad \vdots \qquad \vdots \qquad\qquad \vdots \\
x_{n+m} &= b_m - (a_{m1}x_1 + \cdots + a_{mn}x_n)
\end{aligned}
\tag{1}
$$

The right-hand side may be written more compactly as $\mathbf{b} - A\mathbf{x}$. This in turn enables us to neatly indicate a one-to-one correspondence between points $\mathbf{x}$ in the feasible set $\mathcal{F}$ of the primal problem $\mathbf{P}$ and points $\mathbf{x}_R$ in the feasible set $\mathcal{F}_R$ of the related problem $\mathbf{P}_R$. The correspondence $\mathbf{x} \leftrightarrow \mathbf{x}_R$ is indicated by

$$
\mathbf{x} = \begin{bmatrix} x_1 \\ \vdots \\ x_n \end{bmatrix} \longleftrightarrow \begin{bmatrix} x_1 \\ \vdots \\ x_n \\ x_{n+1} \\ \vdots \\ x_{n+m} \end{bmatrix} = \begin{bmatrix} \mathbf{x} \\ \mathbf{b} - A\mathbf{x} \end{bmatrix} = \mathbf{x}_R
$$

If two points of $\mathcal{F}$ correspond to two points of $\mathcal{F}_R$, then their midpoints also correspond. Therefore, extreme points in $\mathcal{F}$ correspond to extreme points in $\mathcal{F}_R$. Moreover, an extreme point of $\mathcal{F}_R$ will maximize F if and only if the corresponding extreme point in $\mathcal{F}$ maximizes f.

This correspondence accounts for our interest in the related problem $\mathcal{F}_R$, for while it is difficult to identify extreme points of $\mathcal{F}$, we shall soon see that the extreme points of $\mathcal{F}_R$ are easy to identify. Once $\mathbf{x}_R$ is known to be an extreme point of $\mathcal{F}_R$, the correspondence just described will identify an extreme point $\mathbf{x}$ in $\mathcal{F}$; and all the extreme points of $\mathcal{F}$ can thus be obtained if all the extreme points of $\mathcal{F}_R$ are known.

To describe the extreme points of $\mathcal{F}_R$, a review of some terminology from Section 1-2 is in order. In the equality constraints as stated for the Related Problem $\mathbf{P}_R$, the variables $x_{n+1}, \ldots, x_{n+m}$ appear as basic variables. By pivoting on a nonzero coefficient a_{rs} of a nonbasic variable x_s, we obtain an equivalent system of m equations in which x_s will appear as a basic variable (and, of course, x_{n+r} will not appear as a basic variable). Whenever we have a system of m equations that is equivalent to the system of constraints for $\mathbf{P}_R$ and in which some m variables appear as basic variables, we obtain a **basic solution** to the system by setting the nonbasic variables equal to zero. From (1), for instance, we obtain a basic solution by setting $x_1 = \cdots x_n = 0$.

A point $\mathbf{x}_R = (x_1 \ldots, x_n, x_{n+1}, \ldots, x_{n+m})$ *in* $\mathcal{F}_R$ *is an extreme point of* $\mathcal{F}_R$ *if and only if it is a basic solution to the equality constraints used to determine* $\mathcal{F}_R$.

THEOREM A

The requirement that $\mathbf{x}_R$ be a basic solution to the system of m linearly independent equations in $n + m$ unknowns means, of course, that at least n of the coordinates are 0, and the requirement that $\mathbf{x}_R$ be in $\mathfrak{F}_R$ means that the nonzero coordinates are positive. Before proving the theorem, we shall illustrate its usefulness by solving again the problem already solved geometrically in Example A of Section 7-2.

Example A

Primal Problem P	Related Problem P_R
Maximize	Maximize

$$z = f(\mathbf{x}) \qquad\qquad z = F(\mathbf{x}_R)$$
$$= \tfrac{2}{5}x_1 + x_2 + 1 \qquad = \tfrac{2}{5}x_1 + x_2 + 0x_3 + 0x_4 + 0x_5 + 1$$

subject to	subject to

$$
\begin{aligned}
-2x_1 + x_2 &\le 2 & \qquad -2x_1 + x_2 + x_3 &= 2 \\
x_1 - 2x_2 &\le 3 & x_1 - 2x_2 + x_4 &= 3 \\
x_1 + 2x_2 &\le 11 & x_1 + 2x_2 + x_5 &= 11 \\
\mathbf{x} &\ge 0 & \mathbf{x}_R &\ge 0
\end{aligned}
$$

As stated, the basic variables of the related problem are x_3, x_4, and x_5. Suppose we wish to find an equivalent system with the basic variables x_1, x_3, and x_5. Since x_1 is to enter the list and x_4 is leaving, we select as our pivot element the 1 coefficient of x_1 in the second equation (the equation in which x_4 appears). The pivot operation, exactly as it was described in Section 1-3, leads to the system

(2)
$$
\begin{aligned}
-3x_2 + x_3 + 2x_4 &= 8 \\
x_1 - 2x_2 + x_4 &= 3 \\
4x_2 - x_4 + x_5 &= 8
\end{aligned}
$$

Now if we set the nonbasic variables equal to zero, we obtain $x_1 = 3$, $x_3 = 8$, and $x_5 = 8$. All the nonzero coordinates are positive, so we have identified extreme points, first in $\mathfrak{F}_R$ and then in $\mathfrak{F}$.

$$
\mathbf{x}_R = \begin{bmatrix} 3 \\ 0 \\ 8 \\ 0 \\ 8 \end{bmatrix} \longleftrightarrow \begin{bmatrix} 3 \\ 0 \end{bmatrix} = \mathbf{x}
$$

Either point, substituted in the appropriate objective function, gives $z = \tfrac{11}{5}$.

Though the exact form is not standardized, calculations associated with each pivot are commonly recorded in a **tableau** similar to (3) in which the related problem P_R is displayed.

(3)

	x_1	x_2	x_3	x_4	x_5	
	-2	1	1	0	0	2
	①	-2	0	1	0	3
	1	2	0	0	1	11
z	$-\frac{2}{5}$	-1	0	0	0	1

Constant terms in each constraint appear on the right-hand side of the vertical line. Note that the objective function, which takes the form

$$z - \tfrac{2}{5}x_1 - x_2 - 0x_3 - 0x_4 - 0x_5 = 1$$

when all nonconstant terms are written on the left, has been written just below the constraint equations. Since this function is to be evaluated at the extreme points, it is useful in each pivot to eliminate current basic variables from the objective function as well as from the constraints.

The circled 1 in tableau (3) identifies the pivot that was used in going from the problem $\mathbf{P}_R$ to the equivalent system of constraints (2). The tableau corresponding to (2) is indicated in (4), except that the objective function is included in (4). Note that the last line enables us to conclude immediately that when the nonbasic variables x_2 and x_4 are set equal to zero, $z = \frac{11}{5}$.

(4)

	x_1	x_2	x_3	x_4	x_5	
	0	-3	1	2	0	8
	1	-2	0	1	0	3
	0	4	0	$\boxed{-1}$	1	8
z	0	$-\frac{9}{5}$	0	$\frac{2}{5}$	0	$\frac{11}{5}$

Using the circled entry in (4) as a pivot, you should check your understanding by obtaining tableau (5) without peeking ahead at the answer.

(5)

Basic variables x_1, x_3, x_4

	x_1	x_2	x_3	x_4	x_5	
	0	⑤	1	0	2	24
	1	2	0	0	1	11
	0	-4	0	1	-1	-8
z	0	$-\frac{1}{5}$	0	0	$\frac{2}{5}$	$\frac{27}{5}$

Try it once more, using the circled 5 in tableau (5) to obtain tableau (6).

(6)

Basic variables x_1, x_2, x_4

	x_1	x_2	x_3	x_4	x_5	
	0	1	$\frac{1}{5}$	0	$\frac{2}{5}$	$\frac{24}{5}$
	1	0	$-\frac{2}{5}$	0	$\frac{1}{5}$	$\frac{7}{5}$
	0	0	$\frac{4}{5}$	1	$\frac{3}{5}$	$\frac{56}{5}$
z	0	0	$\frac{1}{25}$	0	$\frac{12}{5}$	$\frac{159}{5}$

Note from tableau (5) that setting the nonbasic variables equal to zero does not give an extreme point because not all the coordinates of $(11, 0, 24, -8, 0)$ are nonnegative. But with tableau (6) we return again to an extreme point, namely $(\frac{7}{5}, \frac{24}{5}, 0, \frac{56}{5}, 0)$. We have thus illustrated 4 of the possible $\binom{5}{3} = 10$ ways of choosing the basic unknowns. A complete tabulation of the results follows:

Basic Variables	Basic Solution	Extreme Points of $\mathcal{F}_R$	Extreme Points of $\mathcal{F}$	Value of Objective Function
1, 2, 3	$(7, 2, 14)$	$(7, 2, 14, 0, 0)$	$(7, 2)$	$\frac{29}{5}$
1, 2, 4	$(\frac{7}{5}, \frac{24}{5}, \frac{56}{5})$	$(\frac{7}{5}, \frac{24}{5}, 0, \frac{56}{5}, 0)$	$(\frac{7}{5}, \frac{24}{5})$	$\frac{159}{25}$
1, 2, 5	$(-\frac{7}{3}, -\frac{8}{3}, \frac{56}{3})$			
1, 3, 4	$(11, 24, -8)$			
1, 3, 5	$(3, 8, 8)$	$(3, 0, 8, 0, 8)$	$(3, 0)$	$\frac{11}{5}$
1, 4, 5	$(-1, 4, 12)$			
2, 3, 4	$(\frac{11}{2}, -\frac{7}{2}, 14)$			
2, 3, 5	$(-\frac{3}{2}, -\frac{7}{2}, 14)$			
2, 4, 5	$(2, 7, 7)$	$(0, 2, 0, 7, 7)$	$(0, 2)$	3
3, 4, 5	$(2, 3, 11)$	$(0, 0, 2, 3, 11)$	$(0, 0)$	1

Compare the extreme points of $\mathcal{F}$ as we just found them with Figure 7-1. The maximum of $z = \frac{159}{25}$ occurs at $(\frac{7}{5}, \frac{24}{5})$. $\square$

(The thought of solving a problem with more than 2 or 3 variables by making all possible pivots makes one dizzy. But it is precisely when we get beyond 2 or 3 variables that geometric methods fail. We clearly need, therefore, to develop some more efficient procedure. Toward this end, we shall introduce the **simplex method.**)

(Since a particular solution leads to an extreme point if and only if the entries in the rightmost column (the column of constants) are nonnegative, getting a system into this form is the first goal (Phase I) of the simplex method. It often happens (as it did in our example above) that when a problem is expressed in canonical form, all the constants are nonnegative. Whether this happy state arises from the statement of the problem or from a prudent pivot, its occurrence means that we are ready for Phase II of the simplex method.

Once ready for Phase II, it is inefficient to move (as we did in tableau (5) above) to a system having one or more negative entries in the column of constants. Therefore, pains are taken, once we have entered Phase II of the simplex method, to see that we do not fall back to Phase I again.

It turns out to be easiest to describe Phase II of the simplex method first, then to talk about Phase I. Thus, we shall begin Section 7-4 with a discussion of a problem in which all the constants are nonnegative.

THE PROOF OF THEOREM A (OPTIONAL)

To wrap everything up before moving on, we shall now prove Theorem A.

Proof: Let us first go in the direction of assuming that $\mathbf{r}$ is a basic solution to the equality constraints of $\mathbf{P}_R$. Then it has of course been obtained from some equivalent system in which m variables, say $x_{i_1}, \ldots, x_{i_m}$, appear as basic variables. This equivalent system can be written in the form

$$
\begin{aligned}
x_{i_1} &= r_{i_1} - (a_{1j_1}x_{j_1} + \cdots + a_{1j_n}x_{j_n}) \\
&\vdots \quad\quad \vdots \quad\quad\quad\quad \vdots \quad\quad\quad\quad \vdots \\
x_{i_m} &= r_{i_m} - (a_{mj_1}x_{j_1} + \cdots + a_{mj_n}x_{j_n})
\end{aligned}
\tag{7}
$$

The nonzero coordinates of $\mathbf{r}$, which must be positive since $\mathbf{r}$ is in $\mathfrak{F}_R$, are obtained from (7) by setting the nonbasic variables equal to zero.

If $\mathbf{r}$ is not extreme in $\mathfrak{F}_R$, then there must be two other points, say $\mathbf{s}$ and $\mathbf{t}$, in $\mathfrak{F}_R$ such that $\mathbf{r} = \frac{1}{2}(\mathbf{s} + \mathbf{t})$. And since all the coordinates of both $\mathbf{s}$ and $\mathbf{t}$ are nonnegative, $\mathbf{s}$ and $\mathbf{t}$ (like $\mathbf{r}$) must have 0 coordinates in positions $j_1, \ldots, j_n$. Moreover, both $\mathbf{s}$ and $\mathbf{t}$ must satisfy the system (7) since it is equivalent to the given linear system satisfied by all members of $\mathfrak{F}_R$. It follows that $s_{i_1} = t_{i_1} = r_{i_1}$, and so on; that is, $\mathbf{s} = \mathbf{t} = \mathbf{r}$.

Now suppose that $\mathbf{r}$ in $\mathfrak{F}_R$ is an extreme point of $\mathfrak{F}_R$. Let $\mathbf{r}$ have k nonzero (therefore positive) coordinates which, for notational convenience, we take to be the first k coordinates.

$$
\mathbf{r} = (r_1, \ldots, r_k, 0, \ldots, 0)
$$

Since $\mathbf{r}$ is in $\mathfrak{F}_R$,

$$a_{11}r_1 + \cdots + a_{1k}r_k = b_1$$
$$\vdots \qquad \vdots \qquad \vdots$$
$$a_{m1}r_1 + \cdots + a_{mk}r_k = b_m$$

We now claim that the column vectors

$$\mathbf{A}_{\cdot 1} = \begin{bmatrix} a_{11} \\ \vdots \\ a_{m1} \end{bmatrix}, \ldots, \mathbf{A}_{\cdot k} = \begin{bmatrix} a_{1k} \\ \vdots \\ a_{mk} \end{bmatrix}$$

are linearly independent. If they aren't, then there are k numbers $u_1, \ldots, u_k$, not all zero (but not necessarily all positive either), such that

$$u_1\mathbf{A}_{\cdot 1} + \cdots + u_k\mathbf{A}_{\cdot k} = \mathbf{0}$$

Now we can choose $\alpha > 0$ so that both

$$\mathbf{s} = (r_1 + \alpha u_1, \ldots, r_k + \alpha u_k, 0, \ldots, 0)$$
$$\mathbf{t} = (r_1 - \alpha u_1, \ldots, r_k - \alpha u_k, 0, \ldots, 0)$$

have nonnegative coordinates. Moreover, both $\mathbf{s}$ and $\mathbf{t}$ are in $\mathfrak{F}_R$. Finally, $\mathbf{r} = \frac{1}{2}(\mathbf{s} + \mathbf{t})$, contradicting the fact that $\mathbf{r}$ is an extreme point. We conclude that the vectors $\mathbf{A}_{\cdot 1}, \ldots, \mathbf{A}_{\cdot k}$ are indeed linearly independent, and hence (since the column rank cannot exceed the row rank) that $k \leq m$. Thus, $\mathbf{r}$ is a basic solution to the system of linear constraints to $\mathbf{P}_R$.) $\square$

PROBLEM
SET 7-3

1. If, after writing the constraints of Problem 5 of Problem Set 7-2 in canonical form, we introduce slack variables, we get

$$-4x_1 - 9x_2 + x_3 \qquad\qquad = -18$$
$$6x_1 - 3x_2 \qquad + x_4 \qquad = 16$$
$$2x_1 + 3x_2 \qquad\qquad + x_5 = 16$$

Find equivalent systems having as basic variables the sets indicated below. When the basic solution identifies an extreme point in R^5, identify the corresponding extreme point in R^2. Compare your results with the picture you sketched in solving Problem 5 of Problem Set 7-2.
(a) Basic variables x_1, x_2, and x_3.
(b) Basic variables x_1, x_2, and x_4.
(c) Basic variables x_1, x_3, and x_5.

2. Follow the instructions for Problem 1, using the constraints for Problem 6 of Problem Set 7-2 which, in the form of equalities, becomes

$$x_1 + 2x_2 + x_3 \qquad\qquad = 18$$
$$3x_1 + 2x_2 \qquad + x_4 \qquad = 26$$
$$-x_1 - 4x_2 \qquad\qquad + x_5 = -3$$

3. If, after writing the constraints of Problem 7 of Problem Set 7-2 in canonical form, we introduce slack variables, we get

$$
\begin{aligned}
x_1 + 2x_2 + x_3 + x_4 &= 12 \\
-2x_1 - 4x_2 + x_3 \quad\quad + x_5 &= -8 \\
-x_1 + x_2 - 2x_3 \quad\quad\quad + x_6 &= -4
\end{aligned}
$$

Find equivalent systems having as basic variables the sets indicated below. When extreme points are identified in R^6, identify the corresponding points in R^3.

(a) Basic variables x_1, x_2, and x_3.
(b) Basic variables x_1, x_2, and x_4.
(c) Basic variables x_1, x_3, and x_5.

4. Follow the instructions for Problem 3, using the constraints for Problem 8 of Problem Set 7-2 which, in the form of equalities, become

$$
\begin{aligned}
2x_1 + 4x_2 + 3x_3 + x_4 &= 24 \\
-x_1 + 3x_2 - 9x_3 \quad\quad + x_5 &= -27 \\
x_1 + x_2 \quad\quad\quad\quad + x_6 &= 6
\end{aligned}
$$

THE SIMPLEX METHOD, PHASE II $\quad$ 7-4

We assume throughout this section that the initial tableau has all non-negative constants in the right-hand column, a possible exception being in the last row where we keep track not of constraints but of the objective function. Part of our responsibility in choosing the next pivot element is to see that this condition of nonnegative constants is not destroyed.

Our choice of a pivot element is based in part on an examination of the coefficients of the objective function as they are found in the last row of the tableau. We begin, therefore, by examining more carefully the expressions found for the objective function in the erratic sequence of pivots used for illustrative purposes in Section 7-3. Skipping tableau (5) of that section where we violated the condition of keeping the right-hand side nonnegative, we summarize results from the other tableaus as follows:

Tableau	Extreme Point	Expression for Objective Function
(3)	$(0, 0, 2, 3, 11)$	$z = \frac{2}{5}x_1 + x_2 + 1$
(4)	$(3, 0, 8, 0, 8)$	$z = \frac{9}{5}x_2 - \frac{2}{5}x_4 + \frac{11}{5}$
(6)	$(\frac{7}{5}, \frac{24}{5}, 0, \frac{56}{5}, 0)$	$z = -\frac{1}{25}x_3 - \frac{12}{25}x_5 + \frac{159}{25}$

It is to be emphasized that the various expressions for the objective function are only a matter of convenience—convenience owing to the fact that, in each case, the variables that appear are exactly the non-basic variables that are set equal to 0. Any one of the expressions, however, will give the same value for z at a particular extreme point. Thus, using the first listed extreme point $(0, 0, 2, 3, 11)$, the various expressions for z give

$$z = \tfrac{2}{5}(0) + \quad 0 \quad + 1 = 1 \quad \text{(the convenient form to}$$
$$z = \tfrac{9}{5}(0) - \tfrac{2}{5}(3) + \tfrac{11}{5} = 1 \quad \text{use with this extreme point)}$$
$$z = -\tfrac{1}{25}(2) - \tfrac{12}{25}(11) + \tfrac{159}{25} = 1$$

(Let us say it this way: If a particular expression for z is evaluated at an extreme point $\mathbf{r}$, then any other expression for z evaluated at $\mathbf{r}$ will give the same value.) This being the case, it is instructive to look at the expression for z given in tableau (6). This expression gives $z = \tfrac{159}{25}$ when $x_3 = x_5 = 0$, and since we always have the constraints $x_3 \geq 0$ and $x_5 \geq 0$, it is clear that this expression with its negative coefficients for x_3 and x_5 will never exceed $\tfrac{159}{25}$. But then no other expression for z can ever exceed $\tfrac{159}{25}$ either. Evidently the desired maximum occurs at $(\tfrac{7}{5}, \tfrac{24}{5}, 0, \tfrac{56}{5}, 0)$ where $z = \tfrac{159}{25}$.

More important than the answer is the lesson. Suppose that in some tableau all the constant terms in the right-hand column (except possibly the last row) are nonnegative, so that a feasible point is determined by setting the nonbasic variables equal to zero. If at the same time all the coefficients *of the variables* in the last (objective function) row are nonnegative (so as to be nonpositive when the expression is solved for z), then the constant in this row gives the maximum value for z. Before any pivot step is taken, therefore, we should always check to see whether an optimal point has already been determined.

Test for optimality. *Are all coefficients in the last row, excluding the rightmost constant, nonnegative?*

If the answer is yes, then the maximum value for z is attained at the extreme point determined by that tableau, and this maximum value appears as the rightmost entry in the last row. If, however, the answer is no (that is, there are negative coefficients in the last row), then another pivot is called for. In this case, one's inclination to go after the worst offender seems to be the best plan of attack.

Choose the pivot column. *Select as pivot column the one having the most negative coefficient in the last row, breaking ties by choosing arbitrarily.*

Now suppose that the column s has been selected as the pivot column. There will be m columns, corresponding to m basic variables, that each consist of one 1 with all other entries 0. This will not be the case in column s, however, since x_5 is at this point a nonbasic variable. We now wish to investigate the possibility that all the entries in column s might be nonpositive. The type of situation we have in mind is illustrated by tableau (1). *(1)*

$$s = 3$$

x_1	x_2	x_3	x_4	x_5	
0	1	-5	0	1	3
0	-2	-1	1	0	9
1	-1	-2	0	0	2
0	2	-4	0	0	7

If in this situation we set $x_3 = r > 0$, $x_2 = 0$, and use the three constraint equations to determine

$$x_1 = 2 + 2r, \qquad x_4 = 9 + r, \qquad x_5 = 3 + 5r$$

then the point $(2 + 2r, 0, r, 9 + r, 3 + 5r)$ is feasible for all $r > 0$; and for this point, $z = 7 + 4r$. Since there is no upper bound on how r can be chosen, z grows without bound; and so the maximization problem grows without bound.

The same argument would apply to any problem in which all the entries in a chosen column are negative. These observations give us a test for an unbounded objective function.

Test for an unbounded objective function. *If all entries in the selected pivot column are nonpositive, the objective function is unbounded and the maximization problem has no solution.*

Thus, we are obliged to get on with the business of selecting a pivot row only if column S has some positive entries. It is in the selection of this row that we take precautions to retain the nonnegative feature of the last column. To see how to do this, imagine that row R has been chosen so that $a_{RS} > 0$. Then examine what happens to row i, $i \neq R$, if we use a_{RS} as a pivot. Nothing happens, of course, if $a_{iS} = 0$, but if $a_{iS} \neq 0$, we multiply row R by $-a_{iS}/a_{RS}$ and add it to row i. In the column of constants on the right, this gives us the new element.

$$b_i - \frac{a_{iS}}{a_{RS}} b_R = a_{iS}\left(\frac{b_i}{a_{iS}} - \frac{b_R}{a_{RS}}\right) \tag{2}$$

which we want to keep nonnegative. If $a_{iS} < 0$, then both factors on the right-hand side of (2) will be negative, making (2) nonnegative. If $a_{iS} > 0$, then (2) will be nonnegative if and only if

$$\frac{b_i}{a_{iS}} \geq \frac{b_R}{a_{RS}}$$

This indicates how R should be chosen.

Choose the pivot row. *For all rows having a positive entry a_{iS} in column S, choose the one for which b_i/a_{iS} is a minimum, breaking ties arbitrarily.*

It is with no small pleasure that we note that the value of z obtained after this pivot about a_{RS} will be

$$d - \frac{c_S}{a_{RS}} b_R \geq d$$

the inequality holding because $c_S < 0$, $a_{RS} > 0$, $b_R \geq 0$.

While there is a phenomenon called **cycling** in which one encounters an infinite sequence of values b_R which are 0, this does not happen in practical problems.) We will not concern ourselves with it here, but will turn instead to using the method just described on two problems that are by now quite familiar.

Example A

Maximize

$$f(\mathbf{x}) = \tfrac{2}{5}x_1 + x_2 + 1$$

subject to

$$-2x_1 + x_2 \leq 2$$
$$x_1 - 2x_2 \leq 3$$
$$x_1 + 2x_2 \leq 11$$
$$x_1 \geq 0, \qquad x_2 \geq 0$$

Original tableau (nonbasic variables x_1, x_2 set equal to 0)

	x_1	x_2	x_3	x_4	x_5		Ratios	Extreme point
	-2	①	1	0	0	2	$2/1 \longleftarrow$	$(x_1, x_2) = (0,0)$
	1	-2	0	1	0	3		
	1	2	0	0	1	11	$11/2$	$z = 1$
z	$-\tfrac{2}{5}$	-1	0	0	0	1		

$\uparrow$

The original tableau does not identify an optimal point since there are negative coefficients in the last row. The most negative of these, identified with the vertical arrow, gives us our pivot column. The ratios of constant terms to positive coefficients in this column are indicated, with an arrow pointing to the smallest, thus completing our choice of the first pivot element (circled). The next tableau shows the result of the first pivot, and again we see a negative coefficient in the last row, meaning that we have not yet identified an optimal point.

First pivot (nonbasic variables x_1, x_3 set equal to 0)

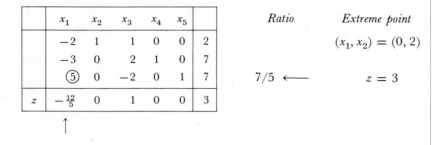

	x_1	x_2	x_3	x_4	x_5		Ratio	Extreme point
	-2	1	1	0	0	2		$(x_1, x_2) = (0, 2)$
	-3	0	2	1	0	7		
	⑤	0	-2	0	1	7		$7/5 \longleftarrow$	$z = 3$
z	$-\frac{12}{5}$	0	1	0	0	3		

$\uparrow$

The rules are followed to obtain our next pivot element (circled), and then to obtain the tableau resulting from this second part.

Second pivot (nonbasic variables x_3, x_5 set equal to 0)

	x_1	x_2	x_3	x_4	x_5		Extreme point
	0	1	$\frac{1}{5}$	0	$\frac{2}{5}$	$\frac{24}{5}$	$(x_1, x_2) = (\frac{7}{5}, \frac{24}{5})$
	0	0	$\frac{4}{5}$	1	$\frac{3}{5}$	$\frac{56}{5}$	
	1	0	$-\frac{2}{5}$	0	$\frac{1}{5}$	$\frac{7}{5}$	$z = \frac{159}{25}$
z	0	0	$\frac{1}{25}$	0	$\frac{12}{5}$	$\frac{159}{25}$	

Having reached a tableau with all nonnegative coefficients in the last row, our theory tells us that an optimal point has been located; so, of course, does comparison with previous solutions of this problem.

Note, with one eye on Figure 7-1, that our series of pivots, in taking us from $(0, 0)$ to $(0, 2)$ to $(\frac{7}{5}, \frac{24}{5})$, have taken us on a tour of successive extreme points moving around the boundary of the feasible set. □

Maximize

$$f(\mathbf{x}) = 4x_1 + 6x_2 + 7x_3$$

Example B

subject to

$$2x_1 + 5x_2 + 3x_3 \leq 120$$
$$x_1 + x_2 + 3x_3 \leq 60$$
$$3x_1 + 4x_2 + 2x_3 \leq 120$$
$$x_1 \geq 0, \qquad x_2 \geq 0, \qquad x_3 \geq 0$$

Original tableau (nonbasic variables x_1, x_2, x_3 set equal to 0)

x_1	x_2	x_3	x_4	x_5	x_6		Ratios	Extreme point
2	5	3	1	0	0	120	$120/3 = 40$	$(0, 0, 0)$
1	1	③	0	1	0	60	$60/3 = 20$ ⟵	$z = 0$
3	4	2	0	0	1	120	$120/2 = 60$	

z	-4	-6	-7	0	0	0	0

↑

The most negative coefficient in the last row gives us our pivot column, identified with a vertical arrow. The ratios of constant terms to positive coefficients in this column are indicated, and an arrow again identifies the smallest ratio. The (circled) pivot element is thus identified, and we obtain the tableau of our first pivot.

First pivot (nonbasic variables x_1, x_2, x_5 set equal to 0)

x_1	x_2	x_3	x_4	x_5	x_6		Ratios	Extreme point
1	④	0	0	-1	0	60	$60/4 = 15$ ⟵	
$\frac{1}{3}$	$\frac{1}{3}$	1	0	$\frac{1}{3}$	0	20	$20/\frac{1}{3} = 60$	$V_4(0, 0, 20)$
$\frac{7}{3}$	$\frac{10}{3}$	0	0	$-\frac{2}{3}$	1	80	$80/\frac{10}{3} = 24$	$z = 140$

z	$-\frac{5}{3}$	$-\frac{11}{3}$	0	0	$\frac{7}{3}$	0	140

↑

The negative coefficients in the last row tell us that we do not have an optimal point. (The designation of the extreme point that we do have by V_4 corresponds with notation on Figure 7-4.) We proceed as before.

Second pivot (nonbasic variables x_1, x_4, x_5 set equal to 0)

x_1	x_2	x_3	x_4	x_5	x_6		Ratios	Extreme point
$\frac{1}{4}$	1	0	$\frac{1}{4}$	$-\frac{1}{4}$	0	15	$15/\frac{1}{4} = 60$	
$\frac{1}{4}$	0	1	$-\frac{1}{12}$	$\frac{5}{12}$	0	15	$15/\frac{1}{4} = 60$	$V_5(0, 15, 15)$
ⓧ$\frac{3}{2}$	0	0	$-\frac{10}{12}$	$\frac{2}{12}$	1	30	$30/\frac{3}{2} = 20$ ⟵	$z = 195$

z	$-\frac{3}{4}$	0	0	$\frac{11}{12}$	$\frac{17}{12}$	0	195

↑

Shall we swing through a third pivot?

Third pivot (nonbasic variables x_4, x_5, x_6 set equal to 0)

	x_1	x_2	x_3	x_4	x_5	x_6		
	0	1	0	$\frac{14}{36}$	$-\frac{10}{36}$	$-\frac{1}{6}$	10	
	0	0	1	$\frac{2}{36}$	$\frac{14}{36}$	$-\frac{1}{6}$	10	
	1	0	0	$-\frac{5}{9}$	$\frac{1}{9}$	$\frac{2}{3}$	20	
z	0	0	0	$\frac{18}{36}$	$\frac{48}{36}$	$\frac{1}{2}$	210	

Extreme point

$V_3(20, 10, 10)$

$z = 210$

Finally, all the coefficients in the last row are nonnegative. We have found an optimal solution. Again we are cheered to notice that it is the same solution that was determined in Example B of Section 7-2 where we examined the value of the objective function at all eight extreme points. □

The solution just obtained can also be viewed geometrically as a tour of successive extreme points (Figure 7-4). It is a guided tour, however. With each pivot we have gone to an extreme point at which the objective function assumed a greater value.

In Problems 1 and 2, use Phase II of the simplex method to find the required maximum. Sketch the feasible set and identify the extreme point that corresponds to each stage of the solution.

PROBLEM
SET 7-4

1. Maximize $z = 3x_1 - x_2$ subject to

$$2x_1 - x_2 \leq 3$$
$$x_1 + x_2 \leq 18$$
$$3x_1 - 2x_2 \leq 6$$
$$3x_1 - x_2 \leq 6$$

2. Maximize $z = 2x_1 + 5x_2$ subject to

$$x_1 + 2x_2 \leq 12$$
$$x_1 - x_2 \leq 6$$
$$2x_1 + x_2 \leq 12$$
$$2x_1 - 4x_2 \leq 6$$

Solve the following linear programming problems using Phase II of the simplex method.

3. Maximize $z = 2x_1 - x_2 + x_3$ subject to

$$-x_1 + 2x_2 + 3x_3 \leq 8$$
$$3x_1 + 5x_2 - x_3 \leq 12$$
$$2x_1 - 3x_2 + 2x_3 \leq 9$$
$$x_1 \geq 0, \qquad x_2 \geq 0, \qquad x_3 \geq 0$$

4. Maximize $z = x_1 + 2x_2 - x_3$ subject to

$$x_1 - x_2 + 2x_3 \leq 2$$
$$2x_1 + 3x_2 - x_3 \leq 8$$
$$3x_1 - x_2 + 2x_3 \leq 5$$
$$x_1 \geq 0, \quad x_2 \geq 0, \quad x_3 \geq 0$$

5. Minimize $z = x_1 - 2x_2 - x_3$ subject to

$$2x_1 - 3x_2 + x_3 \leq 2$$
$$3x_1 + 5x_2 - 2x_3 \leq 6$$
$$x_1 - 4x_2 + 3x_3 \leq 4$$
$$x_1 \geq 0, \quad x_2 \geq 0, \quad x_3 \geq 0$$

6. Minimize $z = 2x_1 - 3x_2 + x_3$ subject to

$$x_1 - 2x_2 - x_3 \leq 0$$
$$2x_1 - x_2 - 2x_3 \leq 4$$
$$x_1 + 3x_2 + x_3 \leq 9$$
$$x_1 \geq 0, \quad x_2 \geq 0, \quad x_3 \geq 0$$

7-5 THE SIMPLEX METHOD, PHASE I

We began Section 7-4 with the assumption that the initial tableau had all nonnegative constants in the right-hand column. Phase II of the simplex method was then described, analytically as a way to find a solution if one existed, geometrically as a guided tour of extreme points of the feasible set.

We now address the problem of what is to be done if the initial tableau includes some negative constants in the right-hand column. If this happens, the negative constants must be tended to before Phase II can begin. For this purpose, we now describe Phase I. (Geometrically, Phase I may be thought of as an effort to locate a first extreme point, from which we can begin the tour of Phase II.

The problem to be addressed here, therefore, is:

Maximize $\qquad z = f(\mathbf{x}) = c_1 x_1 + \cdots + c_n x_n$

(1)   subject to $a_{11}x_1 + \cdots + a_{1n}x_n + x_{n+1} \qquad\qquad = b_1$
$$\vdots \qquad\qquad \vdots \qquad\qquad\qquad \vdots$$
$$a_{m1}x_1 + \cdots + a_{mn}x_n \qquad\quad + x_{n+m} = b_m$$
$$x_1 \geq 0, \ldots, x_{m+n} \geq 0$$

where some b_i are negative. We then consider what we call the

Augmented problem.   Maximize $\widehat{z} = -x_0$

subject to
$$a_{11}x_1 + \cdots + a_{1n}x_n + x_{n+1} \qquad\qquad = b_1 + x_0$$
$$\vdots \qquad\qquad \vdots \qquad\qquad\qquad \vdots$$
$$a_{m1}x_1 + \cdots + a_{mn}x_n \qquad\qquad + x_{n+m} = b_m + x_0 \qquad\qquad (2)$$
$$x_0 \geq 0, \qquad x_1 \geq 0, \qquad \ldots, \qquad x_{n+m} \geq 0$$

We note several things about the augmented problem. First, its feasible set is nonempty; let $x_0 \geq \max\{-b_1, \ldots, -b_m\}$, choose $x_1 = \cdots = x_n = 0$, and determine $x_{n+1}, \ldots, x_{n+m}$. Second, since the objective function of the augmented problem satisfies $\widehat{z} \leq 0$, it clearly is bounded; Theorem B of Section 7-2 tells us that the augmented problem does have an optimal solution. Third, the constraints (2) reduce to the constraints (1) when $x_0 = 0$. Thus, the original constraints are satisfied if and only if the maximum value of $\widehat{z} = -x_0$ is 0. In particular, if the maximum value of $\widehat{z}$ is less than 0, the original problem has an empty feasible set. Finally, the augmented problem has the very attractive feature of being manageable, for while some of the constant terms are negative, one well-chosen pivot turns the tableau to a position where the constants are nonnegative. Let us rewrite (2) in the form of a tableau:

$$(3)$$

	x_0	x_1	$\cdots$	x_n	x_{n+1}	$\cdots$	x_{n+m}	
	-1	a_{11}	$\cdots$	a_{1n}	1	$\cdots$	0	b_1
		$\vdots$		$\vdots$				$\vdots$
	(-1)	a_{R1}	$\cdots$	a_{Rn}		1	$\cdots$	b_R
		$\vdots$		$\vdots$				$\vdots$
	-1	a_{m1}	$\cdots$	a_{mn}	0	$\cdots$	1	b_m
$\widehat{z}$	1	0		0	0	$\cdots$	0	0

Now if row R is chosen so that $b_R \leq b_i$ for $i = 1, \ldots, m$, and if we use the (circled) -1 in the first column as a pivot, we obtain the desired tableau:

	x_0	x_1	$\cdots$	x_n	x_{n+1}	$\cdots$	x_{n+m}	
	0	$a_{11} - a_{R1}$	$\cdots$	$a_{1n} - a_{Rn}$	1	$\cdots$ -1 $\cdots$	0	$b_1 - b_R$
	$\vdots$	$\vdots$		$\vdots$	$\vdots$	$\vdots$	$\vdots$	$\vdots$
	1	$-a_{R1}$	$\cdots$	a_{Rm}	0	$\cdots$ -1 $\cdots$	0	$-b_R$
	$\vdots$	$\vdots$		$\vdots$	$\vdots$	$\vdots$	$\vdots$	$\vdots$
	0	$a_{m1} - a_{R1}$	$\cdots$	$a_{mn} - a_{Rm}$	0	$\cdots$ 1 $\cdots$	1	$b_m - b_R$
$\widehat{z}$	0	a_{R1}	$\cdots$	a_{Rn}	0	$\cdots$ 1 $\cdots$	0	b_R

All the constants in this tableau are nonnegative, so we may now proceed with Phase II of the simplex method to obtain the solution that we know exists. We have already established that this solution $\widehat{z} = d$ satisfies $d \leq 0$, and that the original problem has a solution if and only if $d = 0$. Thus, if $d < 0$, we may quit, concluding that the feasible set of the original problem is empty.

If $d = 0$, we next check to see if x_0 appears in our final tableau as a nonbasic or a basic variable. If it is a basic variable, another pivot is called for to replace x_0 with some other (arbitrarily chosen) variable, thus guaranteeing that, along with the other nonbasic variables, we will set $x_0 = 0$. With $x_0 = 0$, the remaining coordinates will determine an extreme point for the original problem.

We are now in a position where we can shift our attention entirely to the original problem. The columns corresponding to x_0 and the row giving the coefficients of $\widehat{z}$ can be ignored. We would be even better situated at this point if our tableau included a row giving the coefficients of z in terms of the now current nonbasic variables. In anticipation of this, it is therefore common to include in the tableau for the augmented problem a line for keeping a current expression for z. Note the following example.

Example A Maximize

$$z = f(\mathbf{x}) = 3x_1 - 2x_2$$

subject to

$$4x_1 + x_2 \leq 16$$
$$-x_1 - x_2 \leq -7$$
$$2x_1 - x_2 \leq -4$$
$$x_1 \geq 0, \qquad x_2 \geq 0$$

The related problem asks us to maximize

$$z = 3x_1 - 2x_2$$

subject to

$$4x_1 + x_2 + x_3 = 16$$
$$-x_1 - x_2 + x_4 = -7$$
$$2x_1 - x_2 + x_5 = -4$$
$$x_1 \geq 0, \qquad x_2 \geq 0, \qquad x_3 \geq 0, \qquad x_4 \geq 0, \qquad x_5 \geq 0$$

The augmented problem then is to maximize

$$\widehat{z} = -x_0$$

subject to

$$4x_1 + x_2 + x_3 \qquad\qquad = 16 + x_0$$
$$-x_1 - x_2 \quad\; + x_4 \quad\;\; = -7 + x_0$$
$$2x_1 - x_2 \qquad\quad + x_5 = -4 + x_0$$
$$x_1 \geq 0, \qquad x_2 \geq 0, \qquad x_3 \geq 0, \qquad x_4 \geq 0, \qquad x_5 \geq 0$$

The tableau that corresponds to (1), with the addition of a line to keep track of z, is

	x_0	x_1	x_2	x_3	x_4	x_5		
	-1	4	1	1	0	0	16	
	$\boxed{-1}$	-1	-1	0	1	0	-7	←
	-1	2	-1	0	0	1	-4	
z	0	-3	2	0	0	0	0	
$\widehat{z}$	1	0	0	0	0	0	0	

Choosing the row with the least value of b (arrow), we pivot about the corresponding (circled) -1 in the first column.

First pivot (nonbasic variables x_1, x_2, x_4 set equal to 0)

	x_0	x_1	x_2	x_3	x_4	x_5		Ratio
	0	5	2	1	-1	0	23	23/2
	1	1	①	0	-1	0	7	7/1 ←
	0	3	0	0	-1	1	3	
z	0	-3	2	0	0	0	0	
$\widehat{z}$	0	-1	-1	0	1	0	-7	

↑

Since all constants are now nonnegative, we may move to Phase II in our effort to solve the augmented problem, bringing the objective function of the related problem along for the ride. Choosing arbitrarily between columns x_1 and x_2 (in our search for the most negative coefficient in the row for $\widehat{z}$), we have used x_2 and computed the necessary ratios for the positive coefficients in this column. The smallest is $7/1$, determining our second pivot.

Second pivot (nonbasic variables x_0, x_1, x_4 set equal to 0)

	x_0	x_1	x_2	x_3	x_4	x_5		Ratio
	-2	3	0	1	1	0	9	9/3
	1	1	1	0	-1	0	7	7/1
	0	③	0	0	-1	1	3	3/3 ⟵
z	-2	-5	0	0	2	0	-14	
$\widehat{z}$	1	0	0	0	0	0	0	

↑

With this pivot, since all coefficients in the $\widehat{z}$ (last) row are nonnegative, we have found the optimal solution to the augmented problem. Moreover, this optimal maximum value of $\widehat{z}$ is 0, meaning that we can proceed with the related problem. In fact, with the (shaded) row for $\widehat{z}$ and column for x_0 now deleted, we are left with a tableau for the related problem that exhibits the extreme point

$$(x_1, x_2, x_3, x_4, x_5) = (0, 7, 9, 0, 3)$$

We are now ready to use Phase II on the related problem. Noting that -5 is the most (in fact the only) negative coefficient in the objective row for z, we compute the required ratios for this column. The resulting pivot element is indicated (circled) and we perform our third pivot operation.

Third pivot (nonbasic variables x_4 and x_5 set equal to 0)

	x_1	x_2	x_3	x_4	x_5	
	0	0	1	2	-1	6
	0	1	0	$-\frac{2}{3}$	$-\frac{1}{3}$	6
	1	0	0	$-\frac{1}{3}$	$\frac{1}{3}$	1
z	0	0	0	$\frac{1}{3}$	$\frac{5}{3}$	-9

Since all the entries in the last row are nonnegative, we have determined an extreme point

$$(x_1, x_2, x_3, x_4, x_5) = (1, 6, 6, 0, 0)$$

for the related problem at which the objective function takes the value of -9. Or, in terms of the original problem, the extreme point is

$$(x_1, x_2) = (1, 6)$$

at which the objective function $z = 3x_1 - 2x_2$ takes the value $3 - 2(6) = -9.$ $\square$

1–4. In Problems 7–10 of Problem Set 7-2, you were given four problems to solve by geometric methods. Solve these problems again, using the simplex method.

5. Maximize $z = 2x_1 + x_2 - x_3$ subject to the constraints

$$x_1 + x_2 + 3x_3 \leq 9$$
$$2x_1 - x_2 + 3x_3 \geq 2$$
$$3x_1 + x_2 - 2x_3 \geq 5$$

6. Maximize $z = 3x_1 - x_2 + 2x_3$ subject to the constraints

$$2x_1 - x_2 - x_3 \leq 2$$
$$x_1 + x_2 + 2x_3 \geq 2$$
$$-x_1 + 2x_2 + 3x_3 \geq 5$$

All questions refer to the following problem:

$$\text{Maximize} \quad z = 4x_1 + 3x_2 \quad \text{subject to}$$
$$2x_1 - 2x_2 \leq 1$$
$$6x_1 - 2x_2 \leq 13$$
$$x_1 - 2x_2 \geq -7$$
$$2x_1 + 3x_2 \geq 2$$
$$x_1 \geq 0, \quad x_2 \geq 0$$

1. By drawing a graph, locate all extreme points of the feasible set.

2. Solve the problem graphically. Let Q be the extreme point at which the maximum occurs.

3. Write out the related problem.

4. Write out the augmented problem.

5. Following the argument in Section 7-5, find a point in the feasible set of the augmented problem.

6. Solving the augmented problem by the procedure outlined in Section 7-5, you find one of the extreme points listed in Problem 1. Call it A. What are the coordinates of A?

7. You know that a pivot in Phase II will move you from A to an adjacent extreme point.
   (a) Looking at your graph of the feasible set, the point A, and the known solution at Q, to which extreme point do you hope to be moved by the first pivot of Phase II?
   (b) Perform the first pivot of Phase II. What extreme point do you obtain?

8. Find the solution to the given problem by completing Phase II.

9. By removing just one constraint from the given problem, illustrate that an objective function may be unbounded. Notice that your choice of the constraint to be removed is unique; you must choose carefully.

10. Change the function to be maximized to $z = x_1 - 4x_2$. Keeping $x_1 \geq 0$ and $x_2 \geq 0$ as usual, show that the resulting problem has a solution no matter which other single constraint is removed.

MARKOV CHAINS 8

INTRODUCTION | 8-1

A semiretired couple have their home in Illinois, a condominium in Florida, and a winterized cabin in Wisconsin. If they are at their Illinois home on a given weekend, the probability of their being there the next weekend is $\frac{2}{3}$; but if they are not in Illinois on the next weekend, the chances are equally likely that they will have gone to Florida or Wisconsin (so the probability of being in either of these states is $\frac{1}{6}$). Let us designate the three states by E_1 (Illinois), E_2 (Florida), and E_3 (Wisconsin). The information just given may be summarized as follows:

If on a given weekend
they are in state

we are told that the
probabilities of their spending
the next weekend in states

E_1

$E_1 \quad E_2 \quad E_3$

are given by

$$[\begin{matrix} \frac{2}{3} & \frac{1}{6} & \frac{1}{6} \end{matrix}]$$

The vector showing the various probabilities is called a **probability vector.** Probability vectors are characterized by the fact that all entries are nonnegative, and that the sum of the entries is 1.

A complete description of the habits of this peripatetic couple would give us a probability vector for each of the three possible starting states. It could be conveniently summarized as a matrix:

$$\begin{array}{c} \textit{Probability of spending the} \\ \textit{next weekend in state} \\ \begin{array}{ccc} E_1 & E_2 & E_3 \end{array} \end{array}$$

$$\begin{array}{c} \textit{State on a} \\ \textit{given weekend} \end{array} \quad \begin{array}{c} E_1 \\ E_2 \\ E_3 \end{array} \left[\begin{array}{ccc} \frac{2}{3} & \frac{1}{6} & \frac{1}{6} \\ \frac{3}{4} & \frac{1}{12} & \frac{1}{6} \\ \frac{1}{2} & \frac{1}{6} & \frac{1}{3} \end{array} \right]$$

This is called a **stochastic matrix.** Stochastic matrices are square matrices in which each row is a probability vector.

Now, suppose it is known that on a particular Memorial Day weekend it is equally probable that the couple will be in any of three states. What is the probability of their being in Wisconsin for the 4th of July weekend (5 weekends later)? For the Labor Day weekend (14 weekends later)? If they pursue these habits through the remaining years of their life together, what is the probability that the first one to die will die in Wisconsin?

Your intuition might suggest that the answer to the last question— where they will be years hence when one of them dies—should be almost independent of their probable location on the weekend when we first happen to observe them. We will see in Section 8-2 that such suspicions are correct.

The process just described is called a **Markov chain** or a **Markov process.** To be a little more formal, a finite Markov chain or process is a system characterized by the following properties:

1. The system being observed is always in one of the states E_1, $E_2, \ldots, E_n$.
2. A step is taken (an interval of time elapses, a change occurs) after which the system may be in the state in which it started, or in any of the other states.
3. The probability of going from state E_i to state E_j in one step, designated by P_{ij}, depends only on i and j; that is, the state of the process before reaching E_i or after leaving E_j is of no consequence.

It is convenient to display as a stochastic matrix the probabilities of going from state E_i to state E_j in a single step.

$$\begin{array}{c} \textit{State after one step} \\ \begin{array}{cccc} E_1 & E_2 & \cdots & E_n \end{array} \end{array}$$

$$\begin{array}{c} \textit{Beginning} \\ \textit{state} \end{array} \quad \begin{array}{c} E_1 \\ E_2 \\ \vdots \\ E_n \end{array} \left[\begin{array}{cccc} p_{11} & p_{12} & \cdots & p_{1n} \\ p_{21} & p_{22} & \cdots & p_{2n} \\ \vdots & \vdots & & \vdots \\ p_{n1} & p_{n2} & \cdots & p_{nn} \end{array} \right]$$

This matrix is called the **transition matrix** for the Markov chain.

The usefulness of Markov chains can be suggested by describing specific transition matrices that have appeared in recent research articles. In a study to guide the purchase of multiple copies of books in a library, p_{ij} represented for books in a certain category the probability that a book checked out i times in a year would be checked out j times the following year. An analysis of zip code effectiveness in the post office uses the probability that a letter being sorted at level i (levels being a sort by state, by city, by substation, by individual carrier) will next go to level j. The same notion of an item proceeding from one area to another is used in studying the routing of items in the manufacture of specialized jewelry (class rings, company pins, fraternity pins); this time the areas are for plating, painting, etching, backing, and so on. A study of the no-fault concept for automobile insurance uses p_{ij} to represent the probability that a driver involved in i accidents over a 5-year period will be in j accidents during the next 5 years. A study of repeating criminal offenders uses the probability that a person convicted of a crime in class i (1–murder, 2–rape, 3–robbery, 4–assault, 5–burglary, 6–larceny, 7–auto theft) will subsequently be convicted of a crime in class j. Finally, we mention that many of the systems studied in queuing theory are Markov processes in which there may be i customers waiting for service in one time frame, j customers in the next. "Customers" may be airplanes waiting for clearance to land at an airport, calls coming in to a large police station dispatching center, or jobs waiting for processing in a time-sharing computer center. It is fair to say that Markov chains enter into a wide variety of applications.

Markov chains obviously involve matrices, the properties of which should be fresh in your mind. They also involve some probability theory which, though simple, is probably not so fresh. The properties that will be needed can be summarized as follows:

4. If the probability of one event is p_1 and the probability of a second event, independent of the first event, is p_2, then the probability of both events occurring is $p_1 p_2$.
5. If the probability of one event is p_1 and the probability of a second event is p_2, and if it is not possible for both events to happen, then the probability of one or the other happening is $p_1 + p_2$.
6. If the probability of an event in each of n trials is p, then the expected number of occurrences of the event in n trials is np.

8-2 | *THE POWERS OF THE TRANSITION MATRIX*

Suppose that a given Markov process has as its transition matrix the 3×3 matrix

$$P = \begin{bmatrix} p_{11} & p_{12} & p_{13} \\ p_{21} & p_{22} & p_{23} \\ p_{31} & p_{32} & p_{33} \end{bmatrix}$$

Suppose further that when the process first begins (or when we first observe it), the probability of being in state E_1 is $q_1^{(0)}$. Then after Step 1, according to Property 4 of the last section, the probability of still being in state E_1 will be $q_1^{(0)} p_{11}$. Similarly, if the probability of starting in state E_2 is $q_2^{(0)}$, then the probability of being in state E_1 after the first step will be $q_2^{(0)} p_{21}$; and finally if the probability of starting in E_3 is $q_3^{(0)}$, then the probability of being in state E_1 after the first step is $q_3^{(0)} p_{31}$. Since we must begin in one of the three states, and since we can't begin in more than one of them, we see by an extension of Property 5 to three mutually exclusive events that the probability of being in state E_1 after the first step is

(1)

$$q_1^{(0)} p_{11} + q_2^{(0)} p_{21} + q_3^{(0)} p_{31}$$

$$= [q_1^{(0)} \quad q_2^{(0)} \quad q_3^{(0)}] \begin{bmatrix} p_{11} \\ p_{21} \\ p_{31} \end{bmatrix} = \mathbf{q}^{(0)} \mathbf{P}_{\cdot 1}$$

Note that $\mathbf{q}^{(0)}$ has been used to represent the *row* vector $[q_1^{(0)} \quad q_2^{(0)} \quad q_3^{(0)}]$. We will similarly use $\mathbf{q}^{(1)}$ to represent the *row* vector which gives the probability of being in state E_i after Step 1. The reasoning that led to (1) for the probability of being in state E_1 after Step 1 also gives us the probability after Step 1 of being in state

$$
\begin{array}{ccc}
E_1 & E_2 & E_3 \\
[\mathbf{q}^{(0)} \mathbf{P}_{\cdot 1} & \mathbf{q}^{(0)} \mathbf{P}_{\cdot 2} & \mathbf{q}^{(0)} \mathbf{P}_{\cdot 3}]
\end{array}
$$

$$= [q_1^{(0)} \quad q_2^{(0)} \quad q_3^{(0)}] \begin{bmatrix} p_{11} & p_{12} & p_{13} \\ p_{21} & p_{22} & p_{23} \\ p_{31} & p_{32} & p_{33} \end{bmatrix}$$

In more succinct notation,

$$\mathbf{q}^{(1)} = \mathbf{q}^{(0)} P$$

Using $\mathbf{q}^{(2)}$ to represent the row vector that gives the probabilities of being in states E_1, E_2, or E_3 after Step 2, we see that the transition matrix applied to the probability vector after Step 1 gives

$$\mathbf{q}^{(2)} = \mathbf{q}^{(1)} P = [\mathbf{q}^{(0)} P] P = \mathbf{q}^{(0)} P^2$$

The result after k steps is obvious, and the argument given here for a Markov chain with three states applies equally well to one with n states. We have, then, the following theorem:

Suppose a Markov chain with transition matrix P is initially in one of the states $E_1, \ldots, E_n$ with the respective probabilities $\mathbf{q}^{(0)} = [q_1^{(0)} \cdots q_n^{(0)}]$. Then the probabilities of being in states $E_1, \ldots, E_n$ after k steps are given by

$$\mathbf{q}^{(k)} = \mathbf{q}^{(0)} P^k$$

If we let $p_{ij}^{(k)}$ denote the probability of going from state E_i to state E_j in k steps, then the k-step transition matrix becomes

$$P^{(k)} = \begin{bmatrix} p_{11}^{(k)} & \cdots & p_{1n}^{(k)} \\ \vdots & & \vdots \\ p_{n1}^{(k)} & \cdots & p_{nn}^{(k)} \end{bmatrix}$$

This matrix is easily determined in terms of P as an immediate consequence of Theorem A.

$$P^{(k)} = P^k$$

Proof. If we start in state E_i, then $\mathbf{q}^{(0)}$ is the vector $[0 \cdots 1 \cdots 0]$ having a 1 in the ith position as its only nonzero entry. The product $\mathbf{q}^{(0)} P^{(k)}$ gives us the ith row vector $\mathbf{P}_{i\cdot}^{(k)}$. The jth component of this vector is the entry of $\mathbf{q}^{(k)}$ which tells us the probability of being in state E_j after the kth step. □

BACK TO OUR ILLUSTRATION

In Section 8-1 we considered a couple that spent weekends in states E_1, E_2, and E_3, following a Markov process having transition matrix

$$P = \begin{bmatrix} \frac{2}{3} & \frac{1}{6} & \frac{1}{6} \\ \frac{3}{4} & \frac{1}{12} & \frac{1}{6} \\ \frac{1}{2} & \frac{1}{6} & \frac{1}{3} \end{bmatrix}$$

We were told that on a particular Memorial Day weekend, it was equally likely that they were in any one of the states, and we were asked to give probabilities that they were in E_3 after various steps (weekends).

The initial probability vector, corresponding to the fact that any state is equally likely, is $\mathbf{q}^{(0)} = [\frac{1}{3} \quad \frac{1}{3} \quad \frac{1}{3}]$. After one weekend, the probabilities are

$$[\tfrac{1}{3} \quad \tfrac{1}{3} \quad \tfrac{1}{3}] \begin{bmatrix} \frac{2}{3} & \frac{1}{6} & \frac{1}{6} \\ \frac{3}{4} & \frac{1}{12} & \frac{1}{6} \\ \frac{1}{2} & \frac{1}{6} & \frac{1}{3} \end{bmatrix} = [\tfrac{23}{36} \quad \tfrac{5}{36} \quad \tfrac{8}{36}] = \mathbf{q}^{(1)}$$

After two weekends, the probabilities are

$$[\tfrac{23}{36} \quad \tfrac{5}{36} \quad \tfrac{8}{36}] \begin{bmatrix} \tfrac{2}{3} & \tfrac{1}{6} & \tfrac{1}{6} \\ \tfrac{3}{4} & \tfrac{1}{12} & \tfrac{1}{6} \\ \tfrac{1}{2} & \tfrac{1}{6} & \tfrac{1}{3} \end{bmatrix} = [\tfrac{277}{432} \quad \tfrac{67}{432} \quad \tfrac{88}{432}] = \mathbf{q}^{(2)}$$

Expressed in decimal form, the probability vectors for the first five weekends (up through the 4th of July) are

(2)
$$\mathbf{q}^{(1)} = \mathbf{q}^{(0)}P = [0.638889 \quad 0.138889 \quad 0.222222]$$
$$\mathbf{q}^{(2)} = \mathbf{q}^{(0)}P^2 = [0.641204 \quad 0.155092 \quad 0.203704]$$
$$\mathbf{q}^{(3)} = \mathbf{q}^{(0)}P^3 = [0.645640 \quad 0.153742 \quad 0.200617]$$
$$\mathbf{q}^{(4)} = \mathbf{q}^{(0)}P^4 = [0.646042 \quad 0.153855 \quad 0.200102]$$
$$\mathbf{q}^{(5)} = \mathbf{q}^{(0)}P^5 = [0.646137 \quad 0.153845 \quad 0.200017]$$

To find these vectors for 14 weekends by direct computation would give *us* a labor day. On the other hand, the apparent convergence of just the first five vectors does suggest that there might be an answer to the question of what happens to $\mathbf{q}^{(k)}$ as k becomes large.

A more subtle approach suggests itself if we recall from Theorem D of Section 6-3 that the computing of powers of P can be greatly simplified if P is diagonalizable. Accordingly, we solve the characteristic equation

$$\begin{vmatrix} \tfrac{2}{3} - r & \tfrac{1}{6} & \tfrac{1}{6} \\ \tfrac{3}{4} & \tfrac{1}{12} - r & \tfrac{1}{6} \\ \tfrac{1}{2} & \tfrac{1}{6} & \tfrac{1}{3} - r \end{vmatrix} = 0$$

After some algebra, we find the characteristic roots (eigenvalues) to be $r_1 = 1, r_2 = \tfrac{1}{6}, r_3 = -\tfrac{1}{12}$. Since they are all distinct, we know that there exists an invertible matrix Q such that

$$Q^{-1}PQ = \begin{bmatrix} 1 & 0 & 0 \\ 0 & \tfrac{1}{6} & 0 \\ 0 & 0 & -\tfrac{1}{12} \end{bmatrix}$$

It follows (from Theorem D of Section 6-3) that

$$P^k = Q \begin{bmatrix} 1 & 0 & 0 \\ 0 & (\tfrac{1}{6})^k & 0 \\ 0 & 0 & (-\tfrac{1}{12})^k \end{bmatrix} Q^{-1}$$

and since

$$\lim_{k \to \infty} (\tfrac{1}{6})^k = \lim_{k \to \infty} (-\tfrac{1}{12})^k = 0$$

we see that

$$\lim_{k \to \infty} P^k = QDQ^{-1} \quad \text{where} \quad D = \begin{bmatrix} 1 & 0 & 0 \\ 0 & 0 & 0 \\ 0 & 0 & 0 \end{bmatrix}$$

The eigenvectors corresponding to the eigenvalues $r_1 = 1$, $r_2 = \frac{1}{6}$, and $r_3 = -\frac{1}{12}$ can be found in the usual way; they are $\mathbf{v}_1 = \begin{bmatrix} 1 & 1 & 1 \end{bmatrix}$, $\mathbf{v}_2 = \begin{bmatrix} 1 & 1 & -4 \end{bmatrix}$, and $\mathbf{v}_3 = \begin{bmatrix} 2 & -11 & 2 \end{bmatrix}$, respectively. Using these as column vectors to form Q and then finding Q^{-1}, we get a matrix W given by

$$W = QDQ^{-1} = \begin{bmatrix} 1 & 1 & 2 \\ 1 & 1 & -11 \\ 1 & -4 & 2 \end{bmatrix} \begin{bmatrix} 1 & 0 & 0 \\ 0 & 0 & 0 \\ 0 & 0 & 0 \end{bmatrix} \begin{bmatrix} \frac{42}{65} & \frac{10}{65} & \frac{13}{65} \\ \frac{13}{65} & 0 & -\frac{13}{65} \\ \frac{5}{65} & -\frac{5}{65} & 0 \end{bmatrix}$$

$$W = \begin{bmatrix} \frac{42}{65} & \frac{10}{65} & \frac{13}{65} \\ \frac{42}{65} & \frac{10}{65} & \frac{13}{65} \\ \frac{42}{65} & \frac{10}{65} & \frac{13}{65} \end{bmatrix}$$

The special form of matrix W, in which the same probability vector $\mathbf{w} = \begin{bmatrix} \frac{42}{65} & \frac{10}{65} & \frac{13}{65} \end{bmatrix}$ appears in all three rows, illustrates a theorem we shall prove in the next section. Beyond that, it seems to substantiate the intuition of those who felt that the probability of being in a particular state after a long time should be almost independent of the probabilities given for the weekend on which we first observed this couple. For let $\mathbf{q}^{(0)}$ be any initial probability vector, then

$$\mathbf{q}^{(0)}W = \begin{bmatrix} q_1^{(0)}\frac{42}{65} + q_2^{(0)}\frac{42}{65} + q_3^{(0)}\frac{42}{65} & \cdots & q_1^{(0)}\frac{13}{65} + q_2^{(0)}\frac{13}{65} + q_3^{(0)}\frac{13}{65} \end{bmatrix}$$

and since $q_1^{(0)} + q_2^{(0)} + q_3^{(0)} = 1$,

$$\mathbf{q}^{(0)}W = \mathbf{w} \qquad\qquad (3)$$

No matter what the initial probabilities, the probabilities of being in the respective states after a long time tend toward

$$\mathbf{w} = \begin{bmatrix} \frac{42}{65} & \frac{10}{65} & \frac{13}{65} \end{bmatrix} = \begin{bmatrix} 0.646154 & 0.153846 & 0.200000 \end{bmatrix}$$

We saw this trend developing in the sequence of five steps worked out in (2).

EIGENVALUES OF STOCHASTIC MATRICES

One of the eigenvalues found in our illustrative problem was 1, and the corresponding eigenvector had all entries equal to 1. Note that any stochastic matrix must have at least one eigenvalue of 1 since it will always be true that

$$\begin{bmatrix} p_{11} & \cdots & p_{1n} \\ \vdots & & \vdots \\ p_{n1} & \cdots & p_{nn} \end{bmatrix} \begin{bmatrix} 1 \\ \vdots \\ 1 \end{bmatrix} = \begin{bmatrix} p_{11} + \cdots + p_{1n} \\ \vdots \\ p_{n1} + \cdots + p_{nn} \end{bmatrix} = \begin{bmatrix} 1 \\ \vdots \\ 1 \end{bmatrix}$$

The special form of W in our illustrative example above hinges on two key facts: (1) all the eigenvalues were, in absolute value, less than or equal to 1; and (2) the eigenvalue of 1 appeared with multiplicity one. In Problem 17 at the end of this section, we outline a simple argument which shows that any eigenvalue r of a stochastic matrix satisfies $|r| \leq 1$. Thus, the success of the procedure depends on having exactly one eigenvalue of 1; that is, $(x - 1)$ appears only once as a factor of the characteristic equation. This is not true for all stochastic matrices, but it does happen to be true for stochastic matrices in which all entries are positive. (See Karlin, S. 1975. *First course in stochastic processes.* 2d ed. New York: Academic Press.) Although we shall not pursue this further, it is one explanation for our interest in studying in Section 8-3 those stochastic matrices having all positive (no zero) entries.

PROBLEM SET 8-2

In Problems 1–6, part (c) asks for the probable state after a long time. The procedure illustrated in this section—diagonalizing the transition matrix and then raising to powers—can be used. We will see in the next section that there are more direct methods.

1. A woman who intends to jog daily doesn't quite make it. If she jogs one day, she is 75% likely to jog the next, but if she misses a day, she is just as likely as not to miss the next day also.
   (a) Write the transition matrix for this Markov process of two states s_1 (jogged), s_2 (didn't jog).
   (b) If she jogs on Monday, what is the probability that she will jog on Friday?
   (c) As time marches on (we are tempted to say over the long run), what is the probability of her jogging on a particular day?

2. The Alumni Office notes that if an alumnus is a donor in a given year, there is a 70% chance that this person will be a donor the next year. A nondonor in a given year is only 20% likely, however, to be a donor the following year.
   (a) Write the transition matrix for this Markov process of two states s_1 (donates), s_2 (doesn't donate).
   (b) If a new graduate is a 50-50 possibility to be a donor in the first year as an alumnus, what are the probabilities of being a donor in one's fourth year as an alumnus?
   (c) What is the probability that an alumnus of many years standing will be a donor?

3. A certain young woman always buys a Ford or a Chevrolet when she buys a new car. If she owns a Ford, her next car is never a Ford, but if she owns a Chevrolet, she is equally likely to buy either a Ford or a Chevrolet.
   (a) Find the transition matrix for this two-state Markov process.
   (b) If her first car was a Chevrolet, what is the probability that her fourth car was a Ford?
   (c) What is the probability that she will own a Chevrolet when she retires (a long time hence)?

4. Two noncommercial FM stations compete for that audience in their area which is attracted to programming focused on classical music and public affairs. If a listener is tuned to one of the stations, there is a 75% chance

that he will not change his dial to the other station at the half-hour station breaks.

(a) Find the transition matrix for this two-station Markov process.

(b) If the listener is equally likely to turn on s_1 or s_2 at 8:00 A.M., what is the probability that he will turn to s_2 at 10:00 A.M.?

(c) If he listens all day to one of the two stations, what is the probability that he will be listening at bedtime to the station first tuned in that morning?

5. Jane, Dick, and Spot are playing with a frisbee. Dick always throws to Jane, but she is equally likely to throw to Dick or the dog. The dog brings it back to Dick about twice as often as to Jane.

(a) Write the transition matrix for this three-state Markov process.

(b) What is the probability of Spot receiving the third toss if Dick has the frisbee to start with?

(c) In the long run, what is the probability that this Frisbee will go to the dog?

6. A farmer plants one of three crops, A, B, or C, in a certain field, never planting the same crop two years in succession. He prefers crop C and plants it twice as often as the others, but he is equally likely to choose A or B when not choosing C.

(a) Write the transition matrix for this three-state Markov process.

(b) What is the probability that crop C is planted in the fourth year if it is planted in the first year?

(c) If you visit this farmer many years hence, what is the probability that you will see crop C in his field?

7. Suppose the peripatetic couple discussed earlier in the text spent the Memorial Day weekend in Wisconsin. What then is the probability vector that gives their likely state on the Fourth of July? Compare this probability vector with the vector $\mathbf{q}^{(5)}$ obtained in (2).

8. Answer Problem 7 if the Memorial Day weekend is spent in Illinois.

9. A security guard moves between various stations on a campus, remaining about 15 minutes, then moving on to an adjacent station (Figure 8-1). To

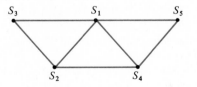

Figure 8-1

make his appearance a little less predictable, it is his practice to pick the adjacent station in a random way (meaning he is equally likely to move to any adjacent station).

(a) Considering this as a Markov process with five states, form the transition matrix.

(b) If the guard begins at s_1, find the vector giving his probable locations an hour later (after four moves).

10. Suppose we have two urns (antique vases now found mostly in mathematics books) each containing three marbles. Three of the marbles are red,

three are white, and the system consists of four states corresponding to the number of red marbles in urn A. A move consists of drawing a marble from each vase and replacing each in the vase opposite from which it was drawn.

(a) Write the transition matrix.

(b) Describe the probable state of urn A after four moves if it initially contained three red marbles.

Problems 11–16 deal with the concept of a random walk. Variations of the random walk are useful in modeling a variety of phenomena. For further discussion, see Kemeny, J., et al., Finite mathematical structures, Prentice-Hall, 1959, pp. 423–434.

11. A pointer indicates one of the positions s_1 through s_6 (Figure 8-2, which may, for example, be thought of as a radio dial with six stations that can be tuned in). If at s_1, it always moves at the next step to s_2; if at s_6, it

Figure 8-2

always moves at the next step to s_5. If at any other s_i, it moves to s_{i-1} (to the left) with probability $\frac{3}{5}$, or to s_{i+1} (to the right) with probability $\frac{2}{5}$.

(a) Write the transition matrix for this six-state Markov process.

(b) If the marker starts at s_4, use a probability vector to indicate the probability of it being in each of the states after three steps.

12. Answer part (b) of Problem 11 for the case in which the marker starts at s_5.

13. Modify Problem 11 so that if the marker reaches s_1 or s_6, the process terminates. Then answer the same questions.

14. For the case in which the marker starts at s_5, answer part (b) of Problem 11 if the transition matrix has been modified as in Problem 13.

15. A player has \$3 and plays a game of chance with a gambler who starts with \$2. She bets \$1 at a time and loses $\frac{3}{5}$ of the time. The game continues until one of the participants has all the money. Show that this problem is equivalent to Problem 13.

16. Show that if the player of Problem 15 begins with \$4 and the gambler with \$1, analysis of the game is equivalent to Problem 14.

17. Follow the outline provided to prove that if r is an eigenvalue of a stochastic matrix P, then $|r| \leq 1$.

(a) Let $\mathbf{x}$ be the eigenvector belonging to r, and let

$$m = \max\{|x_1|, \ldots, |x_n|\} = |x_k|$$

Consider row k of the matrix equation $r\mathbf{x} = P\mathbf{x}$ to show that

$$|r|m = |p_{k1}x_1 + \cdots + p_{kn}x_n|$$

(b) Use the triangle inequality, extended to n terms, to show that $|r|m \leq 1m$, hence that $|r| \leq 1$.

18. Let a and b be chosen, both between 0 and 1 inclusive, in such a way that the determinant of the stochastic matrix

$$P = \begin{bmatrix} a & 1-a \\ 1-b & b \end{bmatrix}$$

is nonzero.

(a) Find the eigenvalues. Use det $P \neq 0$ to show that both eigenvalues r_1 and r_2 satisfy $|r_i| \leq 1$.

(b) Find the eigenvectors $\mathbf{v}_1$ and $\mathbf{v}_2$ corresponding to r_1 and r_2. Use them to form Q.

(c) Express P in the form $P = QDQ^{-1}$ where D is the diagonal matrix equivalent to P.

(d) Find $\lim\limits_{k \to \infty} P^k$.

REGULAR MARKOV CHAINS   8-3

We have already seen that there is great interest in the powers of a transition matrix. In this section we focus attention on a class of matrices that have particularly nice powers. Let us call a stochastic matrix P a **regular matrix** if all the entries of P^k are positive for some k. Thus, the matrix

$$P = \begin{bmatrix} \frac{1}{2} & 0 & \frac{1}{2} \\ \frac{1}{3} & \frac{1}{3} & \frac{1}{3} \\ \frac{1}{4} & \frac{1}{2} & \frac{1}{4} \end{bmatrix}$$

turns out to be regular because, although P itself does not contain all positive entries,

$$P^2 = \begin{bmatrix} \frac{3}{8} & \frac{2}{8} & \frac{3}{8} \\ \frac{13}{36} & \frac{10}{36} & \frac{13}{36} \\ \frac{17}{48} & \frac{14}{48} & \frac{17}{48} \end{bmatrix}$$

Predictably, a Markov chain is said to be a **regular Markov chain** if and only if its transition matrix is regular.

Consider the $n \times n$ regular matrix P, and recall that $p_{ij}^{(k)}$ is used to denote the entry in row i, column j of the k-step transition matrix $P^{(k)} = P^k$. Since

$$P^{(k+1)} = PP^k = \begin{bmatrix} \cdots\cdots\cdots\cdots\cdots \\ p_{i1} & p_{i2} & \cdots & p_{in} \\ \cdots\cdots\cdots\cdots\cdots \end{bmatrix} \begin{bmatrix} \cdots & p_{ij}^{(k)} & \cdots \\ \vdots & \vdots & \vdots \\ \cdots & p_{nj}^{(k)} & \cdots \end{bmatrix}$$

it follows that, for any k,

$$p_{ij}^{(k+1)} = p_{i1}p_{1j}^{(k)} + p_{i2}p_{2j}^{(k)} + \cdots + p_{in}p_{nj}^{(k)} \tag{1}$$

We now focus our attention on some fixed column j, and we let m_k and M_k be the smallest and the largest values respectively in column j of the

k-step transition matrix $P^{(k)}$. Since some power of P has all positive entries, we will begin our argument at that point. Thus, without loss of generality, we assume that the regular matrix P with which we begin has as its smallest entry some $\varepsilon > 0$. For later reference, we note here that $\varepsilon \leq 1/n$, so except for the trivial case of a 1×1 transition matrix, $\varepsilon \leq \frac{1}{2}$.

Now arbitrarily choose two rows, say r and s. Of course $m_k = p_{mj}^{(k)}$ and $M_k = P_{Mj}^{(k)}$ for some entries in column j, and by setting $i = r$ and then $i = s$ in (1), we can write

$$p_{rj}^{(k+1)} = p_{rm}p_{mj}^{(k)} + \sum_{v \neq m} p_{rv}p_{vj}^{(k)}$$

$$p_{sj}^{(k+1)} = p_{sM}p_{Mj}^{(k)} + \sum_{v \neq M} p_{sv}p_{vj}^{(k)}$$

It follows, since $p_{vj}^{(k)} \leq M_k$ and $p_{vj}^{(k)} \geq m_k$ for all v, and by our choice of $p_{mj}^{(k)}$ and $p_{Mj}^{(k)}$, that

$$p_{rj}^{(k+1)} \leq p_{rm}m_k + \sum_{v \neq m} p_{rv}M_k$$

$$p_{sj}^{(k+1)} \geq p_{sM}M_k + \sum_{v \neq M} p_{sv}m_k$$

Note that

$$\sum_{v \neq m} p_{rv}M_k = M_k(p_{r1} + \cdots + p_{rn})$$

where p_{rm} is missing from the sum in parentheses. Since entries in the row vectors of the transition matrix P add up to 1, the sum in the parentheses is $1 - p_{rm}$.

$$p_{rj}^{(k+1)} \leq p_{rm}m_k + (1 - p_{rm})M_k = M_k - p_{rm}(M_k - m_k)$$

and similarly,

$$p_{sj}^{(k+1)} \geq p_{sM}M_k + (1 - p_{sM})m_k = m_k + p_{sM}(M_k - m_k)$$

Of course, $p_{sM} \geq \varepsilon$ and $p_{rm} \geq \varepsilon$, so

$$p_{rj}^{(k+1)} \leq M_k - \varepsilon(M_k - m_k)$$
$$p_{sj}^{(k+1)} \geq m_k + \varepsilon(M_k - m_k)$$

Multiplying the second inequality by -1 and adding gives

(2) $$p_{rj}^{(k+1)} - p_{sj}^{(k+1)} \leq M_k - m_k - 2\varepsilon(M_k - m_k) = (1 - 2\varepsilon)(M_k - m_k)$$

We began by fixing k and j; then r and s were chosen arbitrarily. If (with the same fixed k and j) we choose r so that $p_{rj}^{(k+1)} = M_{k+1}$ (the maximum entry in column j of $P^{(k+1)}$), and choose s so that $P^{(k+1)}_{sj} = m_{k+1}$ (the minimum entry in the same column), then (2) says

$$M_{k+1} - m_{k+1} \le (1 - 2\varepsilon)(M_k - m_k)$$

Choosing $k = 1, 2, \ldots,$

$$M_2 - m_2 \le (1 - 2\varepsilon)(M_1 - m_1)$$
$$M_3 - m_3 \le (1 - 2\varepsilon)(M_2 - m_2) \le (1 - 2\varepsilon)^2 (M_1 - m_1)$$
$$\vdots$$
$$M_k - m_k \le (1 - 2\varepsilon)^{k-1}(M_1 - m_1)$$

Previous observations about ε guarantee that $0 \le 1 - 2\varepsilon < 1$. Thus,

$$\lim_{k \to \infty} (M_k - m_k) = 0$$

In words, we have proved that in an arbitrary column j, the difference between the maximum entry and the minimum entry of P^k goes to zero as k increases. All the entries in the jth column must approach some fixed number, say w_j. We have proved a theorem which shows that the long-term experience of the peripatetic couple discussed in the last section was not accidental.

Let P be a regular matrix. Then   *THEOREM A*

(i) $\displaystyle \lim_{k \to \infty} P^k = W$

(ii) *W consists of n identical row vectors* **w**.

The matrix W, being the limit of a product of stochastic matrices, is again stochastic. Thus, the vector **w** is a probability vector, a fact that we use in proving our next result.

Let P be a regular matrix with **w** *and W as described in Theorem A. Then*   *THEOREM B*

(i) *If* **q** *is a probability row vector,* $\displaystyle \lim_{k \to \infty} \mathbf{q}P^k = \mathbf{w}$.

(ii) $PW = WP = W$

(iii) **w** *is the unique row vector such that* $\mathbf{w}P = \mathbf{w}$.

Proof. By Theorem A,

$$\lim_{k \to \infty} \mathbf{q}P = \mathbf{q}W = \mathbf{w}$$

The last equality follows directly from the fact that **q** is a probability vector and is illustrated in equation (3) of Section 8-2.

To prove (ii), simply note that

$$PW = P \lim_{k \to \infty} P^k = \lim_{k \to \infty} P^{k+1} = (\lim_{k \to \infty} P^k) P = WP$$

and the middle limit is W itself. From $WP = W$, we get

$$\begin{bmatrix} \mathbf{w} \\ \vdots \\ \mathbf{w} \end{bmatrix} P = \begin{bmatrix} \mathbf{w} \\ \vdots \\ \mathbf{w} \end{bmatrix}$$

The first row of this matrix product gives $\mathbf{w}P = \mathbf{w}$.

Finally, to show that $\mathbf{w}$ is the unique vector satisfying $\mathbf{w}P = \mathbf{w}$, suppose there is a second such vector $\mathbf{u}$; then $\mathbf{u}P = \mathbf{u}$. By part (i),

$$\lim_{k \to \infty} \mathbf{u}P^k = \mathbf{w}$$

But

$$\mathbf{u}P^k = (\mathbf{u}P)P^{k-1} = \mathbf{u}P^{k-1} = (\mathbf{u}P)P^{k-2} = \mathbf{u}P^{k-2} = \cdots = \mathbf{u}$$

Hence, $\mathbf{u} = \mathbf{w}$. $\square$

The strength and beauty of Theorem B deserves comment. Part (i) says that for the peripatetic couple whose (regular) transition matrix was given in Section 8-2, the probability vector for the weekend in which we first observed them is of no special consequence when we ask for the long-term probabilities. We have previously noted that this agrees with intuition.

Part (iii) shows us the most direct path to find $\mathbf{w}$. If we write $\mathbf{w}$ as a column vector, then (iii) takes the form $\mathbf{w}^t P = \mathbf{w}^t$. Taking transposes, $P^t \mathbf{w} = \mathbf{w}$, from which it is apparent that P^t has an eigenvalue of 1, and that $\mathbf{w}$ is the eigenvector that belongs to 1.

Example A

Use Theorem B to find the long-term probability vector $\mathbf{w}$ for the peripatetic couple.

Certain that 1 is an eigenvalue of the transpose of the given transition matrix

$$P = \begin{bmatrix} \frac{2}{3} & \frac{1}{6} & \frac{1}{6} \\ \frac{3}{4} & \frac{1}{12} & \frac{1}{6} \\ \frac{1}{2} & \frac{1}{6} & \frac{1}{3} \end{bmatrix}$$

and that the desired vector $\mathbf{w}$ is the corresponding eigenvector, we simply solve in the usual way the system $(P^t - rI)\mathbf{w} = \mathbf{0}$ where $r = 1$. This gives

(3)

$$[P^t - I]\mathbf{w} = \begin{bmatrix} -\frac{1}{3} & \frac{3}{4} & \frac{1}{2} \\ \frac{1}{6} & -\frac{11}{12} & \frac{1}{6} \\ \frac{1}{6} & \frac{1}{6} & -\frac{2}{3} \end{bmatrix} \begin{bmatrix} w_1 \\ w_2 \\ w_3 \end{bmatrix} = \begin{bmatrix} 0 \\ 0 \\ 0 \end{bmatrix}$$

It is short work (at least shorter than what we went through at the end of Section 8-2) to discover that $\mathbf{w}^t = [\frac{42}{65} \quad \frac{10}{65} \quad \frac{13}{65}]$. $\square$

Given our inclination to avoid fractions, it is possible that the eigenvector obtained as a solution to (3) would be [42 10 13]. The required **w** is obtained by dividing this answer by $65 = 42 + 10 + 13$ so as to obtain a probability vector. This is permissible because any scalar multiple of an eigenvector is also an eigenvector.

Find $\lim_{k \to \infty} P^k$ for the stochastic matrix

Example B

$$P = \begin{bmatrix} \frac{1}{2} & 0 & \frac{1}{2} \\ \frac{1}{3} & \frac{1}{3} & \frac{1}{3} \\ \frac{1}{4} & \frac{1}{2} & \frac{1}{4} \end{bmatrix}$$

which was shown at the beginning of this section to be regular.

We begin by finding **w,** the probability vector solution to

$$[P^t - I]\mathbf{w} = \begin{bmatrix} -\frac{1}{2} & \frac{1}{3} & \frac{1}{4} \\ 0 & -\frac{2}{3} & \frac{1}{2} \\ \frac{1}{2} & \frac{1}{3} & -\frac{3}{4} \end{bmatrix} \begin{bmatrix} w_1 \\ w_2 \\ w_3 \end{bmatrix} = \begin{bmatrix} 0 \\ 0 \\ 0 \end{bmatrix}$$

The equivalent system with the coefficient matrix in row echelon form is

$$\begin{bmatrix} 1 & 0 & -1 \\ 0 & 1 & -\frac{3}{4} \\ 0 & 0 & 0 \end{bmatrix} \begin{bmatrix} w_1 \\ w_2 \\ w_3 \end{bmatrix} = \begin{bmatrix} 0 \\ 0 \\ 0 \end{bmatrix}$$

from which we obtain in succession the following forms of a solution:

$$[w_3 \quad \tfrac{3}{4}w_3 \quad w_3] = \tfrac{1}{4}w_3[4 \quad 3 \quad 4] = \tfrac{11}{4}w_3[\tfrac{4}{11} \quad \tfrac{3}{11} \quad \tfrac{4}{11}]$$

$$\lim_{k \to \infty} P^k = W$$

where W has three identical row vectors of $[\tfrac{4}{11} \quad \tfrac{3}{11} \quad \tfrac{4}{11}] = [0.3636$ 0.2727   0.3636]. $\square$

We are finished with Example B in the sense of having found the desired limit. It is interesting (the author thinks it is interesting) to note, since we have already computed P^2, that

$$P^2 = \begin{bmatrix} \frac{3}{8} & \frac{2}{8} & \frac{3}{8} \\ \frac{13}{36} & \frac{10}{36} & \frac{13}{36} \\ \frac{17}{48} & \frac{14}{48} & \frac{17}{48} \end{bmatrix} = \begin{bmatrix} 0.3750 & 0.2500 & 0.3750 \\ 0.3611 & 0.2778 & 0.3611 \\ 0.3542 & 0.2917 & 0.3542 \end{bmatrix}$$

Another multiplication yields

$$P^3 = \begin{bmatrix} \frac{630}{1728} & \frac{468}{1728} & \frac{630}{1728} \\ \frac{628}{1728} & \frac{472}{1728} & \frac{628}{1728} \\ \frac{627}{1728} & \frac{474}{1728} & \frac{627}{1728} \end{bmatrix} = \begin{bmatrix} 0.3646 & 0.2708 & 0.3646 \\ 0.3634 & 0.2731 & 0.3634 \\ 0.3628 & 0.2743 & 0.3628 \end{bmatrix}$$

Even the author is not interested in going beyond this!

1–6. In Problem Set 8-2, the first six problems all had a part (c) which can be easily answered using the methods of this section. Do so.

7. Which of the following matrices are regular?

(a) $\begin{bmatrix} \frac{1}{2} & \frac{1}{8} & \frac{1}{4} & \frac{1}{8} \\ 0 & \frac{1}{2} & 0 & \frac{1}{2} \\ \frac{1}{4} & \frac{1}{2} & 0 & \frac{1}{4} \\ 0 & \frac{1}{3} & 0 & \frac{2}{3} \end{bmatrix}$
(b) $\begin{bmatrix} \frac{1}{3} & 0 & \frac{2}{3} & 0 \\ 1 & 0 & 0 & 0 \\ 0 & \frac{1}{2} & 0 & \frac{1}{2} \\ 0 & 0 & 1 & 0 \end{bmatrix}$

8. Which of the following matrices are regular?

(a) $\begin{bmatrix} 0 & \frac{1}{2} & \frac{1}{2} & 0 \\ 1 & 0 & 0 & 0 \\ 0 & 0 & \frac{1}{3} & \frac{2}{3} \\ \frac{1}{2} & \frac{1}{2} & 0 & 0 \end{bmatrix}$
(b) $\begin{bmatrix} \frac{3}{4} & 0 & \frac{1}{4} & 0 \\ \frac{1}{8} & \frac{1}{4} & \frac{1}{2} & \frac{1}{8} \\ \frac{3}{8} & 0 & \frac{5}{8} & 0 \\ \frac{1}{3} & \frac{1}{6} & \frac{1}{6} & \frac{1}{3} \end{bmatrix}$

9. If the campus guard whose routine is described in Problem 9 of the previous problem set is close to the end of an eight-hour tour of duty, what is the probability of finding him at s_1?

10. If the transfer of marbles between urns as described in Problem 10 of the previous problem set is continued indefinitely, what will be the probability over a long period of time of finding two red marbles in urn A?

11. Beginning with the regular stochastic matrix

$$P = \begin{bmatrix} \frac{1}{2} & \frac{1}{4} & \frac{1}{4} \\ \frac{1}{8} & \frac{1}{2} & \frac{3}{8} \\ \frac{1}{4} & \frac{1}{4} & \frac{1}{2} \end{bmatrix}$$

note that all the terms are positive and that, as determined in our proof of Theorem B, $\varepsilon = \frac{1}{8}$. Choosing column $j = 2$, observe that $M_1 = \frac{1}{2}$, $m_1 = \frac{1}{4}$. Find P^3, M_3, m_3, and verify that

$$M_3 - m_3 \le (1 - 2\varepsilon)^2 (M_1 - m_1)$$

12. Follow the instructions for Problem 11 using column $j = 3$.

13. Find $\lim_{k \to \infty} \begin{bmatrix} 0 & 1 & 0 \\ \frac{2}{3} & 0 & \frac{1}{3} \\ \frac{2}{3} & \frac{1}{3} & 0 \end{bmatrix}^k$

14. Find $\lim_{k \to \infty} \begin{bmatrix} 0 & \frac{1}{2} & \frac{1}{2} \\ \frac{1}{2} & \frac{1}{2} & 0 \\ 0 & 1 & 0 \end{bmatrix}^k$

8-4   ABSORBING MARKOV CHAINS

In the problem of the peripatetic couple which has now become a regular problem for us, we assumed that the couple would be granted a good many years together in their earthly pilgrimage. Asking for the probabilities of their being in a given state when one of them died thus amounted to asking for $\lim_{k \to \infty} P^k$, a question we have been able to answer since this is, as we said, a *regular* problem.

Suppose, however, that we wish to consider the quesiton of death if we know that the couple is in poor health. In this case we might think not of states of the union, but of states of health: fair, poor, and rapidly fading. To these we might add E_0, a kind of lying in state from which one never returns. If all probabilities remain as they were before, except that we allow a probability of $\frac{1}{12}$ for going from the state E_3 of rapidly failing health to the state of death E_0 (and make the appropriate decrease in the probability of going from E_3 to E_3), the transition matrix takes the form

$$
\begin{array}{c}
\textit{Probability of spending} \\
\textit{the next weekend} \\
\textit{in state}
\end{array}
\qquad (1)
$$

$$
\begin{array}{cc}
 & \begin{array}{cccc} E_0 & E_1 & E_2 & E_3 \end{array} \\
\begin{array}{cc} & E_0 \\ \textit{State on a} & E_1 \\ \textit{given weekend} & E_2 \\ & E_3 \end{array} &
\left[\begin{array}{cccc}
1 & 0 & 0 & 0 \\
0 & \frac{2}{3} & \frac{1}{6} & \frac{1}{6} \\
0 & \frac{3}{4} & \frac{1}{12} & \frac{1}{6} \\
\frac{1}{12} & \frac{1}{2} & \frac{1}{6} & \frac{1}{4}
\end{array}\right] = P
\end{array}
$$

In any Markov chain, a state is said to be an **absorbing state** if, once it is entered, it cannot be left. The chain itself is said to be an **absorbing chain** if it has at least one absorbing state *and* if from every nonabsorbing state it is possible to reach some absorbing state. This latter condition should be noted. With this requirement in mind, one might wonder whether, for instance, the matrix (1) defines an absorbing chain. For although the state E_0 is clearly absorbing, it is not clear whether one can get to E_0 from either E_1 or E_2. The 2-step transition matrix is

$$
P^2 = \left[\begin{array}{cccc}
1 & 0 & 0 & 0 \\
\frac{1}{72} & \frac{141}{216} & \frac{11}{72} & \frac{13}{72} \\
\frac{1}{72} & \frac{93}{144} & \frac{23}{144} & \frac{13}{72} \\
\frac{5}{48} & \frac{7}{12} & \frac{10}{72} & \frac{25}{144}
\end{array}\right]
$$

however, so the chain is indeed absorbing. In two steps, the couple may from any state pass (away) into E_0.

If E_i is an absorbing state, then $p_{ii} = 1$ and all other entries in row i are 0. If a chain has m absorbing states, it is common to choose notation that designates these absorbing states as $E_1, \ldots, E_m$. Then the transition matrix will be in what is called the **canonical form.**

(2)

$$
\begin{array}{cc}
 & \begin{matrix} E_1 & \cdots & E_m & E_{m+1} & \cdots & E_n \end{matrix} \\
\begin{matrix} E_1 \\ \vdots \\ E_m \\ \\ E_{m+1} \\ \vdots \\ E_n \end{matrix} &
\left[\begin{array}{c|c} I & Q \\ \hline R & S \end{array}\right]
\end{array}
$$

I is the $m \times m$ identity matrix, Q is an $m \times (n - m)$ matrix consisting entirely of 0 entries, R is an $(n - m) \times m$ matrix giving the probabilities of going from a nonabsorbing to an absorbing state, and S is an $(n - m) \times (n - m)$ matrix giving the probabilities of going from one nonabsorbing state to another.

To have an example to look at now and again later, a Markov chain with two absorbing states E_1 and E_2 and three nonabsorbing states E_3, E_4 and E_5 would have in canonical form the transition matrix

(3)

$$
P = \left[\begin{array}{cc|ccc}
1 & 0 & 0 & 0 & 0 \\
0 & 1 & 0 & 0 & 0 \\
\hline
P_{31} & P_{32} & P_{33} & P_{34} & P_{35} \\
P_{41} & P_{42} & P_{43} & P_{44} & P_{45} \\
P_{51} & P_{52} & P_{53} & P_{54} & P_{55}
\end{array}\right]
$$

You may use the matrix (or review Problem 33 of Problem Set 3-2 that deals with partitioned matrices) to verify that for the canonical form (2),

$$
P^2 = \left[\begin{array}{c|c} I & Q \\ \hline R & S \end{array}\right]\left[\begin{array}{c|c} I & Q \\ \hline R & S \end{array}\right] = \left[\begin{array}{c|c} I & Q \\ \hline R + SR & S^2 \end{array}\right]
$$

Indeed, for any power k,

(4)

$$
P^k = \left[\begin{array}{c|c} I & Q \\ \hline A_k & S^k \end{array}\right] \quad \text{where} \quad A_k = R + SR + \cdots + S^{k-1}R
$$

Note that all entries in S and R, like those of P, are nonnegative. Thus, since $A_k = A_{k-1} + S^{k-1}R$, the entries in A_k can never be smaller than entries in the corresponding positions of A_{k-1}. The proof of Theorem A below makes use of this fact.

The first thing to know about an absorbing Markov chain is that, no matter what the initial state, the probability of being absorbed in some absorbing state as the process goes on is 1.

$(I - S)^{-1}$ exists, we can multiply both sides of (5) on the left by this expression. We get

$$I + S + \cdots + S^{k-1} = (I - S)^{-1}(I - S^k)$$

As k increases, the factor $I - S^k$ gets closer and closer to I, meaning that the sum on the left gets closer and closer to $N = (I - S)^{-1}$. Understanding an infinite series of matrices, just as we do an infinite sum of numbers, to be the limiting value of the sequence of partial sums, and defining $S^0 = I$, we have

$$N = \sum_{k=0}^{\infty} S^k \tag{6}$$

The $(n - m) \times (n - m)$ matrix N is called the **fundamental matrix** for the absorbing Markov chain. For continuity of notation, it is assumed that the rows and columns of N, like those of S, are numbered from $m + 1$ through n. Thus,

$$N = \begin{bmatrix} \eta_{m+1,m+1} & \cdots & \eta_{m+1,n} \\ \vdots & & \vdots \\ \eta_{n,m+1} & \cdots & \eta_{n,n} \end{bmatrix}$$

This matrix is of critical importance in answering the questions posed above, as we now show by answering question 1.

An absorbing Markov chain in nonabsorbing state E_i can be expected to be in the nonabsorbing state E_j a total of η_{ij} times before absorption.   THEOREM B

Proof: The general proof, with greater notational complications, follows the proof we shall give for the absorbing Markov chain (3) which, it will be recalled, has absorbing states E_1, E_2 and nonabsorbing states E_3, E_4, E_5.

Let v_{ij} be the expected number of times that the process, starting in the state E_i, visits E_j before absorption. To further simplify notation, let us take E_3 to be our initial state. Now the expected number of occurrences of an event in n trials (see Property 6 of Section 8-1) is the probability of that event multiplied by the number of trials. Thus, starting in E_3, the expected number of times the process will go from E_3 to E_4 and then return to E_3 is $p_{34}v_{43}$. Similar reasoning applied to the other ways to leave and then return to E_3 gives us

$$p_{33}v_{33} + p_{34}v_{43} + p_{35}v_{53}$$

the expected number of times that the process which starts in E_3 will return to E_3. If we count the fact that the process starts in E_3, then the total number of times that the process can be expected to visit E_3 before absorption, which is v_{33} by definition, is given by

$$v_{33} = 1 + p_{33}v_{33} + p_{34}v_{43} + p_{35}v_{53}$$

Still taking E_3 as the initial state, similar reasoning gives us

$$v_{34} = p_{33}v_{34} + p_{34}v_{44} + p_{35}v_{54}$$
$$v_{35} = p_{33}v_{35} + p_{34}v_{45} + p_{35}v_{55}$$

These expressions may be written in the form

(7)

$$[v_{33} \quad v_{34} \quad v_{35}] = [1 \quad 0 \quad 0] + [p_{33} \quad p_{34} \quad p_{35}]\begin{bmatrix} v_{33} & v_{34} & v_{35} \\ v_{43} & v_{44} & v_{45} \\ v_{53} & v_{54} & v_{55} \end{bmatrix}$$

If we now write the three equations obtained by taking E_4 as the initial state, and then do the same thing using E_5 as the initial state, the nine equations would, in analogy with (7), take the form

$$V = I + SV$$

Rewriting,

$$(I - S)V = I$$

We know (from the lemma) that $(I - S)^{-1}$ exists, so

$$V = (I - S)^{-1}I$$

But $(I - S)^{-1} = N$, so $V = N$; the expected number of visits from E_i to E_j before absorption is $v_{ij} = \eta_{ij}$. □

Since we know the expected number of times the process will be in each of the nonabsorbing states, it is clear that if we begin in state E_i, the expected number of steps until absorption is

$$\eta_{i1} + \eta_{i2} + \cdots + \eta_{i,n-m}$$

Since this easily obtained result answers question 2, we state it as our next theorem.

THEOREM C

The expected number of steps before absorption of an absorbing Markov chain starting in the nonabsorbing state E_i is given by the sum of the entries in the ith row of the fundamental matrix N.

Finally, we find a_{ij}, the probability that a Markov process originating in the nonabsorbing state E_i will be absorbed in state E_j.

THEOREM D

Let the absorbing Markov chain have the canonical transition matrix (2) above. Then a_{ij} are the entries of the $(n - m) \times m$ matrix

$$A = NR$$

Proof: Once again we merely indicate the general proof by giving a proof for the five-state chain with transition matrix (3). We consider first the probability a_{31} of the process beginning in E_3 and finally being absorbed in E_1. There are a number of ways in which this can happen. It may

(a) Go directly in the first step (which it does with probability p_{31}).
(b) Go first to some nonabsorbing state E_k (which it does with probability p_{3k}, $k = 3, 4, 5$) and then eventually from E_k to E_1 (which it does with probability a_{k1}, which is one of the probabilities we seek).

The probability of going from E_3 to E_1 by way of the nonabsorbing state E_k is $p_{3k}a_{k1}$, and because the various ways we have mentioned are mutually exclusive (see Property 5 again), the probability we seek is given by

$$a_{31} = p_{31} + p_{33}a_{31} + p_{34}a_{41} + p_{35}a_{51}$$

Similar expressions may be written for all the a_{ij},

$$\begin{bmatrix} a_{31} & a_{32} \\ a_{41} & a_{42} \\ a_{51} & a_{52} \end{bmatrix} = \begin{bmatrix} p_{31} & p_{32} \\ p_{41} & p_{42} \\ p_{51} & p_{52} \end{bmatrix} + \begin{bmatrix} p_{33} & p_{34} & p_{35} \\ p_{43} & p_{44} & p_{45} \\ p_{53} & p_{54} & p_{55} \end{bmatrix} \begin{bmatrix} a_{31} & a_{32} \\ a_{41} & a_{42} \\ a_{51} & a_{52} \end{bmatrix}$$

or, by using the notation of our canonical form,

$$A = R + SA$$

Rewriting and then multiplying by $(I - S)^{-1} = N$,

$$(I - S)A = R$$
$$A = NR \quad \square$$

A FINAL VISIT

We conclude this chapter as we began it, discussing our peripatetic couple, presumed in the revised transition matrix (1) introduced in this section to be visiting states that correspond to various states of poor health. Introduction of the state E_0 into the transition matrix not only gives us an example of an absorbing chain, but brings the illustration into line with human experience; the probability of the ultimate absorption is 1.

For this example,

$$S = \begin{bmatrix} \frac{2}{3} & \frac{1}{6} & \frac{1}{6} \\ \frac{3}{4} & \frac{1}{12} & \frac{1}{6} \\ \frac{1}{2} & \frac{1}{6} & \frac{1}{4} \end{bmatrix}, \qquad N = (I - S)^{-1} = \begin{bmatrix} \frac{570}{13} & \frac{132}{13} & 12 \\ \frac{558}{13} & \frac{144}{13} & 12 \\ \frac{504}{13} & \frac{120}{13} & 12 \end{bmatrix}$$

According to Theorem B, if the couple is initially in E_1, then they can be expected to spend

$$\eta_{11} = \tfrac{570}{13} \approx 44 \text{ weekends in } E_1$$
$$\eta_{12} = \tfrac{132}{13} \approx 10 \text{ weekends in } E_2$$
$$\eta_{13} = 12 \text{ weekends in } E_3$$

If they begin in state E_2, or E_3, then their expected number of weekends in the various states are read from lines 2 or 3 accordingly. Theorem C merely states the obvious. Depending upon their initial states, one simply adds the figures in lines 1, 2, or 3 to find the total expected number of weekends left before absorption. Finally, corresponding to the fact that there exists only one state of absorption in this problem, we note that

$$NR = \begin{bmatrix} \frac{570}{13} & \frac{132}{13} & 12 \\ \frac{558}{13} & \frac{144}{13} & 12 \\ \frac{504}{13} & \frac{120}{13} & 12 \end{bmatrix} \begin{bmatrix} 0 \\ 0 \\ \frac{1}{12} \end{bmatrix} = \begin{bmatrix} 1 \\ 1 \\ 1 \end{bmatrix}$$

No matter where they start, the probability of being absorbed in E_0 is 1.

We conclude with a disclaimer. Since no one knows the coefficients in a particular individual's transition matrix, it should be clear that the theory of Markov chains is not useful in predicting anyone's life span. The play on words afforded by a couple visiting different states is intended to fix the concepts in your mind, but beyond that we must look elsewhere for practical applications. That is just as well. To paraphrase one old hymn, if we could see beyond this weekend as God can see, we might quit right now.

1. A matrix in canonical form is

$$P = \begin{bmatrix} 1 & 0 & 0 & 0 & 0 \\ 0 & 1 & 0 & 0 & 0 \\ 0 & 0 & \frac{1}{3} & \frac{1}{3} & \frac{1}{3} \\ \frac{1}{3} & \frac{1}{3} & 0 & \frac{1}{3} & 0 \\ \frac{1}{6} & \frac{1}{3} & \frac{1}{6} & \frac{1}{6} & \frac{1}{6} \end{bmatrix}$$

   (a) Find P^3.
   (b) What is the smallest power of P having a nonzero entry in every row of A_r?
   (c) Find S^3.
   (d) Find $A_3 = R + SR + S^2R$.
   (e) Use (a), (c), and (d) to verify equation (4) of this section.
   (f) Find $N = (I - s)^{-1}$.

(g) Find $I + S + S^2 + S^3$. Compare it with N; note equation (6) of this section.

(h) If the process starts in state E_5, how many times might we expect it in E_4 before absorption? In E_3?

(i) If the process starts in E_5, how many steps might we expect before absorption?

(j) If the process starts in E_5, what is the probability that it is absorbed in state E_1?

2. A matrix in canonical form is

$$P = \begin{bmatrix} 1 & 0 & 0 & 0 & 0 \\ 0 & 1 & 0 & 0 & 0 \\ 0 & 0 & \frac{1}{3} & 0 & \frac{2}{3} \\ \frac{1}{2} & 0 & 0 & \frac{1}{2} & 0 \\ 0 & \frac{1}{4} & \frac{1}{2} & 0 & \frac{1}{4} \end{bmatrix}$$

Follow the instructions for Problem 1.

3. The gambling game described in Problem 15 of Problem Set 8-2 has two absorbing states.

(a) Rewrite the transition matrix in canonical form.

(b) What does Theorem A say about this problem?

(c) How many times might the player expect to have $2?

(d) What is the probability of the player wining all the money?

4. Follow the instructions of Problem 3 for the gambling game described in Problem 16 or Problem Set 8-2.

5. A resort transports guests to one of two entry points, s_5 or s_6, of a system of cross-country ski trails that lead back to the lodge. At each of the trail intersections shown in Figure 8-3, we have indicated the probability of a skier going in the various possible directions. (Obviously some energetic skiers will not want to take the shortest route back.)

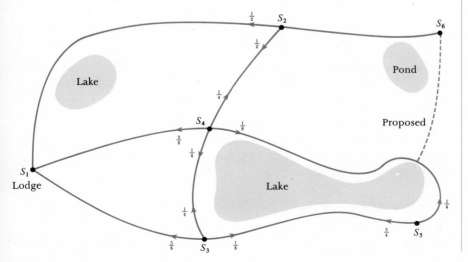

Figure 8-3

(a) Set up the transition matrix for this six-state absorbing chain in canonical form.

(b) How many times is a guest who begins at s_5 likely to pass through s_4 before returning to the lodge? How about one who begins at s_6?

6. Suppose that the ski trails described in Problem 5 are augmented by the proposed trail joining s_6 to a new intersection s_7 between s_4 and s_5. Skiers at s_6 are equally likely to take the old or the new trail. Half the skiers at s_7 go to s_5; the rest divide equally between going to s_4 or s_6. Finally, skiers at s_2 divide equally between going to s_1 or s_4 except for $\frac{1}{8}$ of them who go to s_6. Answer the questions asked in (b) of Problem 5 for this new system.

1. Define the following terms:
   (a) Markov chain
   (b) regular Markov chain
   (c) absorbing Markov chain

2. A Markov chain with four possible states $E_1, \ldots, E_4$ has the transition matrix

$$P = \begin{bmatrix} 0 & \frac{1}{2} & \frac{1}{2} & 0 \\ \frac{1}{2} & \frac{1}{4} & 0 & \frac{1}{4} \\ 0 & 0 & 0 & 1 \\ 0 & \frac{1}{2} & 0 & \frac{1}{2} \end{bmatrix}$$

It is equally likely that the process will begin in either state E_1 or E_4. Give the probabilities for the process being in states E_1 or E_4 after a long time.

Problems 3–7 all refer to a Markov chain of three states E_1, E_2, and E_3 having the transition matrix

$$A = \begin{bmatrix} \frac{1}{2} & \frac{1}{6} & \frac{1}{3} \\ \frac{1}{2} & \frac{1}{2} & 0 \\ \frac{1}{2} & 0 & \frac{1}{2} \end{bmatrix}$$

3. Suppose that the probability of beginning in states E_1, E_2, and E_3 are $\frac{1}{3}$, $\frac{1}{6}$, and $\frac{1}{2}$, respectively.
   (a) What is the probability that the process will begin in state E_3 and be in state E_1 after one step?
   (b) What is the probability of starting in either state E_1 or E_3 and being in state E_1 after one step?
   (c) What is the probability of being in state E_1 after one step?

4. Under the conditions described in Problem 3, what is the probability that the process will be in state E_1 after two steps?

5. Under the conditions described in Problem 3, what is the probability that the process will be in state E_1 after six steps?

6. Find $\lim_{k \to \infty} A^k$.

7. Refer to Problem 3. Could we have chosen beginning probabilities b_1, b_2, and b_3 for the states E_1, E_2, and E_3 so that the probabilities of being in these states after Step 1 would again be b_1, b_2, and b_3? If so, what would the values of b_1, b_2, and b_3 have to be?

Problems 8–10 all refer to a Markov chain of five states $E_1, \ldots, E_5$ having the transition matrix

$$P = \begin{bmatrix} 1 & 0 & 0 & 0 & 0 \\ 0 & 1 & 0 & 0 & 0 \\ \frac{1}{6} & \frac{1}{2} & \frac{1}{3} & 0 & 0 \\ 0 & 0 & \frac{1}{2} & \frac{1}{4} & \frac{1}{4} \\ \frac{1}{3} & 0 & \frac{1}{3} & \frac{1}{3} & 0 \end{bmatrix}$$

8. Prove that this Markov chain is absorbing.

9. What is the fundamental matrix of this absorbing Markov chain?

10. Suppose the process begins in state E_3.
    (a) How many times would you expect the process to be in state E_5 before absorption?
    (b) What is the probability that absorption will occur in state E_2?
    (c) How many steps would you expect before absorption?

9 DIFFERENTIAL EQUATIONS

9-1 | INTRODUCTION

Admirable though numbers are for answering most questions that you have encountered in your study of mathematics, they will not serve as solutions to differential equations. The differential equation

(1)
$$y' = x + y$$

has as its solution not a number, but a function $f(x)$, a function having the property that

$$f'(x) = x + f(x)$$

To get a geometric slant on the solution that we seek, think of (1) this way. The graph of $y = f(x)$ should at any point (x, y) have a slope equal to $x + y$. Thus, at $(1, 1)$, the graph should have a slope of 2; at $(2, 1)$, it should have a slope of 3, and so on. A short line segment drawn with the appropriate slope at a point is called a **lineal element.** A generous sprinkling of these elements over the plane (similar to sprinkling iron filings on a paper covering a magnet) is called a **direction field.** An efficient way to draw a direction field is to assign values (slopes) to y' in (1) and draw lineal elements along the corresponding curve. Thus, for equation (1), all the points with slope 2 lie on the line $x + y = 2$; points with slope 1 lie on $x + y = 1$; points with slope 0, -1, and -2 are also indicated in Figure 9-1.

There is obviously something special about the points on $x + y = -1$, and a quick check (just substitute) shows that $y = -x - 1$ is a solution to (1). It also seems obvious that there must be not just one solution to the equation, but a whole family of solutions. Indeed it is so. Later in

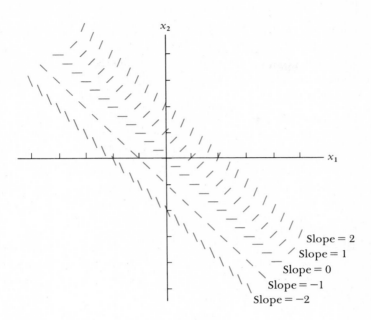

Figure 9-1

this chapter we will develop methods that will lead us to the family of solutions

$$y = Ce^x - x - 1, \qquad C \text{ any constant}$$

Natural questions to ask about a differential equation are:

1. Will a solution always exist?
2. Do solutions always travel together as families?
3. How can we be certain that we have found all the solutions?

For a large and important class of differential equations, these questions are best answered using the concepts and language of linear algebra. This chapter elaborates that theme.

LINEAR HOMOGENEOUS DIFFERENTIAL EQUATIONS WITH CONSTANT COEFFICIENTS

To indicate where differential equations come from, as well as to introduce some terminology, we begin with the following problem:

While leaning over the side of his boat to rinse his hands in the water, a fisherman let a lead sinker slip from his grasp. Describe the motion of the sinker by giving its depth y below the surface x seconds after its release.

We are being asked to model the sinking of the sinker. Until the sinker rests squarely on the bottom, our model will (like all models) rest no less squarely on some assumptions. These will be:

(i) $F = ma$ (force equals mass times acceleration)
(ii) The descending sinker encounters a resistance from the water that is proportional to its rate of descent.

The forces acting on the sinker are its weight W and the resistance R. We shall take the downward direction of motion to be positive, the upward direction of the resistance to be negative. Thus,

$$F = W - R$$

Now $W = mg$, the mass times the (constant) acceleration g due to gravity, and the assumption (ii) is translated into mathematical language by writing $R = ky'$. Hence,

$$F = mg - ky'$$

Finally, since y'' represents the acceleration of a moving body, we have by substituting in (i),

$$mg - ky' = my''$$

If we introduce a new symbol $c = k/m$ for our constant, we get

(1)
$$y'' + cy' = g$$

The function $f(x)$ that tells us how the sinker sinks must satisfy

$$f''(x) + cf'(x) = g \quad \square$$

There is another way to write equation (1) that will be useful to us. It was pointed out in Section 6-1 that the differential operator $Df = f'$ is a linear operator; so, we now point out, is the operator $D^2f = f''$ since, for two functions f_1 and f_2 and two scalars r_1 and r_2,

$$D^2(r_1f_1 + r_2f_2) = r_1D^2f_1 + r_2D^2f_2$$

Thus, we can write (1) in terms of differential operators:

$$(D^2 + cD)y = g$$

Equation (1) is a **differential equation,** a statement of equality involving a function and one or more of its derivatives. It is called **linear** since no term involves more than a first degree expression in y, y', and y''; alternatively, it is linear because it can be written in terms of the linear operator $D^2 + cD$. Since the coefficients of the derivatives (the terms on the *left-hand* side of the equation) are constant, we say that this linear differential equation has **constant coefficients.** Note that this last designation would apply even if the term on the right-hand side of (1) involved some function of x. The term on the right figures into the description of the equation in this way: If the right-hand side (the term not involving y or its derivatives) is 0, the differential equation is said to be **homogeneous.** Thus, (1) is nonhomogeneous. The linear homogeneous differential equation with constant coefficients to which we refer in the next two sections is of the form

$$y^{(n)} + a_{n-1}y^{(n-1)} + \cdots + a_1 y' + a_0 y = 0 \qquad (2)$$

Alternatively, it is of the form

$$(D^n + a_{n-1}D^{n-1} + \cdots + a_1 D + a_0)y = 0$$

Such an equation is said to be of the **nth order.**

SPACES AND SUBSPACES OF FUNCTIONS

We pointed out in Section 2-6 that the set of all functions that have two continuous derivatives on an interval (a, b) form a vector space commonly referred to as the space $C''(a, b)$. Suppose that f and g are two functions (vectors) in this space, both of which satisfy the differential equation

$$y'' - 5y' + 6y = 0 \qquad (3)$$

Then since

$$f''(x) - 5f'(x) + 6f(x) = 0$$
$$g''(x) - 5g'(x) + 6g(x) = 0$$

it follows that for any constant c,

$$f''(x) + cg''(x) - 5[f'(x) + cg'(x)] + 6[f(x) + cg(x)] = 0$$

Thus, $f(x) + cg(x)$ is also a solution to (3), enough to show (Theorem A, Section 2-4) that the solutions to (3) form a subspace of $C''(a, b)$.

A similar result can be proved in the same way for the vector space $C^{(n)}(a, b)$, the space of functions having n continuous derivatives on (a, b).

The collection of functions in $C^{(n)}(a, b)$ which satisfy the homogeneous differential equation

$$y^{(n)} + a_{n-1}y^{(n-1)} + \cdots + a_1 y' + a_0 y = 0$$

for all x in (a, b) forms a subspace of $C^{(n)}(a, b)$.

THEOREM A

There is another way to prove Theorem A that illustrates the economies of setting the study of linear differential equations in the context of linear algebra. Simply observe that solutions to the differential equation of Theorem A are members of the null space of the linear operator $D^n + a_{n-1}D^{n-1} + \cdots + a_1 D + a_0$; hence, they form a subspace.

You have correctly sized up the situation if you now anticipate that the dimension of the solution subspace is related to the order of the equation. But to talk about dimensions of subspaces of functions, we need to think first about the meaning of linear independence for a set of functions.

THE WRONSKIAN

Let f and g be functions in $C'(a, b)$, and suppose there exist constants c_1 and c_2 such that

$$c_1 f(x) + c_2 g(x) = 0 \qquad \text{for } all \ x \text{ in } (a, b)$$

Then of course

$$c_1 f'_1(x) + c_2 g'(x) = 0 \qquad \text{for } all \ x \text{ in } (a, b)$$

Now fix x_0 in (a, b) and consider the system

(4)
$$f(x_0)t_1 + g(x_0)t_2 = 0$$
$$f'(x_0)t_1 + g'(x_0)t_2 = 0$$

This homogeneous system of linear equations in t_1 and t_2 will have a nontrivial solution if and only if the determinant

$$\begin{vmatrix} f(x_0) & g(x_0) \\ f'(x_0) & g'(x_0) \end{vmatrix} = 0$$

Thus, if there is a point x_0 at which this determinant is nonzero, then the only solution to the system (4) is the trivial solution $t_1 = t_2 = 0$. We know, however, that $t_1 = c_1$, $t_2 = c_2$ is a solution to (4), so it must be that $c_1 = c_2 = 0$. The functions f and g are therefore shown to be independent if there is a single x_0 for which the determinant is nonzero. Since this argument obviously extends to n functions, we have the following result:

THEOREM B

Let $f_1(x), \ldots, f_n(x)$ be members of $C^{(n-1)}(a, b)$, and suppose there exists an x_0 in (a, b) such that the determinant

$$W = \begin{vmatrix} f_1(x_0) & \cdots & f_n(x_0) \\ f'_1(x_0) & \cdots & f'_n(x_0) \\ \vdots & & \vdots \\ f_1^{(n-1)}(x_0) & \cdots & f_n^{(n-1)}(x_0) \end{vmatrix} \neq 0$$

Then the functions $f_1(x), \ldots, f_n(x)$ are linearly independent. The crucial determinant in the statement of Theorem B which we happened to designate with a W is generally called the **Wronskian** of the set of functions $f_1(x), \ldots, f_n(x)$.

In Problems 1–6, verify by substitution that the given $f(x)$ and $g(x)$ are both solutions to the given differential equation. Then verify by direct substitution that $h(x) = c_1 f(x) + c_2 g(x)$ is also a solution.

1. $y'' - 2y' = 0$

   $f(x) = e^{2x}, \qquad g(x) = 1$

2. $y'' + y = 0$

   $f(x) = \sin x, \qquad g(x) = \cos x$

3. $y'' - 2y' + y = 0$

   $f(x) = e^x, \qquad g(x) = xe^x$

4. $y'' - y = 0$

   $f(x) = e^x, \qquad g(x) = e^{-x}$

5. $y^{(iv)} + y'' = 0$

   $f(x) = x, \qquad g(x) = \sin x$

6. $y''' - 4y'' + 4y' = 0$

   $f(x) = e^{2x}, \qquad g(x) = xe^{2x}$

For what x are Wronskians of the functions $f(x)$ and $g(x)$, given in Problems 7–12, equal to zero?

7. $f(x) = \sin x$

   $g(x) = \cos x$

8. $f(x) = x^2$

   $g(x) = \sin x$

9. $f(x) = x^2$

   $g(x) = x^3$

10. $f(x) = e^x$

    $g(x) = e^{3x}$

11. $f(x) = e^{3x}$

    $g(x) = xe^{3x}$

12. $f(x) = x \sin x$

    $g(x) = \sin x$

13. Show by an argument not using the Wronskian that $f(x) = e^x$ and $g(x) = xe^x$ are linearly independent. *Hint:* Begin by assuming that $c_1 f(x) + c_2 g(x) = 0$ for *all* x.

14. Show by an argument not using the Wronskian that $f(x) = x^2 + 2$ and $g(x) = x^2 - 3x$ are linearly independent.

A BASIS FOR THE SOLUTION SPACE

9-3

We continue to focus our attention on the nth-order linear homogeneous differential equation

$$y^{(n)} + a_{n-1} y^{(n-1)} + \cdots + a_1 y' + a_0 y = 0$$

in which the coefficients a_i are constant. But as is often the case when trying to get a handle on a problem, mathematical or otherwise, we find it easiest to begin with a simple case. Let us therefore begin with

$$y' + a_0 y = 0$$

Try talking to yourself—out loud—about this problem. You might say something like, "We're supposed to find a function which when multiplied by a constant and added to it's derivative gives 0." Such thinking shortly brings us to consideration of exponential functions, and thus to the function $y = e^{-a_0 x}$ which is seen by substitution to be a solution. Recalling that solutions form subspaces, we next try

$$y = C_1 e^{-a_0 x}$$

and discover that it is a solution for any choice of the constant C_1. (As an alternative to talking to yourself about this problem, see Problem 21 at the end of this section.)

SECOND ORDER

There seems to be little hope, of course, of generalizing a method that depends on listening carefully as you talk to yourself, but something has been accomplished—the form of a solution function has been suggested. Suppose, for example, that we now decide to look for solutions of the form $y = e^{mx}$ to the equation

(1)
$$y'' - 5y' + 6y = 0$$

Derivatives of $y = e^{mx}$ are easily obtained and substituted to get

$$m^2 e^{mx} - 5m e^{mx} + 6e^{mx} = 0$$
$$(m^2 \quad - 5m \quad + 6)e^{mx} = 0$$

Evidently we will obtain solutions to (1) if we choose m so that

$$m^2 - 5m + 6 = 0$$

This is accomplished with modest demands on algebra, leading to $m = 2$ and $m = 3$. The two functions

$$y = e^{2x}, \qquad y = e^{3x}$$

are evidently solutions to (1), and since their Wronskian is

$$\begin{vmatrix} e^{2x} & e^{3x} \\ 2e^{2x} & 3e^{3x} \end{vmatrix} = 3e^{5x} - 2e^{5x} \neq 0 \quad \text{for all } x,$$

these functions are linearly independent. Having used (1) in Section 9-2 to demonstrate that linear combinations of solutions are also solutions, we see that any function of the form

$$y = C_1 e^{2x} + C_2 e^{3x}$$

will also be a solution of (1). Put another way, all members of the 2-dimensional subspace of $C''(-\infty, \infty)$ spanned by the linearly independent functions e^{2x} and e^{3x} are solutions to (1).

We now have a method that seems to hold promise for any second-order equation of the type under consideration. Substitute $y = e^{mx}$ into

$$y'' + a_1 y' + a_0 y = 0$$

obtaining
$$(m^2 + a_1 m + a_0)e^{mx} = 0$$

Then find the roots of the quadratic equation, called the **characteristic equation** of the given differential equation,

$$m^2 + a_1 m + a_0 = 0$$

If the roots are r_1 and r_2, then

$$y = e^{r_1 x}, \qquad y = e^{r_2 x} \qquad (2)$$

will be solutions to the given equation.

When theory like this can be so beautifully developed, it is sometimes wise to move on quickly before multiple and complex questions are raised by those who try to apply the method to specific problems. But these are exactly the kind of questions that will be forthcoming, as you will quickly see if you try the method just described on the equations

$$y'' - 2y' + y = 0 \qquad (3)$$
$$y'' - y' + y = 0 \qquad (4)$$

THE CASE OF MULTIPLE ROOTS

The Wronskian of the solutions given in (2) is

$$\begin{vmatrix} e^{r_1 x} & e^{r_2 x} \\ r_1 e^{r_1 x} & r_2 e^{r_2 x} \end{vmatrix} = (r_2 - r_1)e^{(r_1 + r_2)x}$$

It is nonzero if $r_2 \neq r_1$, but clearly the solutions in (2) are not linearly independent if $r_1 = r_2$. In this case, however, letting $r = r_1 = r_2$, it can be verified that both

$$y = e^{rx}, \qquad y = xe^{rx} \qquad (5)$$

are solutions to the given equation. Moreover, the Wronskian for these functions is

$$\begin{vmatrix} e^{rx} & xe^{rx} \\ re^{rx} & rxe^{rx} + e^{rx} \end{vmatrix} = [(r + 1) - r]e^{2rx} \neq 0$$

so we again have two linearly independent solutions.

Solve the differential equation (3):   *Example A*

$$y'' - 2y' + y = 0$$

Setting $y = e^{mx}$ in the recommended way, we come to

$$(m^2 - 2m + 1)e^{mx} = 0$$

which has 1 as a root of multiplicity two. Naturally $y = e^x$ will be one solution. You should verify (find the required two derivatives and substitute) our assertion that $y = xe^x$ is also a solution. The general solution to the given equation is then given by

$$y = C_1 e^x + C_2 x e^x \quad \square$$

THE CASE OF COMPLEX ROOTS

If $r_1 = a + bi$, then of course $r_2 = a - bi$, but you will be excused for wondering what meaning is to be attached to an expression like

(6)
$$e^{a+bi} = e^a e^{bi} \quad \text{or} \quad e^{a-bi} = e^a e^{-bi}$$

The problem revolves, of course, on the meaning of e^{bi}. It happens that this is a perfectly well-defined and very useful function in the theory of functions of a complex variable, but this is a subject probably not yet studied by most readers of this book. Therefore, rather than digress to a complex subject, we will again proceed by considering simplified cases first (though the interested reader should consider Problems 22 and 23 at the end of the section).

Consider the equation

(7)
$$y'' + b^2 y = 0$$

Substitution of $y = e^{mx}$ leads in the usual way to

$$(m^2 + b^2)e^{mx} = 0$$

and to the roots $r_1 = bi$, $r_2 = -bi$. Formal substitution gives us a somewhat simplified (but no less complex) form of the expressions in (6), namely

$$y = e^{bix}, \quad y = e^{-bix}$$

The simplicity of (7), however, enables us to once again use the approach of talking to ourselves, this time asking for a function which when added to its second derivative gives zero. We soon think of trying either $\sin x$ or $\cos x$. After playing around a bit with the constant b, we discover that two solutions to (7) are given by

$$y = \sin bx, \quad y = \cos bx$$

With these ideas fresh in mind, we return to (6) and try instead the functions

$$y = e^{ax} \cos bx, \quad y = e^{ax} \sin bx$$

These two functions do have the desired property of being linearly independent, since

$$\begin{vmatrix} e^{ax}\cos bx & e^{ax}\sin bx \\ -be^{ax}\sin bx + ae^{ax}\cos bx & be^{ax}\cos bx + ae^{ax}\sin bx \end{vmatrix}$$
$$= e^{2ax}[b\cos^2 bx + a\cos bx\sin bx - (-b\sin^2 bx + a\cos bx\sin bx)]$$
$$= be^{2ax} \neq 0$$

(If b had been zero, we would never have gotten into all this.) Moreover, direct substitution shows that both of these functions are solutions.

Solve the differential equation (4): *Example B*

$$y'' - y' + y = 0$$

Setting $y = e^{mx}$ leads to the characteristic equation

$$m^2 - m + 1 = 0$$

which has roots

$$r_1 = \frac{1 + i\sqrt{3}}{2}, \qquad r_2 = \frac{1 - i\sqrt{3}}{2}$$

The linearly independent functions

$$y = e^{(1/2)x}\cos\frac{\sqrt{3}}{2}x, \qquad y = e^{(1/2)x}\sin\frac{\sqrt{3}}{2}x$$

are both solutions to (4); try them. Thus, any function of the form

$$y = C_1 e^{(1/2)x}\cos\frac{\sqrt{3}}{2}x + C_2 e^{(1/2)x}\sin\frac{\sqrt{3}}{2}x$$

will also be a solution. $\square$

A final observation is in order. For each of the possible cases (distinct real roots, equal roots, or complex roots to the characteristic equation in m obtained by setting $y = e^{mx}$), we have found the Wronskian of the linearly independent solutions, and in each case the Wronskian has been nonzero for *every value* of x.

Let $f_1(x)$ and $f_2(x)$ be two linearly independent solutions to *THEOREM A*

$$y'' + a_1 y' + a_0 y = 0$$

obtained by the methods of this section. Then the Wronskian of these functions is nonzero for all x.

THE GENERAL CASE

We have seen that substitution of $y = e^{mx}$ into the second-order differential equation

$$y'' + a_1 y' + a_0 y = 0$$

leads to the second degree characteristic equation

$$m^2 + a_1 m + a_0 = 0$$

To emphasize how this equation is obtained, we have not yet pointed out a short cut; but you may have noticed that if the given differential equation is written in the form

$$(D^2 + a_1 D + a_0)y = 0$$

the characteristic equation, expressed in terms of D instead of m, is immediately evident.

From the roots of the characteristic equation, we now know how to write down two linearly independent solutions to the given differential equation. It is a fact (which we are not going to prove) that these two linearly independent solutions form a basis for the subspace of all possible solutions. Best of all, our ability to write down linearly independent solutions corresponding to distinct, to multiple, and to complex roots, together with our (theoretical) ability to solve any polynomial equation (according to the fundamental theorem of algebra), now enables us to solve any equation of the form

(8)
$$(D^n + a_{n-1}D^{n-1} + \cdots + a_1 D + a_0)y = 0$$

in which the a_i are real constants. We should note that a root r occurring with multiplicity k will give rise to the k linearly independent functions

$$f_1(x) = e^{rx}, \quad f_2(x) = xe^{rx}, \quad \ldots, \quad f_k(x) = x^{k-1}e^{rx}$$

Example C

Solve the fifth-order differential equation

$$(D + 1)^3(D^2 - 2D + 2)y = 0$$

The roots of the characteristic equation are $-1, -1, -1, 1 + i$, and $1 - i$. The general solution is

$$y = C_1 e^{-x} + C_2 xe^{-x} + C_3 x^2 e^{-x} + C_4 e^x \cos x + C_5 e^x \sin x \quad \square$$

Finally, we have an analog to Theorem A.

THEOREM B

Let $f_1(x), \ldots, f_n(x)$ be linearly independent solutions to the differential equation (8). Then the Wronskian of these functions is nonzero for all x.

Solve the following homogeneous differential equations.

1. $y'' + y' - 2y = 0$

2. $y'' + 4y' + 4y = 0$

3. $y'' - 6y' + 9y = 0$

4. $y'' - y' - 2y = 0$

5. $y'' + 4y = 0$

6. $4y'' + y = 0$

7. $y'' - 4y' + 5y = 0$

8. $y'' - 2y' + 5y = 0$

9. $y''' - 3y'' + y' + 5y = 0$

10. $y''' - 3y'' + 7y' - 5y = 0$

11. $y^{(iv)} - 3y''' + 7y'' - 5y' = 0$

12. $y^{(iv)} - 3y''' + y'' + 5y' = 0$

13. $y''' - 2y'' - 4y' + 8y = 0$

14. $y''' + 2y'' - 4y' - 8y = 0$

15. $2y''' - 4y'' + 3y' - y = 0$

16. $2y''' + 4y'' + 3y' + y = 0$

17. $(D - 2)(D + 3)(D - 1)^2 y = 0$

18. $(D - 3)(D + 2)D^2 y = 0$

19. $(2D^2 + 2D + 1)(D + 3)^2 y = 0$

20. $(2D^2 - 2D + 1)(D + 2)^2 y = 0$

21. The equation $y' + a_0 y = 0$ considered at the beginning of this section may be written in the form

$$\frac{dy}{dx} = -a_0 y$$

Separation of variables gives

$$\frac{dy}{y} = -a_0 \, dx$$

Now obtain $y = C_1 e^{-a_0 x}$ by taking antiderivatives of each side.

22. Substitute $z = bix$ in the series expansion

$$e^z = 1 + z = \frac{z^2}{2!} + \frac{z^3}{3!} + \frac{z^4}{4!} + \cdots$$

to show (formally at least) that

$$e^{bix} = \cos bx + i \sin bx$$

23. Using the definition for e^{bix} suggested in Problem 22, show that

$$A e^{bix} + B e^{-bix} = c_1 \cos bx + c_2 \sin bx$$

THE NONHOMOGENEOUS CASE 9-4

If the constants on the right-hand side of a nonhomogeneous system of linear equations

$$a_{11}x_1 + \cdots + a_{1n}x_n = b_1$$
$$\vdots \qquad \qquad \vdots \quad \vdots$$
$$a_{m1}x_1 + \cdots + a_{mn}x_n = b_m$$

(1)

are all replaced by zeros, we obtain the so-called **corresponding homogeneous system.** Solutions to this homogeneous system form a subspace of R^n, and solutions to the given nonhomogeneous system are correctly described as a translation of this subspace (Section 5-5). That is, if **p** is a particular solution of (1), then any other solution of (1) may be expressed in the form **p** + **v** where **v** is a solution to the corresponding homogeneous system.

Similarly, if the function $q(x)$ on the right-hand side of a nonhomogeneous linear differential equation

(2) $$y^{(n)} + a_{n-1}y^{(n-1)} + \cdots + a_1 y' + a_0 y = q(x)$$

is replaced by zero, we obtain the **corresponding homogeneous differential equation.** Those functions having n continuous derivatives on (a, b) which satisfy the homogeneous equation form a subspace of $C^{(n)}(a, b)$. We shall see that if $p(x)$ is a particular solution to (2), then any other solution to (2) may be expressed in the form $p(x) + f(x)$ where $f(x)$ is a solution to the corresponding homogeneous system. For if $p(x)$ is a particular solution of (2), then

$$(D^n + a_{n-1}D^{n-1} + \cdots + a_1 D + a_0)p(x) = q(x)$$

And if $h(x)$ is any other solution,

$$(D^n + a_{n-1}D^{n-1} + \cdots + a_1 D + a_0)h(x) = q(x)$$

Subtraction of the first equality from the second gives

$$(D^n + a_{n-1}D^{n-1} + \cdots + a_1 D + a_0)(h(x) - p(x)) = 0$$

This says that $f(x) = h(x) - p(x)$ is a solution to the corresponding homogeneous system and, since $h(x) = p(x) + f(x)$, our assertion is proved.

Having learned in the previous section to solve an nth-order homogeneous differential equation, our task in this section is clear: We need to develop methods for finding a particular solution for a nonhomogeneous linear differential equation. We will discuss two methods, perhaps better characterized as one method and one scheme for methodical guesswork. In both cases we simplify notational problems by discussing a second-order equation. Extension of the ideas to equations of any order is easily accomplished in ways that will be obvious.

VARIATION OF PARAMETERS (A METHOD)

Consider the second-order equation

(3) $$y'' + a_1 y' + a_0 y = q(x)$$

having as general solution to the corresponding homogeneous equation the function

(4) $$y = c_1 f_1(x) + c_2 f_2(x)$$

The idea of this method is to choose two functions $c_1(x)$ and $c_2(x)$ so that

$$p(x) = \{c_1(x)f_1(x) + c_2(x)f_2(x)\}$$

will be a particular solution to (3). Begin by noting that

$$p'(x) = \{c_1(x)f_1'(x) + c_2(x)f_2'(x)\} + [c_1'(x)f_1(x) + c_2'(x)f_2(x)]$$

If we now set the bracketed expression on the right equal to 0, computation of $p''(x)$ is simplified to

$$p''(x) = \{c_1(x)f_1''(x) + c_2(x)f_2''(x)\} + c_1'(x)f_1'(x) + c_2'(x)f_2'(x)$$

Consider now those parts of the expressions for $p(x)$, $p'(x)$, and $p''(x)$ that are enclosed in braces. Compare them with the expressions for y, y', and y'' obtained from (4):

$$y = c_1 f_1(x) + c_2 f_2(x)$$
$$y' = c_1 f_1'(x) + c_2 f_2'(x)$$
$$y'' = c_1 f_1''(x) + c_2 f_2''(x)$$

We know that when these expressions are substituted in (3), the left-hand side vanishes identically for all x. In the same way, when our expressions for $p(x), p'(x)$, and $p''(x)$ are substituted in (3), those parts of these expressions that are enclosed in braces will vanish. (Try it.)

We will be left with

$$c_1'(x)f_1'(x) + c_2'(x)f_2'(x) = q(x)$$

Together with our agreement above to set the bracketed terms of $p'(x)$ equal to zero, we therefore seek to choose $c_1'(x)$ and $c_2'(x)$ to satisfy the system

$$f_1(x)c_1'(x) + f_2(x)c_2'(x) = 0$$
$$f_1'(x)c_1'(x) + f_2'(x)c_2'(x) = q(x)$$

This can be solved using Cramer's rule. Note that the coefficient matrix called for in Cramer's rule is just the Wronskian $W(x)$ for the solution functions

$$c_1'(x) = \frac{-q(x)f_2(x)}{W(x)}, \qquad c_2'(x) = \frac{q(x)f_1(x)}{W(x)}$$

Since $f_1(x)$ and $f_2(x)$ are the linearly independent solutions to the homogeneous equation corresponding to (3), we know (Theorem A, Section 9-3) that their Wronskian is nonzero for all x. We can, therefore, be certain of finding $c_1(x)$ and $c_2(x)$, providing only that $q(x)$ is integrable. In particular, with the provision that $q(x)$ be continuous, we have a method that works!

Example A

Solve the nonhomogeneous equation

$$y'' + cy' = g$$

first encountered in Example A of Section 9-2.

The corresponding homogeneous equation is

$$(D^2 + cD)y = 0$$

The roots of the characteristic equation are 0 and $-c$; the general solution to the corresponding homogeneous equation is $y = c_1(1) + c_2 e^{-cx}$. And so, according to the method outlined above, we seek a particular solution of the form

(5)

$$p(x) = \{c_1(x)1 + c_2(x)e^{-cx}\}$$
$$p'(x) = \{c_1(x)(0) - cc_2(x)e^{-cx}\} + [c_1'(x) + c_2'(x)e^{-cx}]$$
$$p''(x) = \{c_1(x)(0) + c^2c_2(x)e^{-cx}\} + c_1'(x)(0) - cc_2'(x)e^{-cx}$$

where we have used the notation of braces and brackets used in the proof above. Actual substitution shows how the terms in braces vanish:

$$\{c_1(x)(0) + c^2c_2(x)e^{-cx}\} + c_1'(x)(0) - cc_2'(x)e^{-cx}$$
$$+ c\{c_1(x)(0) - cc_2(x)e^{-cs}\} = g$$

This equation, together with the bracketed terms of (5) set equal to 0 give the system of equations

$$1c_1'(x) + e^{-cx}c_2'(x) = 0$$
$$(0)c_1'(x) - ce^{-cx}c_2'(x) = g$$

The determinant of the coefficient matrix is

$$\begin{vmatrix} 1 & e^{-cx} \\ 0 & -ce^{-cx} \end{vmatrix} = -ce^{-cx} \neq 0$$

which is, of course, the Wronskian of the linearly independent solutions $f_1(x) = 1$, $f_2(x) = e^{-cx}$ of the corresponding homogeneous system. From Cramer's rule,

$$c_1'(x) = \frac{-ge^{-cx}}{-ce^{-cx}} \qquad c_2'(x) = \frac{g}{-ce^{-cx}}$$

$$c_1(x) = \frac{g}{c}x + k_1 \qquad c_2(x) = \frac{-g}{c^2}e^{cx} + k_2$$

Since we seek any particular solution, the arbitrary constants k_1 and k_2 may be set equal to 0. When we substitute our choices of $c_1(x)$ and $c_2(x)$ into (5),

$$p(x) = \frac{g}{c}x - \left(\frac{g}{c^2}e^{cx} \cdot e^{-cx}\right) = \frac{g}{c}x - \frac{g}{c^2}$$

The general solution to the nonhomogeneous equation may therefore be written as

$$y = c_1 + c_2 e^{-cx} + \frac{g}{c}x$$

(where the $-g/c^2$ has been absorbed into the general constant term). □

UNDETERMINED COEFFICIENTS (A SCHEME FOR GUESSING)

Though the method of variation of parameters always works, it is frequently a lot of work for getting answers that might have been obtained with a little guessing and a little work. Consider again the equation

$$y'' + cy' = g \tag{6}$$

If we ask ourselves what kind of a particular solution might work, there are some obvious things to eliminate. First and second derivatives of a function like

$$p(x) = Ae^x$$

or

$$p(x) = A \cos x + B \sin x$$

will still involve transcendental expressions; linear combinations of them will not reduce to the constant g. The best bet, in fact, would seem to be a polynomial. Moreover, if we set

$$p(x) = A_n x^n + \cdots + A_2 x^2 + A_1 x + A_0$$

then

$$p'(x) = nA_n x^{n-1} + \cdots + 2A_2 x + A_1$$
$$p''(x) = n(n-1)A_n x^{n-2} + \cdots + 2A_2$$

Substitution on the left-hand side of (6) will give us a polynomial of degree $n - 1$, while on the right-hand side our (constant) polynomial is of degree 0. Since these polynomials are equal for all x, they must be identical term by term, and in particular, they are surely both of the same degree; $n - 1 = 0$, $n = 1$.

The polynomial that we try is

$$p(x) = A_1 x + A_0$$
$$p'(x) = A_1$$
$$p''(x) = 0$$

Substitution into $y'' + cy' = g$ gives

$$cA_1 = g$$

With no condition on A_0, we are free to choose it as simply as possible.

Thus, $A_1 = g/c$ and we set $A_0 = 0$. The particular solution thus obtained is (again)

$$p(x) = \frac{g}{c} x$$

This scheme for guessing can be used with reasonable hope of success whenever the term $q(x)$ on the right-hand side has derivatives, all of which are linear combinations of a finite set of familiar functions. In such a situation, the hope is that $p(x)$ might be a linear combination of these same familiar functions. Some examples:

$q(x)$	Trial $p(x)$
$\sin x$	$A \sin x + B \cos x$
xe^{ax}	$Axe^{ax} + Be^{ax}$
polynomial, degree n	$A_n x^n + \cdots + A_1 x + A_0$

Another observation! If a term of the suggested trial function already appears as a solution to the corresponding homogeneous equation, then try multiplying the suggested trial function by x. Thus, in the example above, the rules just given applied to $q(x) = g$, a polynomial of degree zero, suggest as a trial solution $p(x) = A_0$. But as soon as we note that a constant is part of the general solution $y = c_1 + c_2 e^{-cs}$, we multiply our trial function by x, leading us to $p(x) = A_0 x$. This will, in fact, be a more efficient trial function than the one we used, obviating the need for us to discover that the constant term can be set equal to 0.

Example B   Solve the differential equation

$$y'' - 3y' + 2y = e^x + \sin x$$

In this case, it simplifies the work to think of two trial functions corresponding to the two terms on the right. From e^x, we first think of trying $p_1(x) = Ae^x$, but upon noticing that the general solution to $(D^2 - 3D + 2)y = 0$ is $y = c_1 e^x + c_2 e^{2x}$, we modify our trial function to $p_1(x) = Axe^x$. Then from $\sin x$, we think of the trial function $p_2(x) = B \sin x + C \cos x$. From these considerations, we are led to try

$$p(x) = Axe^x + B \sin x + C \cos x$$

Calculation gives

$$p'(x) = Axe^x + Ae^x + B \cos x - C \sin x$$
$$p''(x) = Axe^x + 2Ae^x - B \sin x - C \cos x$$

and substitution gives, after collection of similar terms,

$$-Ae^x + (B + 3C) \sin x + (C - 3B) \cos x = e^x + \sin x$$

Equating coefficients of similar terms gives $A = -1$, $B = \frac{1}{10}$, $C = \frac{3}{10}$. The general solution to the given nonhomogeneous equation is

$$y = c_1 e^x + c_2 e^{2x} - xe^x + \tfrac{1}{10} \sin x + \tfrac{3}{10} \cos x \quad \square$$

In Problems 1 through 4, follow the steps of the method of variation of parameters, identifying the terms in brackets and showing that those in braces do indeed vanish.

1. $y'' + 4y' + 4y = 8 \cos 2x$

2. $y'' + 2y' + y = e^{-2x}$

3. $y'' - y' - 2y = e^x - 2xe^x$

4. $y'' - 2y' + y = 2[x \sin x - \sin x - \cos x]$

5. Find a particular solution to the differential equation of Problem 1 using undetermined coefficients.

6. Find a particular solution to the differential equation of Problem 2 using undetermined coefficients.

7. Solve, using variation of parameters to find a particular solution, the equation

$$y''' + y'' - 5y' + 3y = 9e^{-2x}$$

8. Solve, using variation of parameters to find a particular solution, the equation

$$y''' - 2y'' - y' + 2y = 20 \cos \frac{x}{2} + 5 \sin \frac{x}{2}$$

9. Find a particular solution to the equation of Problem 7 using undetermined coefficients.

10. Find a particular solution to the equation of Problem 8 using undetermined coefficients.

Find complete solutions to the following equations using any method that seems convenient.

11. $y'' + y' - 2y = e^{2x}$

12. $y'' - 2y' - 3y = e^{3x}$

13. $y'' - 2y' + y = xe^x + x^2$

14. $y'' + 4y' + 4y = xe^{-2x} - x$

15. $(D^2 + 1)(D - 2)y = x^2 - 2 \sin x - 4 \cos x$

16. $(D^2 + 1)(D - 1)y = 3x + e^x$

SYSTEMS OF LINEAR DIFFERENTIAL EQUATIONS 9-5

If in the equation

$$y'' - 3y' + 2y = 0 \tag{1}$$

we set $y_1 = y$ and $y_2 = y'$, then

$$y_1' = y' = \qquad y_2$$
$$y_2' = y'' = 3y' - 2y = 3y_2 - 2y_1$$

The original equation is then equivalent to the system

(2)
$$
\begin{aligned}
y_1' &= y_2 \\
y_2' &= -2y_1 + 3y_2
\end{aligned}
$$

In a similar way, a linear nth-order homogeneous differential equation of the type studied in Section 9-3 can be related to a system of n first-order differential equations in n unknowns $y_1, \ldots, y_n$. Such systems of equations also arise directly from certain applications, and they are the object of our attention in this section.

The system we have in mind is of the form

$$
\begin{aligned}
y_1' &= a_{11}y_1 + \cdots + a_{1n}y_n \\
&\ \vdots \qquad\quad \vdots \qquad\qquad \vdots \\
y_n' &= a_{n1}y_1 + \cdots + a_{nn}y_n
\end{aligned}
$$

which may be written, with self-explanatory notation, in the form

(3)
$$
\mathbf{y}' = A\mathbf{y}
$$

As has been our custom in much of this chapter, we again begin by considering a simple case. Suppose

$$
A = \begin{bmatrix} r_1 & & 0 \\ & \ddots & \\ 0 & & r_n \end{bmatrix}
$$

is diagonal, so that the system is of the form

$$
\begin{aligned}
\frac{dy_1}{dx} &= r_1 y_1 \\
\vdots \qquad &\ \vdots \\
\frac{dy_n}{dx} &= r_n y_n
\end{aligned}
$$

Each equation is a first-order linear differential equation in two variables. The solutions are easily obtained.

$$
\begin{aligned}
y_1 &= c_1 e^{r_1 x} \\
\vdots \qquad &\ \vdots \\
y_n &= c_n e^{r_n x}
\end{aligned}
$$

This now points the way to a method for solving (3). If A can be diagonalized, that is, if P exists such that $P^{-1}AP$ is diagonal, then we can solve our system by setting

$$
\mathbf{y} = P\mathbf{z}
$$

For then

$$
\mathbf{y}' = P\mathbf{z}'
$$

and substitution in (3) will give

$$P\mathbf{z}' = AP\mathbf{z}$$

or

$$\mathbf{z}' = P^{-1}AP\mathbf{z} = \begin{bmatrix} r_1 & & 0 \\ & \ddots & \\ 0 & & r_n \end{bmatrix} \mathbf{z}$$

As noted above, the solution in this case is easy:

$$\begin{bmatrix} z_1 \\ \vdots \\ z_n \end{bmatrix} = \begin{bmatrix} c_1 e^{r_1 x} \\ \vdots \\ c_n e^{r_n x} \end{bmatrix}$$

Finally,

$$\begin{bmatrix} y_1 \\ \vdots \\ y_n \end{bmatrix} = [P] \begin{bmatrix} c_1 e^{r_1 x} \\ \vdots \\ c_n e^{r_n x} \end{bmatrix}$$

Solve the system (2) which can be written

Example A

$$\begin{bmatrix} y_1' \\ y_2' \end{bmatrix} = \begin{bmatrix} 0 & 1 \\ -2 & 3 \end{bmatrix} \begin{bmatrix} y_1 \\ y_2 \end{bmatrix}$$

To find whether the coefficient matrix is diagonalizable, we solve the characteristic equation

$$\begin{vmatrix} 0 - r & 1 \\ -2 & 3 - r \end{vmatrix} = -r(3 - r) + 2 = r^2 - 3r + 2 = 0$$

The eigenvalues of $r = 1$ and $r = 2$ are distinct (good news), and their corresponding eigenvectors are [1 1] and [1 2]. We form P and quickly find P^{-1}:

$$P = \begin{bmatrix} 1 & 1 \\ 1 & 2 \end{bmatrix}, \qquad P^{-1} = \begin{bmatrix} 2 & -1 \\ -1 & 1 \end{bmatrix}$$

Setting

$$\begin{bmatrix} y_1 \\ y_2 \end{bmatrix} = \begin{bmatrix} 1 & 1 \\ 1 & 2 \end{bmatrix} \begin{bmatrix} z_1 \\ z_2 \end{bmatrix}$$

leads to

$$\begin{bmatrix} z_1' \\ z_2' \end{bmatrix} = \begin{bmatrix} 1 & 0 \\ 0 & 2 \end{bmatrix} \begin{bmatrix} z_1 \\ z_2 \end{bmatrix}$$

then to

$$\begin{bmatrix} z_1 \\ z_2 \end{bmatrix} = \begin{bmatrix} c_1 e^x \\ c_2 e^{2x} \end{bmatrix}$$

Finally,

$$\begin{bmatrix} y_1 \\ y_2 \end{bmatrix} = \begin{bmatrix} 1 & 1 \\ 1 & 2 \end{bmatrix} \begin{bmatrix} c_1 e^x \\ c_2 e^{2x} \end{bmatrix} = \begin{bmatrix} c_1 e^x + c_2 e^{2x} \\ c_1 e^x + 2c_2 e^{2x} \end{bmatrix} \quad \square$$

Note in particular, since the system (2) was derived from (1) by setting $y_1 = y$, that the result of

$$y = y_1 = c_1 e^x + c_2 e^{2x}$$

is consistent with the solution to (1) that is obtained using the methods of Section 9-3.

None of this speaks, of course, to the situation in which the coefficient matrix is not diagonalizable, or to the situation in which diagonalization can only be accomplished using complex numbers. Having omitted these topics from our treatment of diagonalization in Chapter 6, there is little choice at this point but to omit the treatment of corresponding systems of linear differential equations. The fact is that systems in which the coefficient matrix has some equal or some complex eigenvalues can be solved. This fact by itself underscores the usefulness of developing theory that goes beyond the scope of this book.

We do know enough, however, to see how knowledge about complex eigenvalues can provide useful information about a system of equations. It is appropriate that this book should close with one last application.

STABILITY

Suppose that n vital signs of a patient's health are being monitored in an emergency room, that these readings at time t are given by $\mathbf{x}(t) = [x_1(t) \quad \cdots \quad x_n(t)]$, and that the rate at which any particular reading changes is a function of all the other readings at that time; that is,

(4)

$$\frac{dx_1}{dt} = f_1(x_1(t), \ldots, x_n(t)) = f_1(\mathbf{x}(t))$$

$$\vdots \qquad \qquad \vdots \qquad \qquad \vdots$$

$$\frac{dx_n}{dt} = f_n(x_1(t), \ldots, x_n(t)) = f_n(\mathbf{x}(t))$$

Times of rapid change are periods of crisis for the patient, and at the other extreme we say that the patient has **stabilized** at t_0 with readings $\mathbf{s}_0 = \mathbf{x}(t_0)$ if $f_1(\mathbf{x}(t_0)) = f_n(\mathbf{x}(t_0)) = 0$, since according to (4), none of the readings are then changing.

One naturally wonders at such a time if a slight disturbance affecting one of these readings would trigger another crisis, or whether the system would again return to the steady state s_0. In this latter case, we say the patient is **stable** (in contrast with being stabilized for the moment). Stability is expressed mathematically by writing

$$\lim_{t\to\infty} \mathbf{x}(t) = \mathbf{s}_0$$

To study the behavior of the system (4) as t increases, set

$$x_i(t) = s_i + y_i(t), \qquad i = 1, \ldots, n \tag{5}$$

Then

$$\frac{dx_i}{dt}(t) = \frac{dy_i}{dt}(t)$$

and if notation is simplified by writing y_i for $y_i(t)$, the system (4) can be written in the form

$$\frac{dy_1}{dt} = f_1(s_1 + y_1, \ldots, s_n + y_n) = f_1(\mathbf{s}_0 + \mathbf{y}) \tag{6}$$

$$\vdots \qquad\qquad \vdots \qquad\qquad \vdots$$

$$\frac{dy_n}{dt} = f_n(s_1 + y_1, \ldots, s_n + y_n) = f_n(\mathbf{s}_0 + \mathbf{y})$$

Recall (see the vignette, *The Derivative as a Linear Transformation*, Section 5-3) that for any differentiable function $f_i(\mathbf{x})$,

$$f_i(\mathbf{s}_0 + \mathbf{y}) = f_i(\mathbf{s}_0) + \left[\frac{\partial f_i}{\partial x_1}(\mathbf{s}_0), \ldots, \frac{\partial f_i}{\partial x_n}(\mathbf{s}_0) \right] \cdot [y_1, \ldots, y_n]$$

Since $f_i(\mathbf{s}_0) = 0$, we can write (6) in the form

$$\frac{dy_1}{dt} = \frac{\partial f_1}{\partial x_1}(\mathbf{s}_0)y_1 + \cdots + \frac{\partial f_1}{\partial x_n}(\mathbf{s}_0)y_n$$

$$\vdots \qquad \vdots \qquad\qquad \vdots$$

$$\frac{dy_n}{dt} = \frac{\partial f_n}{\partial x_1}(\mathbf{s}_0)y_1 + \cdots + \frac{\partial f_n}{\partial x_n}(\mathbf{s}_0)y_n$$

or finally, by setting $a_{ij} = \dfrac{\partial f_i}{\partial x_j}(\mathbf{s}_0)$,

$$\mathbf{y}' = A\mathbf{y}$$

exactly the form (3) that has been the subject of this section.

From (5) it is clear that if the patient is to be stable, that is if $\lim_{t\to\infty} \mathbf{x}(t) = \mathbf{s}_0$, then $\lim_{t\to\infty} \mathbf{y}(t) = \mathbf{0}$. This motivates terminology that is used for systems of the form $\mathbf{y}' = A\mathbf{y}$, whether or not the system arises from studying the condition of a patient. Though it is more common to ask what happens if the system is perturbed rather than disturbed, we still say that the system (3) is stable if $\lim_{t\to\infty} \mathbf{y}(t) = \mathbf{0}$.

Up to this point, complex eigenvalues have been nothing but a road-block to further progress, and it must be recognized that the eigenvalues to matrix A in (3) may be complex. Such a possibility need not deter us from stating the fundamental theorem on the stability of a system of linear differential equations.

THEOREM A

The system of differential equations defined by (3) will be stable if and only if all the eigenvalues of A have real parts that are negative.

The interested reader is referred to Bellman, R. 1960. *Introduction to matrix analysis.* New York: McGraw-Hill. Chapter 13.

PROBLEM SET 9-5

Solve the following systems of linear differential equations.

1. $y'_1 = -4y_1 + 6y_2$
   $y'_2 = -y_1 + y_2$

2. $y'_1 = -17y_1 + 30y_2$
   $y'_2 = -9y_1 + 16y_2$

3. $y'_1 = 5y_1 - 6y_2$
   $y'_2 = 2y_1 - 2y_2$

4. $y'_1 = 44y_1 - 18y_2$
   $y'_2 = 105y_1 - 43y_2$

5. $y'_1 = 3y_1 + 4y_3$
   $y'_2 = -2y_1 - 2y_2 - y_3$
   $y'_3 = -2y_1 + 3y_3$

6. $y'_1 = 43y_1 - 105y_3$
   $y'_2 = -2y_1 + 3y_2 + 8y_3$
   $y'_3 = 18y_1 - 44y_3$

CHAPTER 9 SELF-TEST

1. Use the Wronskian to establish the linear independence of the functions $f(x) = e^x$, $g(x) = e^{2x}$, $h(x) = xe^x$.

2. Find general solutions to the following homogeneous differential equations:
   (a) $(D^3 + 3D^2 + 4D + 2)y = 0$
   (b) $(D^3 + 3D^2 - D - 3)y = 0$
   (c) $(D^3 + 3D^2 + 3D + 1)y = 0$

3. Find a particular solution to the nonhomogeneous differential equation $(D^2 + D - 12)y = x - \sin x$
   (a) using the method of undetermined coefficients
   (b) using variation of parameters.

Problems 4–6 refer to the differential equation

$$(D^2 - D - 6)y = q(x)$$

and to the system of equations

$$2x_1 - x_2 + x_4 = b_1$$
$$x_1 - 3x_2 + 2x_3 = b_2$$
$$x_1 + 2x_2 - 2x_3 + x_4 = b_3$$

4. What similarities exist between the general solutions to
   (a) the homogeneous differential equation obtained by setting $q(x) = 0$?
   (b) the homogeneous system of equations obtained by setting $b_1 = b_2 = b_3 = 0$?

5. Find a particular solution to
   (a) the nonhomogeneous differential equation obtained by setting $q(x) = \sin x$
   (b) the nonhomogeneous system of equations obtained by setting $b_1 = 6$, $b_2 = 2$, $b_3 = 4$.

6. What similarities exist between the general solutions of
   (a) the nonhomogeneous differential equation described in Problem 5(a)?
   (b) the nonhomogeneous system of equations described in Problem 5(b)?

Problems 7–10 refer to the differential equation

$$y''' + 2y'' - y' - 2y = 0$$

7. Find a system of first-order linear differential equations that is equivalent to the given equation.

8. Solve the system of linear differential equations obtained in Problem 7.

9. Use the solution obtained in Problem 8 to find a solution to the given third-order differential equation.

10. Solve the given third-order equation using the methods of Section 9-4.

ANSWERS TO ODD-NUMBERED PROBLEMS AND SELF-TESTS

1. $(-1, \frac{1}{2})$ **3.** $k = -\frac{1}{3}$ (horizontal)   **5.** $k = \frac{4}{3}$

$k = \frac{1}{2}$ (vertical)

7. (a) $-3x_1 + x_2 - x_3 = -5$ (b) $\frac{5}{4}x_2 + \frac{5}{2}x_3 = \frac{5}{4}$

$7x_1 \quad - x_3 = 14$ $\qquad \frac{11}{4}x_2 - \frac{5}{2}x_3 = \frac{11}{4}$

$-5x_1 \quad - 5x_3 = -10$ $\qquad x_1 - \frac{3}{4}x_2 - \frac{1}{2}x_3 = \frac{5}{4}$

9. (a) $3x_2 - 4x_3 + x_4 = 3$ (b) $x_1 + \frac{3}{2}x_2 - \frac{5}{2}x_3 + \frac{1}{2}x_4 = 4$

$x_1 \quad - \frac{1}{2}x_3 \quad = \frac{5}{2}$ $\qquad - 6x_2 + 8x_3 - 2x_4 = -6$

$5x_2 - \frac{3}{2}x_3 \quad = -\frac{7}{2}$ $\qquad \frac{1}{2}x_2 + \frac{9}{2}x_3 - \frac{3}{2}x_4 = -8$

$- 2x_2 + \frac{19}{2}x_3 \quad = -\frac{15}{2}$ $\qquad -\frac{19}{2}x_2 + \frac{39}{2}x_3 - \frac{5}{2}x_4 = -15$

9. (c) $\frac{1}{5}x_1 \quad - \frac{16}{5}x_3 + x_4 = \frac{28}{5}$ **11.** $-8x_1 + 3x_2 \quad + x_4 = -17$

$4x_1 \quad - 2x_3 \quad = 10$ $\qquad -2x_1 \quad + x_3 \quad = -5$

$\frac{3}{5}x_1 + x_2 - \frac{3}{5}x_3 \quad = \frac{4}{5}$ $\qquad -3x_1 + 5x_2 \quad = -11$

$\frac{31}{5}x_1 \quad + \frac{29}{5}x_3 \quad = \frac{33}{5}$ $\qquad 19x_1 - 2x_2 \quad = 40$

13. $(\frac{6}{7} - \frac{4}{7}x_3, -\frac{11}{7} + \frac{5}{7}x_3, x_3)$ **19.** $-\frac{1}{2}x_1 + x_2 \quad - \frac{1}{2}x_4 \quad = -\frac{3}{2}$

$(-\frac{2}{5} - \frac{4}{5}x_2, x_2, \frac{11}{5} + \frac{7}{5}x_2)$ $\qquad -\frac{3}{2}x_1 \quad + x_3 + \frac{1}{2}x_4 \quad = \frac{7}{2}$

$(x_1, -\frac{1}{2} - \frac{5}{4}x_1, \frac{3}{2} - \frac{7}{4}x_1)$ $\qquad 2x_1 \quad + x_4 + x_5 = 14$

15. $(1, -1, -\frac{1}{2})$ **17.** (a) $k = \frac{1}{2}$ (b) $k = \frac{5}{3}$ (c) $k = -\frac{1}{2}$

1. $(3, \frac{2}{3}, -2)$ **3.** $(-19 + 10x_4, 11 - 5x_4, x_4 - 3, x_4)$
5. $[x_1, 7 - 7x_1, 8 - 9x_1]$ **7.** No solution   **9.** $(1, -1, 2)$
11. $[x_1, -8x_1, 1 - 11x_1]$ **13.** $(A, -7A, -2A, 7A)$ **15.** $A = 4; B = 0;$
$C = \frac{1}{2}; D = -\frac{1}{2}$ **17.** $A = 1; B = -4; C = 2; D = -3$ **19.** $x_1 = \frac{1}{2};$
$x_2 = -1; x_3 = 1$

PROBLEM SET 1-2

PROBLEM SET 1-3

369

PROBLEM SET 1-4

1. (a) Yes (b) No (c) Yes

3. $\begin{bmatrix} 1 & 0 & 2 & 4 & 0 \\ 0 & 1 & -1 & 1 & 0 \\ 0 & 0 & 0 & 0 & 1 \\ 0 & 0 & 0 & 0 & 0 \end{bmatrix}$

5. $\begin{bmatrix} 1 & 0 & 0 \\ 0 & 1 & 0 \\ 0 & 0 & 1 \\ 0 & 0 & 0 \end{bmatrix}$ **7.** $\begin{bmatrix} 1 & 0 & 0 & 1 \\ 0 & 1 & 0 & -1 \\ 0 & 0 & 1 & 2 \end{bmatrix}$

9. (a) $(-3x_2 - 3x_5, x_2, 0, -x_5, x_5)$ (b) $(3 - 3x_2, x_2, 0, 1)$

11. (a) $(-4x_3 - 9x_4, -x_3 - 4x_4, x_3, x_4)$ (b) $(9 - 4x_3, 4 - x_3, x_3)$

13. $x_1 = \frac{29}{37} + \frac{4}{37}x_4$ **15.** No solution
$x_2 = -\frac{28}{37} + \frac{14}{37}x_4$
$x_3 = \frac{6}{37} + \frac{71}{37}x_4$

17. $(1, 0, 2)$ **19.** A line through the origin and $(1, -\frac{7}{3}, -\frac{8}{3})$

21. There must be (a) one solution and (b) only one solution.

23. $\begin{bmatrix} 1 & 0 & 0 \\ 0 & 1 & 0 \\ 0 & 0 & 1 \end{bmatrix}, \begin{bmatrix} 1 & * & 0 \\ 0 & 0 & 1 \\ 0 & 0 & 0 \end{bmatrix}, \begin{bmatrix} 1 & * & * \\ 0 & 0 & 0 \\ 0 & 0 & 0 \end{bmatrix}$

$\begin{bmatrix} 1 & 0 & 0 \\ 0 & 1 & * \\ 0 & 0 & 0 \end{bmatrix}, \begin{bmatrix} 0 & 1 & 0 \\ 0 & 0 & 1 \\ 0 & 0 & 0 \end{bmatrix}, \begin{bmatrix} 0 & 1 & * \\ 0 & 0 & 0 \\ 0 & 0 & 0 \end{bmatrix}$

25. $(i, -1, 1 + i)$

CHAPTER 1 SELF-TEST

1. See page 5. **2.** $(x_1, x_2, x_3, x_4, x_5) = (x_5 - 2x_2, x_2, -x_5, -2x_5, x_5)$

3. Nonhomogeneous; $(x_1, x_2) = (2, -1)$
Homogeneous; $(x_1, x_2, x_3) = (-2x_3, x_3, x_3)$

4. $\begin{bmatrix} 1 & 0 & 0 & \frac{1}{5} & -1 \\ 0 & 1 & 0 & -\frac{12}{5} & 3 \\ 0 & 0 & 1 & -\frac{3}{5} & 2 \\ 0 & 0 & 0 & 0 & 0 \end{bmatrix}$

5. $(\frac{1}{2}, -1, -\frac{3}{2}, \frac{5}{2})$ **6.** $\frac{10}{3}x_1 \quad - x_3 - 3x_4 = \frac{7}{3}$
$\frac{2}{3}x_1 + x_2 \quad - x_4 = \frac{2}{3}$
$\frac{19}{3}x_1 \quad + 4x_3 - 2x_4 = \frac{10}{3}$

7. $x_1 + x_2 \quad - 2x_4 \quad = 1$
$-x_1 \quad + x_3 + x_4 \quad = 1$
$-3x_1 \quad + 7x_4 + x_5 = -5$

8. $(x_1, x_2, x_3, x_4) = (\frac{49}{10} + \frac{13}{10}x_3, \frac{79}{10} + \frac{33}{10}x_3, x_3, -\frac{47}{10} - \frac{19}{10}x_3)$ **9.** If $k \neq 1$, there are no solutions. If $k = 1$, then $(x_1, x_2, x_3) = (\frac{4}{3}x_3, -1 + \frac{7}{3}x_3, x_3)$.

10. $(r, s, t) = (-1, -2, 4)$

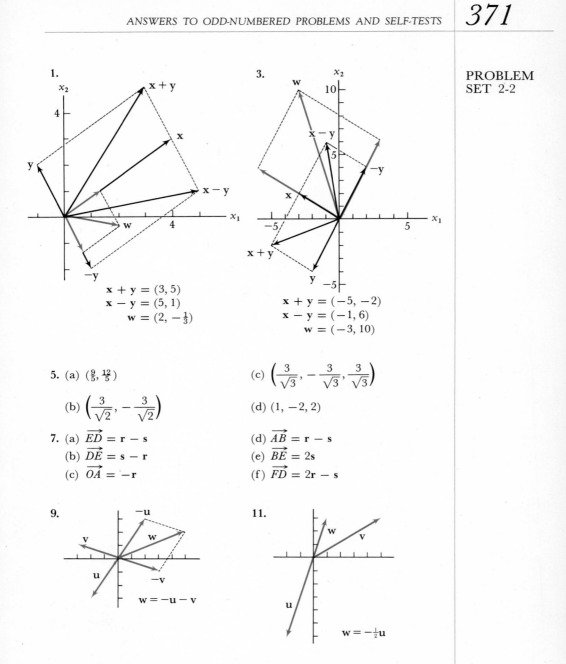

1.

$$\mathbf{x} + \mathbf{y} = (3, 5)$$
$$\mathbf{x} - \mathbf{y} = (5, 1)$$
$$\mathbf{w} = (2, -\tfrac{1}{3})$$

3.

$$\mathbf{x} + \mathbf{y} = (-5, -2)$$
$$\mathbf{x} - \mathbf{y} = (-1, 6)$$
$$\mathbf{w} = (-3, 10)$$

5. (a) $(\tfrac{9}{5}, \tfrac{12}{5})$

(b) $\left(\dfrac{3}{\sqrt{2}}, -\dfrac{3}{\sqrt{2}}\right)$

(c) $\left(\dfrac{3}{\sqrt{3}}, -\dfrac{3}{\sqrt{3}}, \dfrac{3}{\sqrt{3}}\right)$

(d) $(1, -2, 2)$

7. (a) $\overrightarrow{ED} = \mathbf{r} - \mathbf{s}$
(b) $\overrightarrow{DE} = \mathbf{s} - \mathbf{r}$
(c) $\overrightarrow{OA} = -\mathbf{r}$

(d) $\overrightarrow{AB} = \mathbf{r} - \mathbf{s}$
(e) $\overrightarrow{BE} = 2\mathbf{s}$
(f) $\overrightarrow{FD} = 2\mathbf{r} - \mathbf{s}$

9.

$$\mathbf{w} = -\mathbf{u} - \mathbf{v}$$

11.

$$\mathbf{w} = -\tfrac{1}{2}\mathbf{u}$$

13. Linearly dependent
$$3\mathbf{x}_1 + 2\mathbf{x}_2 + \tfrac{1}{2}\mathbf{x}_3 - 4\mathbf{x}_4 = 0$$
17. Linearly dependent
$$2\mathbf{x}_1 + 3\mathbf{x}_2 - \mathbf{x}_3 = 0$$
21. $\mathbf{w} = 2\mathbf{x}_1 + 3\mathbf{x}_2 + \mathbf{x}_3$
$\mathbf{v} = 3\mathbf{x}_1 + 5\mathbf{x}_2 + \mathbf{x}_3$
25. $(\tfrac{7}{3}, \tfrac{2}{3}, -\tfrac{4}{3})$

15. Linearly independent

19. $\mathbf{w} = 3\mathbf{x}_1 + 2\mathbf{x}_2 + 2\mathbf{x}_3$

23. $D(5, 2, -2)$
$R(3, 1, -\tfrac{3}{2})$

PROBLEM SET 2-3

1. $\theta = 45°$ **3.** $\theta = \text{Arccos} \dfrac{\sqrt{2}}{10} = 82°$ **5.** $\text{proj}_{\mathbf{r}}\, \mathbf{s} = \mathbf{r}$; $\text{proj}_{\mathbf{s}}\, \mathbf{r} = \frac{1}{2}\mathbf{s}$

7. 26 square units   **9.** (a) $\overrightarrow{AB} = (1, 1, -1)$, $\overrightarrow{BC} = (-4, 2, 3)$,

$\overrightarrow{OA} = (3, 1, 1)$ (b) $\sqrt{3}$ (c) $\text{Arccos} \dfrac{4}{5\sqrt{5}}$ (d) $\sqrt{114}$

11. $\text{Arccos} \dfrac{3}{5\sqrt{2}}$ **13.** $D(\frac{3}{2}, -\frac{1}{2}, 3)$ **15.** $\mathbf{u} = (-12, -10, 9)$

PROBLEM SET 2-4

1. Yes   **3.** No   **5.** No   **7.** $\mathbf{r}_1 - \mathbf{r}_2 - 3\mathbf{r}_3 - 2\mathbf{r}_3 - 2\mathbf{r}_4 = \mathbf{0}$
9. $\mathbf{r}_1 + 3\mathbf{r}_2 - \mathbf{r}_3 + 2\mathbf{r}_4 = \mathbf{0}$ **11.** $x_2 = x_1 - x_5$, $x_4 = -x_1 + 2x_5$
15. No   **17.** Yes   **21.** (15, 44, 22, 88, 5, 90)

PROBLEM SET 2-5

1. $AB = \begin{bmatrix} 24 & 19 \\ 5 & 4 \end{bmatrix}$, $BA = \begin{bmatrix} 73 & 106 \\ -31 & -45 \end{bmatrix}$

3. $AC = \begin{bmatrix} 71 & -111 \\ 16 & -25 \end{bmatrix}$, $CA = \begin{bmatrix} 29 & 41 \\ 12 & 17 \end{bmatrix}$

5. $A^t B^t = \begin{bmatrix} 73 & -31 \\ 106 & -45 \end{bmatrix}$, $(AB)^t = B^t A^t$

7. $(BA)^t = A^t B^t$

9. (a) $E = \begin{bmatrix} 24 & 19 \\ 5 & 4 \end{bmatrix}$ (b) $F = \begin{bmatrix} 45 & -71 \\ -19 & 30 \end{bmatrix}$

(c) $EC = \begin{bmatrix} 158 & -249 \\ 33 & -52 \end{bmatrix}$ (d) $AF = EC$

11. (a) $R = \begin{bmatrix} 71 & -111 \\ 16 & -25 \end{bmatrix}$ (b) $S = \begin{bmatrix} 1 & -1 \\ 1 & 0 \end{bmatrix}$

(c) $RD = \begin{bmatrix} 22 & -9 \\ 5 & -2 \end{bmatrix}$ (d) $AS = RD$

13. $AI = IA = A$ **15.** $LA = I$ **17.** $PC = I$

19. $M = \begin{bmatrix} 3 & -3 & -1 & 1 \\ 6 & 0 & 2 & 4 \\ -1 & -1 & -1 & -1 \end{bmatrix}$, $N = \begin{bmatrix} -1 & -2 & 5 \\ 1 & -1 & 8 \end{bmatrix}$

$MC = AN = \begin{bmatrix} -4 & -5 & 7 \\ 2 & -2 & 16 \\ -2 & -1 & -3 \end{bmatrix}$

21. $B^t A^t = (AB)^t = M^t$

PROBLEM SET 2-6

11. -6 **13.** $\begin{bmatrix} 0 & 0 & 1 \\ 0 & 1 & 0 \\ 1 & 0 & 0 \end{bmatrix}$ **15.** $\begin{bmatrix} 1 & 0 & 0 \\ 0 & 1 & -2 \\ 0 & 0 & 1 \end{bmatrix}$

17. $\begin{bmatrix} 1 & 0 & 0 & 0 \\ 0 & 1 & 0 & 0 \\ 0 & 0 & 0 & 1 \\ 0 & 0 & 1 & 0 \end{bmatrix}$
19. $\begin{bmatrix} 1 & 0 & 0 & -\frac{1}{2} \\ 0 & 1 & 0 & 0 \\ 0 & 0 & 1 & 0 \\ 0 & 0 & 0 & 1 \end{bmatrix}$

23. No. $(A + B)^2 = A^2 + AB + BA + B^2$

PROBLEM SET 2-7

1. $\begin{bmatrix} 1 & 0 \\ 0 & \frac{1}{3} \end{bmatrix}$
3. $\begin{bmatrix} 1 & 0 & 0 \\ 0 & 1 & 0 \\ 4 & 0 & 1 \end{bmatrix}$
5. $\begin{bmatrix} 0 & 0 & 1 & 0 \\ 0 & 1 & 0 & 0 \\ 1 & 0 & 0 & 0 \\ 0 & 0 & 0 & 1 \end{bmatrix}$

7. $\begin{bmatrix} \frac{1}{8} & -\frac{3}{8} & \frac{1}{8} \\ \frac{3}{8} & -\frac{9}{8} & -\frac{5}{8} \\ \frac{1}{8} & \frac{5}{8} & \frac{1}{8} \end{bmatrix}$ $A = \begin{bmatrix} 1 & 0 & 0 & \frac{2}{8} \\ 0 & 1 & 0 & -\frac{10}{8} \\ 0 & 0 & 1 & \frac{2}{8} \end{bmatrix}$

9. $\begin{bmatrix} -1 & 2 & 1 \\ \frac{1}{2} & -\frac{1}{2} & -\frac{1}{2} \\ \frac{7}{2} & -\frac{11}{2} & -\frac{5}{2} \end{bmatrix}$ $A = \begin{bmatrix} 1 & 0 \\ 0 & 1 \\ 0 & 0 \end{bmatrix}$

11. $\begin{bmatrix} 2 & -1 & -1 & 0 \\ -\frac{2}{3} & \frac{2}{3} & \frac{1}{3} & 0 \\ -\frac{5}{3} & \frac{2}{3} & \frac{1}{3} & 0 \\ 3 & -2 & -1 & 1 \end{bmatrix}$ $A = \begin{bmatrix} 1 & 0 & 0 & -3 \\ 0 & 1 & 0 & \frac{2}{3} \\ 0 & 0 & 1 & \frac{8}{3} \\ 0 & 0 & 0 & 0 \end{bmatrix}$

13. $\begin{bmatrix} -1 & 1 \\ -2 & 3 \end{bmatrix}$
15. $\begin{bmatrix} \frac{4}{14} & \frac{1}{14} \\ -\frac{2}{14} & \frac{3}{14} \end{bmatrix}$
17. No inverse exists.

19. $\begin{bmatrix} -\frac{13}{28} & \frac{7}{28} & \frac{11}{28} \\ -\frac{1}{28} & \frac{7}{28} & \frac{3}{28} \\ -\frac{22}{28} & \frac{14}{28} & \frac{10}{28} \end{bmatrix}$
21. $\begin{bmatrix} -\frac{8}{15} & -\frac{2}{15} & \frac{7}{15} \\ \frac{53}{15} & \frac{2}{15} & -\frac{37}{15} \\ -\frac{26}{15} & \frac{1}{15} & \frac{19}{15} \end{bmatrix}$

23. $\begin{bmatrix} \frac{21}{6} & \frac{9}{6} & -\frac{4}{6} & \frac{7}{6} \\ \frac{3}{6} & \frac{3}{6} & 0 & \frac{3}{6} \\ -\frac{6}{6} & 0 & \frac{2}{6} & -\frac{2}{6} \\ -\frac{9}{6} & -\frac{3}{6} & \frac{4}{6} & -\frac{1}{6} \end{bmatrix}$
35. $y = \frac{117}{107}x + \frac{638}{107}$

PROBLEM SET 2-8

1. row echelon form   column echelon form

$\begin{bmatrix} 1 & 0 & 0 \\ 0 & 1 & 0 \\ 0 & 0 & 1 \\ 0 & 0 & 0 \\ 0 & 0 & 0 \end{bmatrix}$ $\begin{bmatrix} 1 & 0 & 0 \\ 0 & 1 & 0 \\ 0 & 0 & 1 \\ 0 & 2 & 6 \\ -1 & 1 & 3 \end{bmatrix}$

3. row echelon form   column echelon form

$\begin{bmatrix} 1 & 0 & 0 & -\frac{38}{7} \\ 0 & 1 & 0 & -\frac{71}{7} \\ 0 & 0 & 1 & \frac{25}{7} \\ 0 & 0 & 0 & 0 \end{bmatrix}$ $\begin{bmatrix} 1 & 0 & 0 & 0 \\ 0 & 1 & 0 & 0 \\ 0 & 0 & 1 & 0 \\ -1 & -\frac{1}{2} & \frac{1}{2} & 0 \end{bmatrix}$

5. $(-7, 1, -11, -3)$ **7.** $(1, -13, 11, 0, 6)$ **9.** $(-1, 2, 0, 1) +$ $r(3, -7, -28, -12)$ **11.** $\varnothing$

CHAPTER 2
SELF-TEST

1. (a) $D(3, -\frac{1}{2}, \frac{5}{2})$ (b) $\angle A = 45°$ (c) $E(7, 2, 2)$ **2.** (a) Yes; five vectors in R^4 must be linearly dependent.   (b) No; the set is linearly dependent; in fact, $r_1 + r_2 + r_3 - r_4 = 0$.   (c) Yes; $s = 3r_1 + 2r_2 - 4r_3$
3. Subspaces are (d) and (i).   **4.** $\{r_1, r_2, r_3\}$ **6.** $|s_1|^2 + |s_2|^2 = |s_1 + s_2|^2$ **7.** No.

8. (a) $A^{-1} = \begin{bmatrix} 1 & 0 & 0 \\ 0 & 1 & 0 \\ 3 & 0 & 1 \end{bmatrix}$ **9.** $A = \begin{bmatrix} 15 & -2 & 8 \\ -33 & 5 & -18 \\ -9 & 1 & -5 \end{bmatrix}$

(b) $B^{-1} = \begin{bmatrix} 2 & 0 & 1 \\ -13 & 2 & -7 \\ -7 & 1 & -4 \end{bmatrix}$

(c) $(AB)^{-1} = B^{-1}A^{-1} = \begin{bmatrix} 5 & 0 & 1 \\ -34 & 2 & -7 \\ -19 & 1 & -4 \end{bmatrix}$

10. (a) Rank $A = 3$ (b) $(8x_3 + x_5, 10x_3 + 4x_5, x_3, -7x_3 - 3x_5, x_5)$
(c) $(-6 + 8x_3 + x_5, -5 + 10x_3 + 4x_5, x_3, 2 - 7x_3 - 3x_5, x_5)$

PROBLEM
SET 3-2

1. (a) odd; (b) even; (c) even; (d) odd   **3.** (a) 1; (b) 5; (c) -1
5. 14   **7.** -33 **9.** 14   **11.** $-\det A$ **13.** $-2 \det B$
15. $\det A$ **17.** $\det B$ **19.** $(\det A)(\det B) = -66$ **21.** $\dfrac{1}{\det A} = \frac{1}{6}$
23. 12   **25.** 30

PROBLEM
SET 3-3

9. $A^{-1} = \frac{1}{83} \begin{bmatrix} 26 & -8 & 19 \\ -15 & 11 & 5 \\ 9 & 10 & -3 \end{bmatrix}$; $\det A^{-1} = -83$

11. $x = \frac{11}{2}, -3$ **15.** $2x - 10y + 1 = 0$ **17.** A line passing through $(4, -3)$ and $(-2, 5)$

PROBLEM
SET 3-4

1. 81   **3.** 36   **5.** (a) 27; (b) 3; (c) 81
7. $\frac{11}{6}$ **9.** $\frac{11}{6}$ **11.** $x_1 = 3$ **13.** $x_1 = 1$
$x_2 = 2$ $x_2 = 2$
$x_3 = -1$

15. adj $= \begin{bmatrix} 3 & -3 \\ 2 & 4 \end{bmatrix}$; inverse $= \begin{bmatrix} \frac{3}{18} & -\frac{3}{18} \\ \frac{2}{18} & \frac{4}{18} \end{bmatrix}$

17. adj $= \begin{bmatrix} -6 & 9 & 3 \\ 6 & -9 & -3 \\ -12 & 18 & 6 \end{bmatrix}$; no inverse exists

19. $\text{adj} = \begin{bmatrix} 7 & -1 & 5 \\ -3 & 3 & -6 \\ 2 & 1 & 4 \end{bmatrix}$; inverse $= \begin{bmatrix} \frac{7}{9} & -\frac{1}{9} & \frac{5}{9} \\ -\frac{3}{9} & \frac{3}{9} & -\frac{6}{9} \\ \frac{2}{9} & \frac{1}{9} & \frac{4}{9} \end{bmatrix}$

23. When $\begin{vmatrix} a_1 & a_2 & 1 \\ b_1 & b_2 & 1 \\ c_1 & c_2 & 1 \end{vmatrix} = 0$, points A, B, and C are colinear, in which case the "circle" through them degenerates to a line.

25. $\begin{vmatrix} x_1 & x_2 & x_3 & 1 \\ a_1 & a_2 & a_3 & 1 \\ b_1 & b_2 & b_3 & 1 \\ c_1 & c_2 & c_3 & 1 \end{vmatrix} = 0$

31. See W. F. Osgood and W. C. Graustein, *Plane and solid analytic geometry*, Macmillan, 1920.

CHAPTER 3 SELF-TEST

1. Odd   **2.** All terms are 0.

3. $1 \begin{vmatrix} -2 & 0 \\ 3 & 2 \end{vmatrix} - (-1) \begin{vmatrix} 1 & 0 \\ 0 & 2 \end{vmatrix} + 1 \begin{vmatrix} 1 & -2 \\ 0 & 3 \end{vmatrix} = -4 + 2 + 3 = 1$

4. -4 **5.** -8

6. $\text{adj}\, A = \begin{bmatrix} -7 & 7 & -7 \\ -6 & 5 & -3 \\ 4 & -1 & 2 \end{bmatrix}$; $A\,(\text{adj}\, A) = (\det A)I = 7I$

7. $\frac{57}{2}$

PROBLEM SET 4-2

1. (a) Closed; (b) Not closed; (c) Closed; (d) Not closed; (e) Not closed; (f) Not closed; (g) Closed   **3.** Not closed under addition or multiplication by real numbers   **5.** (a) Vector space; (b) Not a vector space; (c) Vector space; (d) Not a vector space   **7.** Yes   **9.** (a) Subspace; (b) Not a subspace; (c) Not a subspace   **17.** L is a plane, M is a line. $L \cup M$ consists of points on either L or M. $L \cap M = \{(0,0,0)\}$. $L + M = R^3$.   **19.** $M\,"+"\,N \neq N\,"+"\,M$ After noting that I acts as the zero element, explain why property (5) fails. $r(M\,"+"\,N) \neq rM\,"+"\,rN$; $(r + s)M \neq rM\,"+"\,sM$

PROBLEM SET 4-3

1. $w = \frac{1}{4}v_1 + \frac{3}{4}v_2 + \frac{1}{4}v_3$ **3.** $w(x) = 2v_1(x) - v_2(x)$
5. $w(x) = v_4(x) - v_3(x)$ **7.** $w(x)$ is not in the span.
9. $w = \frac{1}{2}v_1 + 2v_2$ **11.** Linearly independent   **13.** Linearly dependent $v_1(x) + v_2(x) - 2v_3(x) = 0$ **15.** Linearly independent   **17.** Linearly dependent $2v_1(x) - v_2(x) = 0$ **19.** Linearly independent   **21.** No; $x_1 + x_2 + x_3 - x_4 = 0$ **23.** No; $2x_1 - 3x_2 - x_3 = 0$ **25.** Not a basis $5v_1 - 6v_2 + v_3 = 0$ $w = 2v_1 - v_2$ **27.** Not a basis $4v_1 + 3v_2 + 2v_3 = 0$; w is not in the span.   **29.** (a) True; (b) False; (c) True; (d) False

PROBLEM SET 4-4

1. 3   3. 3   5. 9   7. 4   9. Infinite   11. 2   13. Yes
15. No   17. No   19. 3

23. $r_1(x) = \dfrac{2}{(x_1 - x_2)^2}\left[x - \dfrac{x_1 + x_2}{2}\right](x - x_2)$

$r_2(x) = \dfrac{-4}{(x_1 - x_2)^2}(x - x_1)(x - x_2)$

$r_3(x) = \dfrac{2}{(x_1 - x_2)^2}\left(x - \dfrac{x_1 + x_2}{2}\right)(x - x_1)$

PROBLEM SET 4-5

1. $(5, 3)_B$ 3. $(-\frac{5}{2}, 2)_B$ 5. $(5, -3, 4)_B$ 7. $(1, 0, 2)_B$

9. (a) $(-3, -5, 2)_{B_1}$ 11. $\mathbf{w} = -\frac{32}{7}\mathbf{v}_1 + \frac{12}{7}\mathbf{v}_2$

(b) $P = \begin{bmatrix} 0 & 1 & 0 \\ -\frac{1}{2} & 0 & \frac{1}{2} \\ \frac{1}{2} & -1 & \frac{1}{2} \end{bmatrix}$ $P = \begin{bmatrix} \frac{2}{7} & -\frac{3}{7} \\ \frac{1}{7} & \frac{2}{7} \end{bmatrix}$

13. $P = Q^{-1}$ 15. $\mathbf{v}_1 = (2, 0, 0)$

$\mathbf{v}_2 = (\sqrt{2}, \sqrt{2}, 0)$

$\mathbf{v}_3 = \left(\dfrac{3}{2}, \dfrac{3\sqrt{2} - 3}{2}, \dfrac{3}{\sqrt[4]{2}}\right)$

PROBLEM SET 4-6

1. (a) No; (b) Yes; (c) No   3. (a) Yes; (b) No   5. (a) Yes; (b) No
7. (a) Yes; (b) No   11. $\dfrac{4}{3\sqrt{35}}$

PROBLEM SET 4-7

1. $\mathbf{w} = -\dfrac{1}{\sqrt{5}}\mathbf{u}_1 - \dfrac{13}{\sqrt{5}}\mathbf{u}_2 = -\dfrac{27}{5}\mathbf{v}_1 - \dfrac{11}{5}\mathbf{v}_2$

3. $\mathbf{w} = \dfrac{1}{\sqrt{5}}\mathbf{u}_1 + \dfrac{18}{\sqrt{5}}\mathbf{u}_2 = \dfrac{37}{5}\mathbf{v}_1 + \dfrac{16}{5}\mathbf{v}_2$

5. $\mathbf{w} = -\mathbf{u}_1 - \dfrac{3}{2}\mathbf{u}_2 + \dfrac{7}{\sqrt{2}}\mathbf{u}_3$ 7. $\mathbf{w} = \mathbf{u}_1 + \dfrac{7}{\sqrt{2}}\mathbf{u}_2 + \dfrac{5}{\sqrt{2}}\mathbf{u}_3$

9. $\text{proj}_w\,\mathbf{u} = \dfrac{3}{\sqrt{2}}\mathbf{u}_1 + \dfrac{1}{\sqrt{6}}\mathbf{u}_2 + \dfrac{1}{\sqrt{3}}\mathbf{u}_3 = \begin{bmatrix} 2 & 0 \\ 0 & -1 \end{bmatrix}$

11. $\text{proj}_w\,\mathbf{u}(x) = \dfrac{8\sqrt{6}}{15}u_1(x) = \dfrac{8}{5}x$ 13. $\text{proj}_w\,\mathbf{u}(x) = \dfrac{\sqrt{\pi}}{2}u_0(x) = \dfrac{1}{2}$

15. $\mathbf{w}_1 = \left[\dfrac{1}{3}, \dfrac{2}{3}, 0, -\dfrac{2}{3}\right]$ 17. $\mathbf{w}_1 = \left[0, \dfrac{1}{\sqrt{2}}, 0, \dfrac{1}{\sqrt{2}}\right]$

$\mathbf{w}_2 = \dfrac{1}{\sqrt{693}}[20, 4, -9, 14]$ $\mathbf{w}_2 = \sqrt{\dfrac{2}{3}}\left[1, \dfrac{1}{2}, 0, -\dfrac{1}{2}\right]$

$\mathbf{w}_3 = \sqrt{\dfrac{3}{28}}\left[\dfrac{4}{3}, -\dfrac{4}{3}, 2, \dfrac{4}{3}\right]$

19. $\mathbf{v}_4 = (1, -3, -1, 1)$

1. (a) The set is not closed with respect to addition.   (b) $(1, 1, 1, \ldots)$ has no additive inverse.   (c) There is no zero element.   (d) There is no zero element.   **3.** (a) True; (b) False; (c) False; (d) False   **4.** 5
6. The dimension is 12; so $\{v_1, v_2, v_3\}$ is not a basis.   **7.** They form a subspace of dimension 8.   **8.** The set is linearly dependent;
$6u_1 - 2u_2 - u_3 - u_4 = 0$.

10. $P = \begin{bmatrix} 1 & 1 \\ 1 & -1 \end{bmatrix}$; $w_{B_1} = (-1, 7)$.

Check: $-1u_1 + 7u_2 = (-8, 5) = 3v_1 - 4v_2$.

1. No   **3.** Yes   **5.** No   **7.** No   **9 and 11.** (a) Yes; (b) Yes; (c) Yes   **13 and 15.** (a) No; (b) Yes (though it degenerates to a line segment in 15); (c) Intersection of the diagonals of the image parallelogram   **17.** Yes; $T(u) = T(rv) = rT(v)$ **19.** Yes   **21.** Yes
23. Yes

1. $\begin{bmatrix} 5 & -1 & -3 \\ 3 & 2 & -2 \end{bmatrix} \begin{bmatrix} -1 \\ 3 \\ -2 \end{bmatrix} = \begin{bmatrix} -2 \\ 7 \end{bmatrix}$ **3.** $\begin{bmatrix} 1 & -1 \\ 0 & 4 \\ -1 & 3 \\ 2 & 0 \end{bmatrix} \begin{bmatrix} -2 \\ 7 \end{bmatrix} = \begin{bmatrix} -9 \\ 28 \\ 23 \\ -4 \end{bmatrix}$

5. $\begin{bmatrix} 3 & -1 \\ -4 & 2 \end{bmatrix} \begin{bmatrix} 2 \\ 1 \end{bmatrix} = \begin{bmatrix} 5 \\ -6 \end{bmatrix}$ **7.** $\begin{bmatrix} \frac{9}{2} & -\frac{1}{4} \\ -1 & \frac{1}{2} \end{bmatrix} \begin{bmatrix} 5 \\ 6 \end{bmatrix} = \begin{bmatrix} 21 \\ -2 \end{bmatrix}$

9. $[-1, 10, 2, -2]$ **11.** $[-7, 6, 19]$

13. $\begin{bmatrix} 2 & -3 & -1 \\ 12 & 8 & -8 \\ 4 & 7 & -3 \\ 10 & -2 & -6 \end{bmatrix} \begin{bmatrix} -1 \\ 3 \\ -2 \end{bmatrix} = \begin{bmatrix} -9 \\ 28 \\ 23 \\ -4 \end{bmatrix}$ **15.** $\begin{bmatrix} 0 & 1 \\ 1 & 0 \end{bmatrix}$

17. $\begin{bmatrix} 1 & 0 & 0 \\ 0 & 1 & 0 \end{bmatrix}$ **19.** $\begin{bmatrix} \frac{2}{3} & -\frac{1}{3} & \frac{2}{3} \\ -\frac{1}{3} & \frac{2}{3} & \frac{2}{3} \\ \frac{2}{3} & \frac{2}{3} & -\frac{1}{3} \end{bmatrix}$

21. $f(2.2, 5.3) \approx 3.117$ **23.** $T(5.3, 1.8) \approx \begin{bmatrix} 8.283 \\ .450 \end{bmatrix}$

25. $\begin{bmatrix} 1 & 1 & \frac{3}{2} & \frac{3}{2} \\ -1 & -1 & \frac{3}{2} & \frac{3}{2} \\ \frac{1}{2} & -\frac{1}{2} & 0 & 0 \\ -\frac{1}{2} & \frac{1}{2} & 0 & 0 \end{bmatrix}$

1. (a) $\begin{bmatrix} -5 & -1 \\ 4 & 5 \end{bmatrix} = \begin{bmatrix} -1 & 1 \\ 2 & 1 \end{bmatrix} \begin{bmatrix} 3 & 2 \\ -2 & 1 \end{bmatrix}$; (b) $w = \begin{bmatrix} 18 \\ -5 \end{bmatrix}_B$;

(c) $T(w) = \begin{bmatrix} 41 \\ 29 \end{bmatrix}_B$; (d) $\begin{bmatrix} \frac{10}{7} & -\frac{19}{7} \\ \frac{13}{7} & \frac{39}{7} \end{bmatrix}$; (e) $T(w) = \begin{bmatrix} -\frac{17}{7} \\ \frac{169}{7} \end{bmatrix}_{B'}$

3. (a) $\begin{bmatrix} 0 & 2 \\ -1 & -5 \end{bmatrix} = \begin{bmatrix} 1 & -1 \\ -2 & 1 \end{bmatrix} \begin{bmatrix} 1 & 3 \\ 1 & 1 \end{bmatrix}$; (b) $\mathbf{w} = \begin{bmatrix} 1 \\ 3 \end{bmatrix}_B$;

(c) $T(\mathbf{w}) = \begin{bmatrix} 11 \\ 6 \end{bmatrix}_B$;

(d) $\begin{bmatrix} \frac{5}{2} & \frac{13}{2} \\ \frac{3}{2} & \frac{7}{2} \end{bmatrix}$; (c) $T(\mathbf{w}) = \begin{bmatrix} \frac{7}{2} \\ \frac{5}{2} \end{bmatrix}_{B'}$

5. (a) $\mathbf{w} = \begin{bmatrix} -23 \\ 31 \end{bmatrix}_S$, $T(\mathbf{w}) = \begin{bmatrix} -12 \\ 111 \end{bmatrix}_S$; (b) $M = \begin{bmatrix} \frac{11}{3} & \frac{7}{3} \\ -\frac{1}{3} & \frac{10}{3} \end{bmatrix}$

7. (a) $\mathbf{w} = \begin{bmatrix} -2 \\ 1 \end{bmatrix}_S$, $T(\mathbf{w}) = \begin{bmatrix} 5 \\ -16 \end{bmatrix}_S$; (b) $M = \begin{bmatrix} -4 & -3 \\ 13 & 10 \end{bmatrix}$

9. (a) $P = \begin{bmatrix} 1 & -1 \\ 3 & -2 \end{bmatrix}$; (b) $Q = \begin{bmatrix} 1 & 2 & 0 \\ 1 & 0 & 1 \\ 1 & -1 & -1 \end{bmatrix}$;

(c) $\mathbf{w} = \begin{bmatrix} 1 \\ 4 \end{bmatrix}_{B_n}$; (d) $T(\mathbf{w}) = \begin{bmatrix} 7 \\ 4 \\ -1 \end{bmatrix}_{B_m}$;

(e) $\begin{bmatrix} \frac{11}{5} & -\frac{9}{5} \\ \frac{7}{5} & -\frac{3}{5} \\ \frac{4}{5} & -\frac{1}{5} \end{bmatrix}$; (f) $T(\mathbf{w}) = \begin{bmatrix} \frac{13}{5} \\ \frac{11}{5} \\ \frac{7}{5} \end{bmatrix}_{B'_m}$

11. (a) $P = \begin{bmatrix} 1 & -1 & 0 \\ 1 & 0 & 1 \\ -1 & 1 & 1 \end{bmatrix}$; (b) $Q = \begin{bmatrix} 3 & 4 \\ 2 & 3 \end{bmatrix}$;

(c) $\mathbf{w} = \begin{bmatrix} -2 \\ 1 \\ 4 \end{bmatrix}_{B_n}$; (d) $T(\mathbf{w}) = \begin{bmatrix} 5 \\ -20 \end{bmatrix}_{B_m}$;

(e) $\begin{bmatrix} -42 & 31 & 11 \\ 31 & -23 & -8 \end{bmatrix}$; (f) $T(\mathbf{w}) = \begin{bmatrix} 95 \\ -70 \end{bmatrix}_{B'_m}$

13. (a) $\mathbf{w} = \begin{bmatrix} -3 \\ 7 \end{bmatrix}_S$ $T(\mathbf{w}) = \begin{bmatrix} 3 \\ 5 \\ 6 \end{bmatrix}_S$;

(b) $M = \begin{bmatrix} -1 & 0 \\ -4 & -1 \\ 5 & 3 \end{bmatrix}$

15. (a) $\mathbf{w} = \begin{bmatrix} 10 \\ -1 \\ -9 \end{bmatrix}_S$ $T(\mathbf{w}) = \begin{bmatrix} 85 \\ 35 \end{bmatrix}$;

(b) $M = \begin{bmatrix} \frac{62}{5} & \frac{69}{5} & \frac{14}{5} \\ 6 & 7 & 2 \end{bmatrix}$

25. Yes

PROBLEM SET 5-5

1. All matrices of the form $\begin{bmatrix} 0 & 0 \\ t & 0 \end{bmatrix}$

3. All functions of the form $f(x) = a_1 x + a_0$ **5.** All polynomials of the form $p(x) = c$ **7.** (a) R^2; (b) $\mathbf{0}$; (c) The subspace spanned by $\mathbf{e}_1$

9. (a) The subspace spanned by $\begin{bmatrix} 1 \\ -2 \end{bmatrix}$; (b) The subspace spanned by

$\begin{bmatrix} 2 \\ 1 \end{bmatrix}$; (c) Same as (a)   **11.** (a) 3; (b) 1; (c) 2   **13.** (a) 5; (b) 3

17. Yes   **19.** (a) $(-7, -5, 3, 0)$, $(-11, -10, 0, 3)$; (b) $(1, 3, 0)$, $(0, 2, -1)$ **21.** (a) $(18, 1, 1, 0, 11)$, $(-2, 0, 1, 1, -1)$; (b) $(1, 0, 0, -1)$, $(0, 1, 0, -3)$, $(0, 0, 1, 1)$

1. Yes   **2.** Yes; kernel $= \{(x_1, x_1)\}$ **3.** e^x, e^{-x}, $\sinh x$, $\cosh x$ are in the kernel

CHAPTER 5 SELF-TEST

4. $\begin{bmatrix} 1 & 0 & 1 \\ 0 & 1 & 1 \\ 1 & 1 & 0 \end{bmatrix}$ **5.** $\begin{bmatrix} 2 & 1 & 0 & 0 \\ 1 & 0 & 1 & 1 \\ 0 & -1 & 0 & -1 \\ 0 & 1 & 2 & 0 \end{bmatrix}$

6. Possible null spaces: the origin, a line through the origin, a plane through the origin, all of R^3.

7. Kernel basis: $\begin{bmatrix} -1 \\ 1 \\ 1 \end{bmatrix}$; range basis: $\begin{bmatrix} 1 \\ 0 \\ -1 \end{bmatrix}$ and $\begin{bmatrix} 0 \\ 1 \\ 1 \end{bmatrix}$

8. $\begin{bmatrix} \frac{3}{2} & \frac{9}{2} \\ -\frac{1}{2} & \frac{1}{2} \end{bmatrix}$

1. $P = \begin{bmatrix} 1 & 1 \\ 1 & 2 \end{bmatrix}$ **3.** $P = \begin{bmatrix} 2 & -3 \\ -3 & 5 \end{bmatrix}$

PROBLEM SET 6-2

$P^{-1}AP = \begin{bmatrix} -1 & 0 \\ 0 & 3 \end{bmatrix}$ $P^{-1}AP = \begin{bmatrix} 2 & 0 \\ 0 & 3 \end{bmatrix}$

5. $P = \begin{bmatrix} 1 & -2 & 0 \\ 0 & -1 & 1 \\ -1 & 3 & 0 \end{bmatrix}$ **7.** $P = \begin{bmatrix} 2 & 0 & 1 \\ -1 & 1 & 2 \\ 1 & 0 & 1 \end{bmatrix}$

$P^{-1}AP = \begin{bmatrix} 2 & 0 & 0 \\ 0 & 1 & 0 \\ 0 & 0 & 3 \end{bmatrix}$ $P^{-1}AP = \begin{bmatrix} 1 & 0 & 0 \\ 0 & 1 & 0 \\ 0 & 0 & 2 \end{bmatrix}$

9. Eigenvalues: $1, 1, -1$ **11.** $P = \begin{bmatrix} -1 & -1 & 1 & 0 \\ -5 & -3 & 3 & -1 \\ -2 & -2 & 3 & 0 \\ -3 & -2 & 2 & -1 \end{bmatrix}$

$P^{-1}AP = \begin{bmatrix} 1 & 0 & 0 & 0 \\ 0 & 1 & 0 & 0 \\ 0 & 0 & 2 & 0 \\ 0 & 0 & 0 & -1 \end{bmatrix}$

13. $B = \{(-1, 1, 0), (2, -5, -2), (-1, 1, 1)\}$, $\mathbf{w}_B^t = (0, -\frac{1}{3}, \frac{1}{3})$

15. $B = \{(-2, 1, 0), (1, 0, 1), (-1, 1, 0)\}$, $\mathbf{w}_B^t = (-4, -3, 3)$

17. $\begin{bmatrix} 0 & 1 \\ 1 & 0 \end{bmatrix}$ has eigenvectors $\begin{bmatrix} 1 \\ 1 \end{bmatrix}$ and $\begin{bmatrix} -1 \\ 1 \end{bmatrix}$.

19. Yes

PROBLEM SET 6-3

1. $2x_1^2 - x_3^2 + x_1x_2 + 2x_2x_3 - 2x_1x_3$

7. $P = \begin{bmatrix} \frac{1}{3} & -\frac{2}{3} & \frac{2}{3} \\ -\frac{2}{3} & \frac{1}{3} & \frac{2}{3} \\ \frac{2}{3} & \frac{2}{3} & \frac{1}{3} \end{bmatrix}$; eigenvalues are $9, -9, 18$.

9. $P = \begin{bmatrix} \dfrac{1}{\sqrt{3}} & \dfrac{1}{\sqrt{2}} & -\dfrac{1}{\sqrt{6}} \\ -\dfrac{1}{\sqrt{3}} & 0 & -\dfrac{2}{\sqrt{6}} \\ -\dfrac{1}{\sqrt{3}} & \dfrac{1}{\sqrt{2}} & \dfrac{1}{\sqrt{6}} \end{bmatrix}$; eigenvalues are $0, 6, -6$.

11. $P = \begin{bmatrix} \dfrac{1}{\sqrt{5}} & 0 & \dfrac{2}{3} \\ 0 & \dfrac{1}{\sqrt{5}} & \dfrac{2}{3} \\ -\dfrac{2}{\sqrt{5}} & -\dfrac{2}{\sqrt{5}} & \dfrac{1}{3} \end{bmatrix}$; eigenvalues are $9, 9, -9$.

13. P is as in Problem 9; eigenvalues are $3, 3, -3$.   **21.** 0

PROBLEM SET 6-4

1. $\frac{11}{2}(x_1')^2 + \frac{1}{2}(x_2')^2 = 15$ (ellipse)

3. $\frac{19}{2}(x_1')^2 - \frac{7}{2}(x_2')^2 = 12$ (hyperbola)   **5.** $25(x_1')^2 + 6x_1' - 8x_2' = 16$

7. $x_1'^2 - x_2'^2 + 2x_3'^2 = 4$ **9.** $x_2'^2 - x_3'^2 = 9$

1. Eigenvalues are 4, 8   **3.** Eigenvalues are 4, 8, 12
5. Eigenvalues are 3, 3, 6   **9.** $(\frac{4}{3}, \frac{1}{3}, 1)$

1. $(0, 1, 0)$ belongs to -1; $(2, 1, -3)$ belongs to 1.   **3.** $(0, 1, 0)$ belongs to 2; $(-1, -5, 2)$ belongs to -1.   **5.** Eigenvalues; $1, -1, 2$ **7.** Eigenvalues; $3, -1, 2$ **9.** $(0, 1, 0)$ belongs to 2; $(0, 0, 1)$ belongs to -3.
11. $(1, 0, 0, 0)$ and $(0, 0, 1, 0)$ belong to 2; $(0, 0, 0, 1)$ belongs to 1.

2. $B = \{(1, 3, 0), (0, -2, 1)\}$ **3.** $B = \{(1, 1), (2, -5)\}$ **4.** $a \pm bi$
5. No   **6.** $3x_2^2 + 6x_3^2 = 9$ has an elliptic cylinder as its graph.

7. $A^5 = \begin{bmatrix} -\frac{4}{7}(\frac{1}{6})^5 + \frac{3}{7} & \frac{4}{7}(\frac{1}{6})^5 + \frac{4}{7} \\ \frac{3}{7}(\frac{1}{6})^5 + \frac{3}{7} & -\frac{3}{7}(\frac{1}{6})^5 + \frac{4}{7} \end{bmatrix}$

1. The maximum of 51 occurs at $(7, 2)$.   **3.** The maximum of 272 occurs at $(4, 8)$.   **5.** The maximum of $\frac{68}{3}$ occurs at $(4, \frac{8}{3})$.   **7.** The maximum of $\frac{26}{3}$ occurs at $(0, \frac{10}{3}, \frac{16}{3})$.   **9.** The maximum of 16 occurs at $(6, 0, 4)$.   **13.** The maximum of 2 occurs at $(4, 0)$, even though the feasible set is unbounded.

1. $(4, \frac{8}{3}, 22, 0, 0) \leftrightarrow (4, \frac{8}{3})$;

   $(15, -\frac{14}{3}, 0, -88, 0)$ is not extreme;

   $(\frac{8}{3}, 0, -\frac{22}{3}, 0, \frac{32}{3})$ is not extreme

3. $(-\frac{20}{9}, \frac{40}{9}, \frac{48}{9}, 0, 0, 0)$ is not extreme;

   $(4, 0, 0, 8, 0, 0) \leftrightarrow (4, 0, 0)$;

   $(20, 0, -8, 0, 40, 0)$ is not extreme

1. The maximum of 6 occurs at $(3, 3)$.   **3.** The maximum of $\frac{69}{8}$ occurs at $(\frac{33}{8}, 0, \frac{3}{8})$.   **5.** The maximum of 1 occurs at $(1, 0, 0)$.

5. The maximum of 18 occurs at $(9, 0, 0)$.

1. $(0, \frac{2}{3}), (0, \frac{7}{2}), (4, \frac{11}{2}), (3, \frac{5}{2}), (\frac{7}{10}, \frac{2}{10})$
2. Maximum of $\frac{65}{2}$ occurs at $Q(4, \frac{11}{2})$.
3. Maximize $4x_1 + 3x_2$ subject to

$$
\begin{aligned}
2x_1 - 2x_2 + x_3 \qquad\qquad\qquad &= 1 \\
6x_1 - 2x_2 \qquad + x_4 \qquad\qquad &= 13 \\
-x_1 + 2x_2 \qquad\qquad + x_5 \qquad &= 7 \\
-2x_1 - 3x_2 \qquad\qquad\qquad + x_6 &= -2
\end{aligned}
$$

All $x_i \geq 0$

4. Minimize $\hat{z} = x_0$ subject to

$$
\begin{aligned}
2x_1 - 2x_2 + x_3 \qquad\qquad\qquad &= 1 + x_0 \\
6x_1 - 2x_2 \qquad + x_4 \qquad\qquad &= 13 + x_0 \\
-x_1 + 2x_2 \qquad\qquad + x_5 \qquad &= 7 \\
-2x_1 - 3x_2 \qquad\qquad\qquad + x_6 &= -2
\end{aligned}
$$

All $x_i \geq 0$

5. $(x_0, x_1, x_2, x_3, x_4, x_5, x_6) = (0, 0, \frac{2}{3}, \frac{7}{3}, \frac{43}{3}, \frac{17}{3}, 0)$ **6.** $A(0, \frac{2}{3})$ **7.** (a) It would be most efficient to move clockwise to $(0, \frac{7}{2})$.   (b) Unfortunately, the simplex algorithm takes us counterclockwise to $(\frac{7}{10}, \frac{2}{10})$ **8.** Same as 2 above.   **9.** Remove $-x_1 + 2x_2 \le 7$ **10.** Argue geometrically, plotting the level curves $x_1 - 4x_2 = k$.

PROBLEM SET 8-2

1. (a) $\begin{bmatrix} \frac{3}{4} & \frac{1}{4} \\ \frac{2}{4} & \frac{2}{4} \end{bmatrix}$; (b) $[\frac{171}{256} \quad \frac{85}{256}]$; (c) $\frac{2}{3}$

3. (a) $\begin{bmatrix} 0 & 1 \\ \frac{1}{2} & \frac{1}{2} \end{bmatrix}$; (b) $[\frac{5}{16} \quad \frac{11}{16}]$; (c) $\frac{1}{3}$

5. (a) $\begin{bmatrix} 0 & \frac{1}{2} & \frac{1}{2} \\ 1 & 0 & 0 \\ \frac{1}{3} & \frac{2}{3} & 0 \end{bmatrix}$; (b) $[\frac{2}{3} \quad \frac{1}{3} \quad 0]$; (c) $\frac{1}{2}$

9. (a) $\begin{bmatrix} 0 & \frac{1}{4} & \frac{1}{4} & \frac{1}{4} & \frac{1}{4} \\ \frac{1}{3} & 0 & \frac{1}{3} & \frac{1}{3} & 0 \\ \frac{1}{2} & \frac{1}{2} & 0 & 0 & 0 \\ \frac{1}{3} & \frac{1}{3} & 0 & 0 & \frac{1}{3} \\ \frac{1}{2} & 0 & 0 & \frac{1}{2} & 0 \end{bmatrix}$; (b) $[\frac{274}{864}, \frac{185}{864}, \frac{110}{864}, \frac{185}{864}, \frac{110}{864}]$

11. (a) $\begin{bmatrix} 1 & 0 & 0 & 0 & 0 & 0 \\ \frac{3}{5} & 0 & \frac{2}{5} & 0 & 0 & 0 \\ 0 & \frac{3}{5} & 0 & \frac{2}{5} & 0 & 0 \\ 0 & 0 & \frac{3}{5} & 0 & \frac{2}{5} & 0 \\ 0 & 0 & 0 & \frac{3}{5} & 0 & \frac{2}{5} \\ 0 & 0 & 0 & 0 & 1 & 0 \end{bmatrix}$; (b) $[0, \frac{27}{125}, 0, \frac{66}{125}, 0, \frac{32}{125}]$

13. (a) $\begin{bmatrix} 0 & 1 & 0 & 0 & 0 & 0 \\ \frac{3}{5} & 0 & \frac{2}{5} & 0 & 0 & 0 \\ 0 & \frac{3}{5} & 0 & \frac{2}{5} & 0 & 0 \\ 0 & 0 & \frac{3}{5} & 0 & \frac{2}{5} & 0 \\ 0 & 0 & 0 & \frac{3}{5} & 0 & \frac{2}{5} \\ 0 & 0 & 0 & 0 & 0 & 1 \end{bmatrix}$; (b) $[0, \frac{27}{125}, 0, \frac{36}{125}, 0, \frac{62}{125}]$

PROBLEM SET 8-3

7. (a) Not regular; (b) Regular   **13.** $\begin{bmatrix} \frac{8}{20} & \frac{9}{20} & \frac{3}{20} \\ \frac{8}{20} & \frac{9}{20} & \frac{3}{20} \\ \frac{8}{20} & \frac{9}{20} & \frac{3}{20} \end{bmatrix}$

PROBLEM SET 8-4

1. (a) $\frac{1}{216}\begin{bmatrix} 1 & 0 & 0 & 0 & 0 \\ 0 & 1 & 0 & 0 & 0 \\ 62 & 80 & 18 & 38 & 18 \\ 104 & 104 & 0 & 8 & 0 \\ 67 & 112 & 9 & 19 & 9 \end{bmatrix}$;

(b) $A_3 = \frac{1}{6^3}\begin{bmatrix} 62 & 80 \\ 104 & 104 \\ 67 & 112 \end{bmatrix}$;

(c) $S^3 = \dfrac{1}{6^3}\begin{bmatrix} 18 & 38 & 18 \\ 0 & 8 & 0 \\ 9 & 19 & 9 \end{bmatrix}$; (f) $N = \begin{bmatrix} \frac{5}{3} & 1 & \frac{2}{3} \\ 0 & \frac{3}{2} & 0 \\ \frac{1}{3} & \frac{1}{2} & \frac{4}{3} \end{bmatrix}$;

(g) $\begin{bmatrix} 1.58 & .79 & .58 \\ 0 & 1.48 & 0 \\ .29 & .39 & 1.29 \end{bmatrix}$;

(h) Expect it in E_4 $\frac{1}{2}$ times, in E_3 $\frac{1}{3}$ times; (i) $\frac{13}{6}$; (j) $\frac{7}{18}$

3. (a) $\begin{bmatrix} 1 & 0 & 0 & 0 & 0 & 0 \\ 0 & 1 & 0 & 0 & 0 & 0 \\ \frac{3}{5} & 0 & 0 & \frac{2}{5} & 0 & 0 \\ 0 & \frac{3}{5} & 0 & 0 & \frac{2}{5} & 0 \\ 0 & 0 & \frac{3}{5} & 0 & 0 & \frac{2}{5} \\ 0 & \frac{2}{5} & 0 & \frac{3}{5} & 0 & 0 \end{bmatrix}$;

(b) The probability of going broke or eventually winning all is 1;
(c) $\frac{60}{101}$; (d) $\frac{56}{101}$ 5. Beginning in S_5, the probability of passing through S_4 is $\frac{112}{171}$. Beginning in S_6, the probability of passing through S_4 is $\frac{116}{171}$.

2. $[\frac{1}{2}\ 0\ 0\ \frac{1}{2}] P^n \to [\frac{2}{11}\ \frac{4}{11}\ \frac{1}{11}\ \frac{4}{11}]$. The probabilities of being in states E_1 or E_4 after a long time are $\frac{2}{11}$ and $\frac{4}{11}$, respectively.   3. (a) $\frac{1}{4}$; (b) $\frac{5}{12}$; (c) $\frac{1}{2}$ 4. $\frac{1}{2}$ 5. $\frac{1}{2}$

CHAPTER 8
SELF-TEST

6. $\begin{bmatrix} \frac{3}{6} & \frac{1}{6} & \frac{2}{6} \\ \frac{3}{6} & \frac{1}{6} & \frac{2}{6} \\ \frac{3}{6} & \frac{1}{6} & \frac{2}{6} \end{bmatrix}$ 7. $[\frac{3}{6}\ \frac{1}{6}\ \frac{2}{6}]$ 8. $P^2 = \begin{bmatrix} 1 & 0 & 0 & 0 & 0 \\ 0 & 1 & 0 & 0 & 0 \\ \frac{2}{9} & \frac{6}{9} & \frac{1}{9} & 0 & 0 \\ \frac{8}{48} & \frac{12}{48} & \frac{18}{48} & \frac{7}{48} & \frac{3}{48} \\ \frac{14}{36} & \frac{6}{36} & \frac{10}{36} & \frac{3}{36} & \frac{3}{36} \end{bmatrix}$

Since all entries in the shaded area are positive, P is a regular matrix.

9. $\begin{bmatrix} \frac{3}{2} & 0 & 0 \\ \frac{21}{16} & \frac{24}{16} & \frac{6}{16} \\ \frac{15}{16} & \frac{8}{16} & \frac{18}{16} \end{bmatrix}$ 10. (a) 0; (b) $\frac{3}{4}$; (c) $\frac{3}{2}$

7. Always nonzero   9. $x = 0$ 11. Always nonzero

PROBLEM
SET 9-2

PROBLEM
SET 9-3

1. $y = c_1 e^x + c_2 e^{-2x}$ 3. $y = c_1 e^{3x} + c_2 x e^{3x}$
5. $y = c_1 \sin 2x + c_2 \cos 2x$ 7. $y = e^{2x}(c_1 \cos x + c_2 \sin x)$
9. $y = c_1 e^{-x} + e^{2x}(c_2 \cos x + c_3 \sin x)$
11. $y = c_1 e^x + c_2 + e^x(c_3 \cos 2x + c_4 \sin 2x)$
13. $y = c_1 e^{2x} + c_2 e^{-2x} + c_3 x e^{2x}$
15. $y = c_1 e^x + e^{x/2}\left(c_2 \cos \dfrac{x}{2} + c_3 \sin \dfrac{x}{2}\right)$
17. $y = c_1 e^{2x} + c_2 e^{-3x} + c_3 e^x + c_4 x e^x$
19. $y = c_1 e^{-3x} + c_2 x e^{-3x} + e^{-x/2}\left(c_4 \cos \dfrac{x}{2} + c_5 \sin \dfrac{x}{2}\right)$

PROBLEM SET 9-4

1. $y = c_1 e^{2x} + c_2 x e^{2x} + 8 \cos 2x$ 3. $y = c_1 e^{2x} + c_2 e^{-x} + x e^x$

7. $y = c_1 e^x + c_2 x e^x + c_3 e^{-3x} + e^{-2x}$ 11. $y = c_1 e^x + c_2 e^{-2x} - \frac{1}{5} e^{2x}$

13. $y = c_1 e^x + c_2 x e^x + x^2 + 4x + 6 + \frac{1}{6} x^3 e^x$

15. $y = c_1 \cos x + c_2 \sin x + c_3 e^{2x} - \frac{1}{2} x^2 - \frac{1}{2} x + \frac{3}{4} + x \sin x$

PROBLEM SET 9-5

1. $y_1 = 2e^{-x} + 3e^{-2x}$ 3. $y_1 = 3e^x + 2e^{2x}$
   $y_2 = e^{-x} + e^{-2x}$ $y_2 = 2e^x + e^{2x}$

5. $y_1 = \qquad 2e^x - e^{-x}$
   $y_2 = e^{-2x} - e^x + e^{-x}$
   $y_3 = \qquad - e^x + e^{-x}$

CHAPTER 9 SELF-TEST

1. $\begin{vmatrix} e^x & e^{2x} & x e^x \\ e^x & 2e^{2x} & x e^x + e^x \\ e^x & 4e^{2x} & x e^x + 2e^x \end{vmatrix} = -e^{4x} \neq 0$

2. (a) $y = C_1 e^{-x} + e^{-x}(C_2 \cos x + C_3 \sin x)$;
   (b) $y = C_1 e^x + C_2 e^{-x} + C_3 e^{-3x}$;
   (c) $y = C_1 e^x + C_2 x e^x + C_3 x^2 e^x$

3. $p(x) = -\frac{1}{12} x + \frac{13}{170} \sin x + \frac{1}{170} \cos x - \frac{1}{144}$

4. In both cases, the solution set is a 2-dimensional subspace of the vector space in which solutions are to be found.

5. (a) $p(x) = -\frac{7}{50} \sin x + \frac{1}{50} \cos x$; (b) $(2, 0, 0, 2)$

6. In each case, the general solution of the nonhomogeneous problem is the sum of a particular solution of that problem and the general solution of the homogeneous problem.

7. $y_1' = \qquad y_2$ 8. $y_1 = C_1 e^x - C_2 e^{-x} + C_3 e^{2x}$
   $y_2' = \qquad y_3$ $y_2 = C_1 e^x + C_2 e^{-x} - 2C_3 e^{2x}$
   $y_3' = 2y_1 + y_2 - 2y_3$ $y_3 = C_1 e^x - C_2 e^{-x} + 4C_3 e^{2x}$

9. $y = C_1 e^x - C_2 e^{-x} + C_3 e^{2x}$

INDEX

A

absorbing
    chain 332–343
    state 333
adjacency matrix 82
adjoint 134
angle between vectors 46–48
area
    under linear transformation 203
    parallelogram 105
    triangle 45, 48, 105, 129
associative 74
augmented matrix 22
augmented problem (linear programming) 310

B

back substitution 13, 16–18
balancing chemical equations 63
barycentric coordinates 157
basic
    equation 8
    solution 297
    variable 8
basis
    change of 169–173, 217–228
    in R^n 58
    standard 59
    in a vector space 156, 160

best approximation 89
bilinear 254

C

$C'(a, b)$ 144
canonical form
    Jordan 274
    linear programming 284
    Markov transition matrix 333
Cauchy-Schwarz-Bunyakovsky inequality 179
Cayley-Hamilton theorem 276
centroid 158
Change
    of coordinates 169–173
    of basis 169–173
    of basis, effect on linear transformation 217
characteristic (matrices)
    equation 241
    value 242
    vector 242
characteristic equation (differential equations) 351
chemical equations 62
closure
    under addition 144
    under scalar multiplication 144
coefficient matrix 22
cofactor 108

385